Perspectives in Formal Induction, Revision and Evolution

Editor-in-Chief

Wei Li, Beijing, China

Series Editors

Jie Luo, New York, NY, USA

Yuefei Sui, Institute of Computing Technology, Chinese Academy of Sciences, Beijing, China

Perspectives in Formal Induction, Revision and Evolution is a book series focusing on the logics used in computer science and artificial intelligence, including but not limited to formal induction, revision and evolution. It covers the fields of formal representation, deduction, and theories or meta-theories of induction, revision and evolution, where the induction is of the first level, the revision is of the second level, and the evolution is of the third level, since the induction is at the formula stratum, the revision is at the theory stratum, and the evolution is at the logic stratum.

In his book "The Logic of Scientific Discovery", Karl Popper argues that a scientific discovery consists of conjecture, theory, refutation, and revision. Some scientific philosophers do not believe that a reasonable conjecture can come from induction. Hence, induction, revision and evolution have become a new territory for formal exploration. Focusing on this challenge, the perspective of this book series differs from that of traditional logics, which concerns concepts and deduction.

The series welcomes proposals for textbooks, research monographs, and edited volumes, and will be useful for all researchers, graduate students, and professionals interested in the field.

More information about this series at http://www.springer.com/series/16789

Wei Li · Yuefei Sui

R-CALCULUS: A Logic of Belief Revision

Science Press
Beijing

Springer

Wei Li
Beihang University
Beijing, China

Yuefei Sui
Institute of Computing Technology
Chinese Academy of Sciences
Beijing, China

ISSN 2731-3689 ISSN 2731-3697 (electronic)
Perspectives in Formal Induction, Revision and Evolution
ISBN 978-981-16-2946-4 ISBN 978-981-16-2944-0 (eBook)
https://doi.org/10.1007/978-981-16-2944-0

Jointly published with Science Press, Beijing, China
The print edition is not for sale in China Mainland. Customers from China Mainland please order the print book from Science Press.

This Springer imprint is published by the registered company Springer Nature Singapore Pte Ltd.
The registered company address is: 152 Beach Road, #21-01/04 Gateway East, Singapore 189721, Singapore

Preface to the Series

Classical mathematical logics (propositional logic, first-order logic, and modal logic) and applied logics (temporal logic, dynamic logic, situation calculus, etc.) concern deduction, a logical process from universal statements to particular statements. Description logics concern concepts which and deduction compose of general logics.

Induction and belief revision are the topics of general logics and philosophical logics, and evolution is a new research area in computer science. Formalizing induction, revision, and evolution is a goal of this series.

Revision is omnipresent in sciences. A new theory usually is a revision of an old theory or several old theories. Copernicus' heliocentric theory is a revision of the Tychonic system; the theory of relativity is a revision of the classical theory of movement; the quantum theory is a revision of classical mechanics; etc. The AGM postulates are a set of conditions a reasonable revision operator should satisfy. Professor Li proposed a calculus for first-order logic, called R-calculus, which is sound and complete with respect to maximal consistent subsets. R-calculus has several variants which can be used in other logics, such as nonmonotonic logics, can propose new problems in the classical logics, and will be used in bigdata.

Popper proposed in his book The Logic of Scientific Discovery that a scientific discovery consists of four aspects: conjecture, theory, refutation, and revision. Some scientific philosophers refuted that a reasonable conjecture should come from induction. Hence, induction, revision, and evolution becomes a new territory to be discovered in a formal way.

An induction process is from particular statements to universal statements. A typical example is the mathematical induction, which is a set of nontrivial axioms in Peano arithmetic which makes Peano arithmetic incomplete with respect to the standard model of Peano arithmetic. A logic for induction is needed to guide data-mining in artificial intelligence. Data-mining is a canonical induction process, which mines rules from data.

In biology, *Evolution is change in the heritable traits of biological populations over successive generations.* In sciences, Darwin's evolution theory is an evolution of intuitive theories of plants and animals. In logic, an evolution is a generating process of combining two logics into a new logic, where the new logic should have the traits of two logics. Hence, we should define the corresponding heritable traits of logics, sets of logics, and sequences of logics. Simply speaking, predicate modal logic is an evolution of propositional modal logic and predicate logic, and there are many new problems in predicate modal logic to be solved, such as the constant domain semantics, the variant domain semantics, etc..

The series should focus on formal representation, deduction, theories or meta-theories of induction, revision, and evolution, where the induction is of the first level, the revision is of the second level, and the evolution is of the third level, because the induction is at the formula stratum, the revision is at the theory stratum, and the evolution is at the logic stratum.

The books in the series differ in level: some are overviews and some highly specialized. Here, the books of overviews are for undergraduated students; and the highly specialized ones are for graduated students and researchers in computer science and mathematical logic.

Beijing, China
October 2016

Li Wei
Luo Jie
Song Fangming
Sui Yuefei
Wang Ju
Zhu Wujia

Preface

There are two kinds of research on belief revision: (1) one is to build a set of postulates that a reasonable belief revision operator should satisfy, and typical examples are AGM postulates for revising a belief base K by a belief A, and DP postulates for iterating revision of a belief base K by a sequence $A_1, ..., A_n$ of beliefs; and (2) another is to define a concrete belief revision operator which satisfies the AGM postulates and the DP postulates.

R-calculus is a Gentzen-typed deduction system which is nonmonotonic, and is a concrete belief revision operator which is proved to satisfy the AGM postulates and the DP postulates. This book is to extend R-calculus

(i) from first-order logic to propositional logic, description logics, modal logic, and logic programming;
(ii) from minimal change semantics to subset minimal change, pseudo-subformula minimal change and deduction-based minimal change (the last two minimal changes are newly defined);

and prove soundness and completeness theorems with respect to these minimal changes in these logics. To make R-calculus computable, we gave an approximate R-calculus which uses finite injury priority method in recursion theory. Moreover, two applications of R-calculus are given to default theory and semantic inheritance networks.

One goal of this book is to give a roadmap for researching in theoretic computer science by showing how a simple idea (R-calculus) is developed into a series of more and more complicated theories and immersed into other theories, such as default logic, semantic inheritance networks, etc. From R-calculus, we deleted the requirement that Δ is a set of atoms and developed **R** into **S** with $\subseteq$ -minimal change; with an idea to preserve as much as possible subformulas of formulas to be revised, we developed **S** into **T** with $\preceq$ -minimal change; after **T** we could develop **U** with deduction-based minimal change and different R-calculi for different logics and purposes.

- R-Calculus
  - decidable logics
    - Propositional logic
    - Description logic
    - Propositional modal logic
    - Logic programming
    - $\mathbf{S}$, $\times\mathbf{T}$, $\boxed{\mathbf{U}}$
  - First-order logic
    - First-order logic
      - $\mathbf{S}^{\mathrm{FOL}}$
      - $\mathbf{T}^{\mathrm{FOL}}$
    - Nonmonotonicity
      - Nonmonotonic PL
      - $\Gamma \not\vdash A$
      - RC vs. Default logic
    - Approximate RC
      - $\mathbf{F}^{\mathrm{app}}$
      - $\mathbf{F}^{\mathrm{rec}}$
  - Applications
    - Default logic
      - $\mathbf{S}^{\mathrm{D}}$
      - $\mathbf{T}^{\mathrm{D}}$
      - $\mathbf{U}^{\mathrm{D}}$
    - Semantic networks
      - $\mathbf{S}^{\mathrm{SN}}$
      - $\mathbf{T}^{\mathrm{SN}}$

Structure of the Book

Beijing, China
October 2016

Wei Li
Yuefei Sui

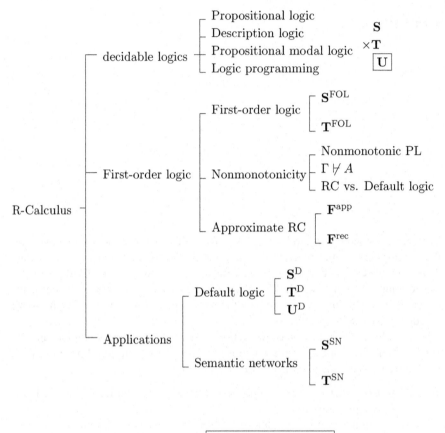

Contents

Chapter 1
Introduction

Belief revision is a topic of philosophy and then of computer science. In computer science, to update databases, truth maintenance systems were proposed by Doyle [6], and belief revision becomes one branch of artificial intelligence; and in philosophy, the logical research of the logic of belief needs consider how human's belief is updated and revised.

1.1 Belief Revision

Belief revision is a process of accepting a new information and updating what we believe.

update: update is an operation of changing the old beliefs to take into account the change;

revision: revision is a process of inserting a new information into a set of old beliefs without generating an inconsistency.[1]

Belief revision requires $K \circ A$ to be closed under the logical inference and consistent, where K is a theory to be revised and A is a formula to revise. $K \circ A$ is equivalent to adding knowledge A in K, delete from $K + A$ the knowledge which makes $K + A$ inconsistent, and revise $K + A$, where

$$\begin{array}{ll} \textit{addition } K + A: & \textit{add } A \textit{ to } K; \\ \textit{deletion } K - A: & \textit{delete } A \textit{ from } K; \\ \textit{revision } K \circ A: & \textit{add } A \textit{ to } K \textit{ and delete contradictory knowledge from } K + A. \end{array}$$

There are basically two kinds of revision schemes: the model-based revision scheme and the syntax-based revision scheme.

[1] From https://www.wikipedia.com/.

W. Li and Y. Sui, *R-CALCULUS: A Logic of Belief Revision*, Perspectives in Formal Induction, Revision and Evolution,
https://doi.org/10.1007/978-981-16-2944-0_1

AGM postulates, proposed by Alchourrón, Gärdenfors and Makinson [1–3], are a set of basic requirements a revision operator should satisfy:

(A1) $K \circ A$ is a belief set;
(A2) $A \in K \circ A$;
(A3) $K \circ A \subseteq K + A = \mathrm{Cn}(K \cup \{A\})$;
(A4) If $\neg A \notin K$ then $K + A \subseteq K \circ A$;
(A5) $K \circ A = \mathrm{Cn}(\bot)$ only if $\models \neg A$;
(A6) If $\models A \leftrightarrow B$ then $K \circ A \equiv K \circ B$;
(A7) $K \circ (A \wedge B) \subseteq (K \circ A) + B$;
(A8) If $\neg A \notin K \circ B$ then $(K \circ A) + B \subseteq K \circ (A \wedge B)$,

where $\mathrm{Cn}(K)$ is the theoretical closure of K.

Darwiche and Pearl [8] proposed the following postulates for iterated revisions:

(DP1) If $A \models B$, then $(K \circ B) \circ A \equiv \Gamma \circ A$;
(DP2) If $A \models \neg B$, then $(K \circ B) \circ A \equiv K \circ A$;
(DP3) If $K \circ A \models B$, then $(K \circ B) \circ A \models B$;
(DP4) If $K \circ A \not\models \neg B$, then $(K \circ B) \circ A \not\models \neg B$.

Similar to the logical principles for logic, there are basic principles for belief revision [5, 7, 15]:

(P1) Principle of consistency: if A is consistent then $K \circ A$ is consistent;
(P2) Principle of success: $A \in K \circ A$;
(P3) Principle of the minimal change: assume that $K \cup \{A\}$ is inconsistent.

(P3.1) In size: let $K \circ A \equiv K' \cup \{A\}$, then $|K - K'|$ is minimal, that is, for any L, if $K \circ A \subseteq L \subseteq K \cup \{A\}$ then either L is inconsistent, or $L = K \circ A$;
(P3.2) In deduction: by the Levy identity:

$$K \circ A \equiv (K - \neg A) \cup \{A\},$$

let $K \circ A \equiv K' \cup \{A\}$, for any L, if $L \cup \{A\}$ is consistent and $K \vdash L \vdash K'$ then $K' \vdash L$, where $K - \neg A$ is the contraction of K by $\neg A$;

(P4) Assume that there is an ordering on the theory to be revised. For iterated belief revision, there is a natural ordering on revising formulas if we assume principle of success. From the logical point of view, we assume as less as possible epistemological except structural assumptions on theories to be revised and revising sentences.

1.2 R-Calculus

The first author [17] developed a Gentzen-typed deduction system in first-order logic [23], called R-calculus, to reduce a configuration $\Delta|\Gamma$ into a consistent theory $\Delta \cup \Theta$, where Θ is a minimal change of Γ by Δ, i.e., a maximal set of Γ which is consistent with Δ, where Δ is a set of atoms, and Γ, Θ are consistent theories of first-order logic. Here, $\Delta|\Gamma$ corresponds to iterating revision $\Gamma \circ \Delta$. Hence, the deduction system is a concrete revision operator which is shown to satisfy AGM postulates.

R-calculus consists of the following deduction rules:

- **Structural rules**:

$$(\text{contraction}^L)\ \Delta, A, A|\Gamma \Rightarrow \Delta, A|\Gamma \qquad (\text{contraction}^R)\ \Delta|A, A, \Gamma \Rightarrow \Delta|A, \Gamma$$
$$(\text{interchange}^L)\ \Delta, A, B|\Gamma \Rightarrow \Delta, B, A|\Gamma\ (\text{interchange}^R)\ \Delta|A, B, \Gamma \Rightarrow \Delta|B, A, \Gamma;$$

- **R-axiom**:

$$\Delta, A|\neg A, \Gamma \Rightarrow \Delta, A|\Gamma;$$

- **R-cut rule**:

$$\frac{\Gamma_1, A \vdash B \quad A \mapsto_T B \quad \Gamma_2, B \vdash C \quad \Delta|C, \Gamma_2 \Rightarrow \Delta|\Gamma_2}{\Delta|\Gamma_1, A, \Gamma_2 \Rightarrow \Delta|\Gamma_1, \Gamma_2};$$

- **R-logical rules**:

$$(R_1^{\wedge})\ \frac{\Delta|A, \Gamma \Rightarrow \Delta|\Gamma}{\Delta|A \wedge B, \Gamma \Rightarrow \Delta|\Gamma}$$
$$(R_2^{\wedge})\ \frac{\Delta|B, \Gamma \Rightarrow \Delta|\Gamma}{\Delta|A \wedge B, \Gamma \Rightarrow \Delta|\Gamma}$$
$$(R^{\vee})\ \frac{\Delta|A, \Gamma \Rightarrow \Delta|\Gamma \quad \Delta|B, \Gamma \Rightarrow \Delta|\Gamma}{\Delta|A \vee B, \Gamma \Rightarrow \Delta|\Gamma}$$
$$(R^{\rightarrow})\ \frac{\Delta|\neg A, \Gamma \Rightarrow \Delta|\Gamma \quad \Delta|B, \Gamma \Rightarrow \Delta|\Gamma}{\Delta|A \rightarrow B, \Gamma \Rightarrow \Delta|\Gamma}$$
$$(R^{\forall})\ \frac{\Delta|A(t), \Gamma \Rightarrow \Delta|\Gamma}{\Delta|\forall x A(x), \Gamma \Rightarrow \Delta|\Gamma}$$
$$(R^{\exists})\ \frac{\Delta|A(x), \Gamma \Rightarrow \Delta|\Gamma}{\Delta|\exists x A(x), \Gamma \Rightarrow \Delta|\Gamma}$$

where t is term and x is a variable not occurring freely in Δ and Γ.

The deduction rules in R-calculus are corresponding to those in Gentzen deduction system of first-order logic [23]. In Gentzen deduction system, a sequent $\Gamma \Rightarrow \Delta$ is reduced to atomic sequents $\Gamma' \Rightarrow \Delta'$ by using the rules of the right-hand side and of the left-hand side, where $\Gamma' \Rightarrow \Delta'$ is atomic if Γ', Δ' are sets of atoms, and is an axiom if and only if $\Gamma' \cap \Delta' \neq \emptyset$. In R-calculus, a configuration $\Delta|A, \Gamma$ is reduced to literal configurations of form $\Delta|l, \Gamma$ by using the deduction rules for logical symbols (logical connectives and quantifiers in first-order logic), where $\Delta|l, \Gamma \Rightarrow \Delta, l|\Gamma$ is an axiom if and only if $\Delta \nvdash \neg l$; otherwise, $\Delta|l, \Gamma \Rightarrow \Delta|\Gamma$, i.e., l is deleted from theory $\{l\} \cup \Gamma$.

1.3 Extending R-Calculus

This book is to extend R-calculus from first-order logic to propositional logic, description logics, modal logic, and logic programming, with respect to from minimal change (maximal consistent subsets) to pseudo-subformula minimal change and deduction-based minimal change. In such a way, we obtain the following R-calculi $\mathbf{Y}^X$ and prove that $\mathbf{Y}^X$ is sound and complete with respect to $\mathbf{Y}$, where

- X is a logic, such as propositional logic, description logics, modal logic, logic programming; and
- $\mathbf{Y}$ is one of subset-minimal changes $\mathbf{S}$, pseudo-subformula-minimal change $\mathbf{T}$ and deduction-based minimal changes $\mathbf{U}$.

For propositional logic, there is such an R-calculus $\mathbf{S}$ in which a configuration $\Delta|\Gamma$ is reduced to a theory $\Delta \cup \Theta$ (denoted by $\vdash_{\mathbf{S}} \Delta|\Gamma \Rightarrow \Delta, \Theta$, i.e., reduction $\Delta|\Gamma \Rightarrow \Delta, \Theta$ is provable in Gentzen deduction system $\mathbf{S}$ for R-calculus) if and only if Θ is a subset-minimal ($\subseteq$-minimal) change of Γ by Δ (denoted by $\models_{\mathbf{S}} \Delta|\Gamma \Rightarrow \Delta, \Theta$, soundness theorem and completeness theorem), that is, Θ is a subtheory of Γ, which is maximal consistent (not *maximally consistent*) with Δ, that is, for any theory Ξ with $\Theta \subset \Xi \subseteq \Gamma$, Ξ is inconsistent with Δ. Here, we use Δ, Θ to denote $\Delta \cup \Theta$. Therefore, R-calculus $\mathbf{S}$ is sound and complete with respect to $\subseteq$-minimal change [8, 9, 15].

There are several other definitions of minimal change:

(i) Pseudo-subformula-minimal ($\preceq$-minimal) change, where $\preceq$ is the pseudo-subformula relation, just as the subformula relation $\leq$, where a formula A is a pseudo-subformula of B if eliminating some subformulas in B results in A.
(ii) Deduction-minimal ($\vdash$-minimal) change, where a theory Θ is a $\vdash$-minimal change of Γ by Δ if Θ is consistent with Δ; $\Gamma \vdash \Theta$, and for any theory Ξ with $\Gamma \vdash \Xi \vdash \Theta$ and $\Theta \nvdash \Xi$, Ξ is inconsistent with Δ. Because the deduction relation $\vdash$ is dense in the set of all the theories, it is impossible to find such a $\vdash$-minimal change. Hence, we consider the following
(iii) Deduction-based minimal ($\vdash_{\preceq}$-minimal) change, where a theory Θ is a $\vdash_{\preceq}$-minimal change of Γ by Δ (denoted by $\models_{\mathbf{U}} \Delta|\Gamma \Rightarrow \Delta, \Theta$), if $\Theta \preceq \Gamma$ is consistent with Δ, and for any theory Ξ with $\Theta \prec \Xi \preceq \Gamma$, either $\Delta, \Xi \vdash \Theta$ and $\Delta, \Theta \vdash \Xi$, or Ξ is inconsistent with Δ.

Corresponding to pseudo-subformula-minimal change and deduction-based minimal change, we have the following R-calculi:

- There is an R-calculus $\mathbf{T}$ in which a configuration $\Delta|\Gamma$ is reduced to a theory $\Delta \cup \Theta$ (denoted by $\vdash_{\mathbf{T}} \Delta|\Gamma \Rightarrow \Delta, \Theta$, i.e., $\Delta|\Gamma \Rightarrow \Delta, \Theta$ is provable in Gentzen deduction system $\mathbf{T}$) if and only if Θ is a $\preceq$-minimal change of Γ by Δ (denoted by $\models_{\mathbf{T}} \Delta|\Gamma \Rightarrow \Delta, \Theta$). This is soundness theorem and completeness theorem for $\mathbf{T}$; and

- There is an R-calculus **U** in which a configuration $\Delta|\Gamma$ is reduced to a theory $\Delta \cup \Theta$ (denoted by $\vdash_{\mathbf{U}} \Delta|\Gamma \Rightarrow \Delta, \Theta$, i.e., $\Delta|\Gamma \Rightarrow \Delta, \Theta$ is provable in Gentzen deduction system **U**) if and only if Θ is a $\vdash_{\preceq}$-minimal change of Γ by Δ (denoted by $\models_{\mathbf{U}} \Delta|\Gamma \Rightarrow \Delta, \Theta$). This is soundness theorem and completeness theorem for **U**.

The same consideration works in description logics [4]. There are corresponding R-calculi $\mathbf{S}^{\mathrm{DL}}$ and $\mathbf{T}^{\mathrm{DL}}$ which are sound and complete with respect to $\subseteq$-minimal change and $\preceq$-minimal change, respectively. Notice that here the minimal changes are not about statements in description logics, but about concepts. A concept set X is a $\subseteq$-minimal change of a concept set Y by a concept set Z. Precisely, we define

(i) Subset-minimal ($\subseteq$-minimal) change for concepts, where for any theories Γ and Δ in a description logic, a theory Θ is a $\subseteq$-minimal change of Γ by Δ, if $\Theta \subseteq \Gamma$ is consistent with Δ, and for any theory Ξ with $\Theta \subset \Xi \subseteq \Gamma$, Ξ is inconsistent with Δ. There is an R-calculus $\mathbf{S}^{\mathrm{DL}}$ such that $\Delta|\Gamma$ is reduced to a theory $\Delta \cup \Theta$ in $\mathbf{S}^{\mathrm{DL}}$ (denoted by $\vdash_{\mathbf{S}^{\mathrm{DL}}} \Delta|\Gamma \Rightarrow \Delta, \Theta$) if and only if Θ is a $\subseteq$-minimal change of Γ by Δ (denoted by $\models_{\mathbf{S}^{\mathrm{DL}}} \Delta|\Gamma \Rightarrow \Delta, \Theta$).

(ii) Pseudo-subconcept-minimal ($\preceq$-minimal) change for concepts, where $\preceq$ is the pseudo-subconcept relation, different from the subconcept relation $\sqsubseteq$ in description logic and also different from the para-subconcept relation (corresponding to the subformula relation in propositional logic), where a concept C is a pseudo-subconcept of D if eliminating some atomic concept symbols in D results in C. Θ is a $\preceq$-minimal change of Γ by Δ (denoted by $\models_{\mathbf{T}^{\mathrm{DL}}} \Delta|\Gamma \Rightarrow \Delta, \Theta$) if Θ is a pseudo-subtheory of Γ consistent with Δ such that for each statement $C(a)$ in Θ, C is a pseudo-subconcept of some concept D with $D(a)$ in Γ (denoted by $\Theta \preceq \Gamma$), and for any consistent theory Ξ with $\Theta \prec \Xi \preceq \Gamma$, Ξ is inconsistent with Δ. There is an R-calculus $\mathbf{T}^{\mathrm{DL}}$ such that $\Delta|\Gamma$ is reduced to a theory $\Delta \cup \Theta$ in $\mathbf{T}^{\mathrm{DL}}$ (denoted by $\vdash_{\mathbf{T}^{\mathrm{DL}}} \Delta|\Gamma \Rightarrow \Delta, \Theta$) if and only if Θ is a $\preceq$-minimal change of Γ by Δ .

(iii) Deduction-based minimal ($\vdash_{\preceq}$-minimal) change for concepts, where a theory Θ is a $\vdash_{\preceq}$-minimal change of Γ by Δ (denoted by $\models_{\mathbf{U}^{\mathrm{DL}}} \Delta|\Gamma \Rightarrow \Delta, \Theta$), if $\Theta \preceq \Gamma$ is consistent with Δ, and for any theory Ξ with $\Theta \prec \Xi \preceq \Gamma$, either $\Delta, \Xi \vdash \Theta$ and $\Delta, \Theta \vdash \Xi$, or Ξ is inconsistent with Δ. There is an R-calculus $\mathbf{U}^{\mathrm{DL}}$ such that $\Delta|\Gamma$ is reduced to a theory $\Delta \cup \Theta$ in $\mathbf{U}^{\mathrm{DL}}$ (denoted by $\vdash_{\mathbf{U}^{\mathrm{DL}}} \Delta|\Gamma \Rightarrow \Delta, \Theta$) if and only if Θ is a $\vdash_{\preceq}$-minimal change of Γ by Δ .

We still do not know whether is a sound and complete R-calculus $\mathbf{U}^{\mathrm{DL}}$ for $\vdash_{\preceq}$-minimal change in description logic.

For propositional modal logic [10], there are R-calculi $\mathbf{S}^{\mathbf{M}}$ and $\mathbf{T}^{\mathbf{M}}$ for modal logic which are sound and complete with respect to $\subseteq$ / $\preceq$-minimal changes in modal logic, respectively. Similar to Hoare's logic [14] in that a program is a modality which changes a state into another, a revising theory Δ in R-calculus changes a theory Γ into another Θ. Therefore, a Hoare-typed R-modal logic $\mathbf{H}_{\mathbf{R}}$ will be given and a possible world semantics will be built so that a formula $\Gamma[\Delta]\Theta$ is provable in $\mathbf{H}_{\mathbf{R}}$ if and only if $\Delta|\Gamma \Rightarrow \Delta, \Theta$ is provable in **S**.

For logic programming [12, 16], there are R-calculi $\mathbf{S}^{\mathrm{LP}}$ and $\mathbf{T}^{\mathrm{LP}}$ which are sound and complete with respect to $\subseteq$ / $\preceq$-minimal changes, respectively.

1.4 Approximate R-Calculus

The above considerations are in these logics in which deduction relation $\vdash$ is decidable [21]. For semi-decidable logics, a configuration $\Delta|A, \Gamma$ may not be reduced to an atomic one $\Delta|l, \Gamma$ in finite steps, when A is consistent with Δ. Because A being inconsistent with Δ is semi-decidable and A being consistent with Δ is undecidable. For such logics, R-calculi are to decompose $\Delta|A, \Gamma$ into $\Delta|l, \Gamma$ when A is inconsistent with Δ, and have an axiom of the following form for A being consistent with Δ :

$$\frac{\Delta \nvdash \neg A}{\Delta|A, \Gamma \Rightarrow \Delta, A|\Gamma}.$$

Therefore, we have the following R-calculus for first-order logic: there are $\subseteq$ / $\preceq$-minimal changes in first-order logic and corresponding R-calculi $\mathbf{S}^{\mathrm{FOL}}$ and $\mathbf{T}^{\mathrm{FOL}}$ such that $\mathbf{S}^{\mathrm{FOL}}$ and $\mathbf{T}^{\mathrm{FOL}}$ are sound and complete with respect to $\subseteq$ / $\preceq$-minimal changes, respectively.

R-calculi are nonmonotonic in nature [13, 20]. Because if $\Delta|A, \Gamma \Rightarrow \Delta, A|\Gamma$ then $\Delta, \neg A|A, \Gamma \Rightarrow \Delta, \neg A|\Gamma$. It will be proved that $\nvdash$ in propositional logic is nonmonotonic, and a sound and complete Gentzen deduction system will be given for sequents $\Gamma \not\Rightarrow \Delta$. We find that each nonmonotonic logic involves relation $\nvdash$, where $\vdash$ is monotonic. For example, in default logic, given a default theory (Δ, D), a consistent theory E is an extension of (Δ, D) if E is minimal such that $E \supseteq \Delta$, and for each normal default $A \leadsto B \in D$ (i.e., $\frac{A:B}{B} \in D$), if $E \vdash A$ and $E \nvdash \neg B$ then $E \vdash B$.

Because $\Delta \nvdash \neg A$ is not decidable and $\Delta \vdash \neg A$ is semidecidable in first-order logic, in terms of the finite injury priority method in recursion theory which was invented independently by two young men Friedberg [11] and Munich [19] to solve Post's problem for recursively enumerable sets [21], we give an approximate R-calculus [18] in which $\Delta|\Gamma$ being reduced to Δ, Θ is a computable procedure with finite injuries. Similar discussions work for the deduction in default logic, where a formula A being deducible from a default theory (Δ, D) is formalized into a Gentzen-typed deduction system $\mathbf{D}$ which is sound and complete with respect to extensions, that is, for any extension E, there is an ordering $\leq$ on D such that $(\Delta, D^{\leq})$ is reduced to E in $\mathbf{D}$; and if (Δ, D) is reduced to a theory E in $\mathbf{D}$ then E is an extension of (Δ, D). The finite injury priority method will be used in default logic to compute the extensions of a default theory, and it will be shown that the infinite injury priority method is needed to compute the strong extensions of a default logic.

1.5 Applications of R-Calculus

R-calculus is to be immersed in nonmonotonic logics. We will give two applications of R-calculus in default logic and semantic inheritance networks.

For default logic, because there is no uniformly defined deduction relation $\vdash$ in default logic, we give deduction systems $\mathbf{S}^{\mathrm{D}}$, $\mathbf{T}^{\mathrm{D}}$ and $\mathbf{U}^{\mathrm{D}}$ to deduce the extensions of a default theory, where three deduction systems may give different extensions of a default theory (Δ, D) under the same ordering of D, and the different extensions have their intuitive meanings in practical application. We will first give Gentzen deduction systems for the sequents consisting of normal defaults which are shown to be sound and complete with respect to the $\subseteq / \preceq / \vdash_{\preceq}$-minimal change.

For example, let (Δ, D) be a default theory, where Δ says that someone is a human (denoted by p) and this man has no arms (denoted by $\neg q$); and D contains one default $\dfrac{p : q \wedge r}{q \wedge r}$, which says that a man defaultly has arms and legs, where r denotes that this man has legs. By default logic, (Δ, D) has an extension $\{p, \neg q\}$, that is, we know that this man is a human and has no arms. In practice, we hope to deduce that this man is a human (p), has no arms, and has legs. Formally,

$$\text{from } p, \neg q, \frac{p : q \wedge r}{q \wedge r}, \text{ defaultly infer } r.$$

If a man has no arms then defaultly this man has legs. In $\mathbf{S}^{\mathrm{D}}$, we have the extension $\{p, \neg q\}$, and in $\mathbf{T}^{\mathrm{D}}$, we have extension $\{p, \neg q, r\}$.

For semantic (inheritance) networks, which are sets of statements of forms

$$C \sqsubseteq D, C \not\sqsubseteq D,$$

there are corresponding R-calculi $\mathbf{S}^{\mathrm{SN}}$ and $\mathbf{T}^{\mathrm{SN}}$ such that for any theories Δ, Γ and Θ of semantic inheritance networks, $\Delta|\Gamma$ is reduced to $\Delta \cup \Theta$ in $\mathbf{S}^{\mathrm{SN}}$, $\mathbf{U}^{\mathrm{SN}}$ (denoted by $\vdash_{\mathbf{S}^{\mathrm{SN}}} \Delta|\Gamma \Rightarrow \Delta, \Theta$, $\vdash_{\mathbf{T}^{\mathrm{SN}}} \Delta|\Gamma \Rightarrow \Delta, \Theta$, respectively) if and only if Θ is a $\subseteq$-/$\preceq$-minimal change of Γ by Δ (denoted by $\models_{\mathbf{S}^{\mathrm{SN}}} \Delta|\Gamma \Rightarrow \Delta, \Theta$, $\models_{\mathbf{T}^{\mathrm{SN}}} \Delta|\Gamma \Rightarrow \Delta, \Theta$, respectively), respectively. That is,

$$\vdash_{\mathbf{S}^{\mathrm{SN}}} \Delta|\Gamma \Rightarrow \Delta, \Theta \text{ iff } \models_{\mathbf{S}^{\mathrm{SN}}} \Delta|\Gamma \Rightarrow \Delta, \Theta,$$
$$\vdash_{\mathbf{T}^{\mathrm{SN}}} \Delta|\Gamma \Rightarrow \Delta, \Theta \text{ iff } \models_{\mathbf{T}^{\mathrm{SN}}} \Delta|\Gamma \Rightarrow \Delta, \Theta,$$

where $\subseteq$-/$\preceq$-minimal change is subset-/pseudo-substatement/deduction-based minimal change, respectively. Before giving R-calculi $\mathbf{S}^{\mathrm{SN}}$ and $\mathbf{T}^{\mathrm{SN}}$, we will give a Gentzen deduction system $\mathbf{G}_4$ for semantic (inheritance) networks, which is proved to be sound and complete with respect to the semantics of semantic (inheritance) network.

We believe that given a logic with deduction relation $\vdash$, there is an R-calculus sound and complete with respect to some minimal change.

A not easy extension of R-calculus is the one to the nonlogical aspects of a logic. In knowledge representation [22], knowledge is classified into the logical stratum, the epistemological stratum, the ontological stratum, the conceptual stratum and the linguistical stratum. The current knowledge representation is at the ontological stratum, that is, we use ontologies to represent knowledge (called the ontology technique). Correspondingly, there is a logic, that is, description logics, and a belief revision (currently the ontology revision). The description logics give logical systems for reasoning ontologies, where the unique logical (in the description logic sense) and nonlogical (in the first-order logical sense) symbol is the subsumption relation $\sqsubseteq$. As a logical symbol, $\sqsubseteq$ should be an important ingredient in designing a Gentzen deduction system. Hence, a Gentzen deduction system for semantic (inheritance) networks is built, in which $\sqsubseteq$ is taken as logical. For ontological revision, there should be some ontological requirements, not just logical requirements, that an ontological revision should satisfy, which is open to come. With these can we consider the logic and R-calculus at the conceptual stratum and the linguistical stratum? where R-calculus at the linguistical stratum should be the belief revision of our daily life.

Our notation is standard and given as needed. We use Δ, Γ, Θ to denote theories; A, B, C formulas in propositional logic and first-order logic, etc., and concepts in description logics (A atomic concepts, C, D, E compound concepts); c, d constants, t, s terms, $\Delta|\Gamma$ a configuration which is equivalent to $\Gamma \circ \Delta$ in AGM postulates. In meta-language, we use $\sim$, &, or, **A**, **E**; and in logical languages, we use $\neg, \wedge, \vee, \forall, \exists$, corresponding to $\sim$, &, or, **A**, **E**, respectively.

References

1. C.E. Alchourrón, P. Gärdenfors, D. Makinson, On the logic of theory change: partial meet contraction and revision functions. J. Symb. Logic **50**, 510–530 (1985)
2. C.E. Alchourrón, D. Makinson, Hierarchies of regulation and their logic, in *New Studies in Deontic Logic*. ed. by R. Hilpinen (D. Reidel Publishing Company, Dordrecht, 1981), pp. 125–148
3. C.E. Alchourrón, D. Makinson, On the logic of theory change: contraction functions and their associated revision functions. Theoria **48**, 14–37 (1982)
4. F. Baader, D. Calvanese, D.L. McGuinness, D. Nardi, P.F. Patel-Schneider, *The Description Logic Handbook: Theory, Implementation, Applications* (Cambridge University Press, Cambridge, 2003)
5. A. Bochman, A foundational theory of belief and belief change. Artif. Intell. **108**, 309–352 (1999)
6. J. Doyle, A truth maintenance system. Artif. Intell. **12**, 231–272 (1979)
7. M. Dalal, Investigations into a theory of knowledge base revision: preliminary report, in *Proceedings of the AAAI-88*, St. Paul, MN (1988), pp. 475–479
8. A. Darwiche, J. Pearl, On the logic of iterated belief revision. Artif. Intell. **89**, 1–29 (1997)
9. E. Fermé, S.O. Hansson, AGM 25 years, twenty-five years of research in belief change. J. Philosoph. Logic **40**, 295–331 (2011)
10. M. Fitting, R.I. Mendelsohn, *First Order Modal Logic* (Kluwer, Dordrecht, 1998)
11. R.M. Friedberg, Two recursively enumerable sets of incomparable degrees of unsolvability. Proc. Natl. Acad. Sci. **43**, 236–238 (1957)

12. P. Gärdenfors, H. Rott, Belief revision, in *Handbook of Logic in Artificial Intelligence and Logic Programming*, vol. 4, ed. by D.M. Gabbay, C.J. Hogger, J.A. Robinson. Epistemic and Temporal Reasoning (Oxford Science Pub, Oxford, 1995), pp. 35–132
13. M.L. Ginsberg (ed.), *Readings in Nonmonotonic Reasoning* (Morgan Kaufmann, San Francisco, 1987)
14. C.A.R. Hoare, An axiomatic basis for computer programming. Commun. ACM **12**, 576–580 (1969)
15. J. Lang, L. van der Torre, From belief change to preference change, in *ECAI 2008 -Proceedings of the 18th European Conference on Artificial Intelligence, Patras, Greece, July 21–25, (Frontiers in Artificial Intelligence and Applications, vol. 178)*, eds. by M. Ghallab, C.D. Spyropoulos, N. Fakotakis, N.M. Avouris (2008), pp. 351–355
16. J.W. Lloyd, *Foundations of Logic Programming*, 2nd edn. (Springer, Berlin, 1987)
17. W. Li, R-calculus: an inference system for belief revision. Comput. J. **50**, 378–390 (2007)
18. W. Li, Y.F. Sui, The R-calculus and the finite injury priority method. J. Comput. **12**, 127–134 (2017)
19. A.A. Muchnik, On the separability of recursively enumerable sets (in Russian). Dokl. Akad. Nauk SSSR, N.S. **109**, 29–32 (1956)
20. R. Reiter, A logic for default reasoning. Artif. Intell. **13**, 81–132 (1980)
21. R.I. Soare, *Recursively Enumerable Sets and Degrees, a Study of Computable Functions and Computably Generated Sets* (Springer, Berlin, 1987)
22. J.F. Sowa, Semantic Networks, in *Encyclopedia of Artificial Intelligence*, ed. by S.C. Shapiro (1987)
23. G. Takeuti, *Proof Theory*, in *Handbook of Mathematical Logic. Studies in Logic and the Foundations of Mathematics*, ed. by J. Barwise (North-Holland, Amsterdam, 1987)

Chapter 2
Preliminaries

A logic consists of a logical language, syntax and semantics. The logical language specifies what symbols can be used in the logic, and the symbols are decomposed into two classes: logical and nonlogical, where the logical symbols are the ones used in each language, and the nonlogical symbols are those which are different for different logical languages; the syntax specifies what strings of symbols are meaningful (formulas) in logic, and the semantics specifies the truth-values of formulas under an assignment (or a model).

In propositional logic, there is no nonlogical symbol, because propositional variables are logical. Hence, there is only one logical language of propositional logic.

In first-order logic, logical connectives and quantifiers are logical, and the constant symbols, function symbols and relation symbols are nonlogical.

In description logics, concept constructors $\bot, \top, \neg, \sqcap, \sqcup, \forall, \exists, \sqsubseteq$ are logical, and the constant symbols, role symbols and atomic concept symbols are nonlogical.

For the sake of self-containedness, this chapter gives basic definitions in propositional logic [1], first-order logic [4, 5, 12, 14] and description logics [9, 13].

2.1 Propositional Logic

There are several complete sets of logical connectives for propositional logic, such as

$$\{\neg, \wedge\}, \{\neg, \vee\}, \{\neg, \to\}, \{\neg, \wedge, \vee\}, \{\neg, \wedge, \vee, \to, \leftrightarrow\}.$$

We choose $\{\neg, \wedge, \vee\}$ as the set of logical connectives, and other connectives are defined as follows:

$$\begin{aligned} &A \to B = \neg A \vee B \\ &A \leftrightarrow B = (A \to B) \wedge (B \to A) = (\neg A \vee B) \wedge (\neg B \vee A). \end{aligned}$$

W. Li and Y. Sui, *R-CALCULUS: A Logic of Belief Revision*, Perspectives in Formal Induction, Revision and Evolution,
https://doi.org/10.1007/978-981-16-2944-0_2

The semantics of propositional logic is defined by assignments, functions from propositional variables to {0, 1}-values.

2.1.1 Syntax and Semantics

A logical language of propositional logic consists of the following symbols:

- propositional variables: $p_0, p_1, \ldots$; and
- logical connectives: $\neg, \wedge, \vee$.

A string A of symbols is a formula if

$$A ::= p|\neg A_1|A_1 \wedge A_2|A_1 \vee A_2,$$

where p or $\neg p$ is called a literal.

The semantics of propositional logic is given by an assignment v, a function from propositional variables to {0, 1}.

Given an assignment v, a formula A is true in v, denoted by $v \models A$, if

$$\begin{cases} v(p) = 1 & \text{if } A = p \\ v \not\models A_1 & \text{if } A = \neg A_1 \\ v \models A_1 \ \& \ v \models A_2 & \text{if } A = A_1 \wedge A_2 \\ v \models A_1 \ or \ v \models A_2 & \text{if } A = A_1 \vee A_2, \end{cases}$$

where $\sim$, &, *or* are symbols used in the meta-language, and correspondingly, $\neg, \wedge, \vee$ are the ones used in the language. Therefore, $v \not\models A_1$ can be represented as $\sim (v \models A_1)$.

A sequent δ is a pair (Γ, Δ), denoted by $\Gamma \Rightarrow \Delta$, where Γ, Δ are sets of formulas.

A sequent $\Gamma \Rightarrow \Delta$ is satisfied in v, denoted by $v \models_{\mathbf{G}_1} \Gamma \Rightarrow \Delta$, if $v \models \Gamma$ implies $v \models \Delta$, where

$$v \models \Gamma \text{ means that for each formula } A \in \Delta, v \models A;$$
$$v \models \Delta \text{ means that for some formula } B \in \Delta, v \models B.$$

A sequent $\Gamma \Rightarrow \Delta$ is valid, denoted by $\models_{\mathbf{G}_1} \Gamma \Rightarrow \Delta$, if for any assignment v,

$$v \models_{\mathbf{G}_1} \Gamma \Rightarrow \Delta.$$

2.1.2 Gentzen Deduction System

Gentzen deduction system $\mathbf{G}_1$ consists of the following axioms and deductions:

- **Axioms**:
$$\frac{\Gamma \cap \Delta \neq \emptyset}{\Gamma \Rightarrow \Delta},$$
where Γ, Δ are sets of propositional variables.
- **Deduction rules**:

$$(\neg^L)\ \frac{\Gamma \Rightarrow A, \Delta}{\Gamma, \neg A \Rightarrow \Delta} \qquad (\neg^R)\ \frac{\Gamma, B \Rightarrow \Delta}{\Gamma \Rightarrow \neg B, \Delta}$$

$$(\wedge_1^L)\ \frac{\Gamma, A_1 \Rightarrow \Delta}{\Gamma, A_1 \wedge A_2 \Rightarrow \Delta} \qquad (\wedge^R)\ \frac{\Gamma \Rightarrow B_1, \Delta \quad \Gamma \Rightarrow B_2, \Delta}{\Gamma \Rightarrow B_1 \wedge B_2, \Delta}$$

$$(\wedge_2^L)\ \frac{\Gamma, A_2 \Rightarrow \Delta}{\Gamma, A_1 \wedge A_2 \Rightarrow \Delta}$$

$$(\vee^L)\ \frac{\Gamma, A_1 \Rightarrow \Delta \quad \Gamma, A_2 \Rightarrow \Delta}{\Gamma, A_1 \vee A_2 \Rightarrow \Delta} \qquad (\vee_1^R)\ \frac{\Gamma \Rightarrow B_1, \Delta}{\Gamma \Rightarrow B_1 \vee B_2, \Delta}$$

$$(\vee_2^R)\ \frac{\Gamma \Rightarrow B_2, \Delta}{\Gamma \Rightarrow B_1 \vee B_2, \Delta}$$

Definition 2.1.1 A sequent $\Gamma \Rightarrow \Delta$ is provable in $\mathbf{G}_1$, denoted by $\vdash_{\mathbf{G}_1} \Gamma \Rightarrow \Delta$, if there is a sequence $\Gamma_1 \Rightarrow \Delta_1, \ldots, \Gamma_n \Rightarrow \Delta_n$ of sequents such that $\Gamma_n \Rightarrow \Delta_n = \Gamma \Rightarrow \Delta$, and for each $1 \leq i \leq n$, $\Gamma_i \Rightarrow \Delta_i$ is an axiom or deduced from the previous sequents by one of the deduction rules in $\mathbf{G}_1$.

2.1.3 Soundness and Completeness Theorem

Theorem 2.1.2 (Soundness theorem) *For any sequent* $\Gamma \Rightarrow \Delta$, *if* $\vdash_{\mathbf{G}_1} \Gamma \Rightarrow \Delta$ *then* $\models_{\mathbf{G}_1} \Gamma \Rightarrow \Delta$.

Proof We prove that each axiom is valid and each deduction rule preserves the validity.

To show the validity of axioms, assume that $\Gamma \cap \Delta \neq \emptyset$, where Γ, Δ are sets of propositional variables. For any assignment v, assume that $v \models \Gamma$. Then, there is a propositional variable p such that $p \in \Gamma \cap \Delta$, and by assumption, $v \models p$, and hence, $v \models \Delta$.

To show that $(\neg_1^L)$ preserves the validity, assume that for any assignment v, $v \models \Gamma \Rightarrow A_1, \Delta$. Then, for any assignment v, if $v \models \Gamma, \neg A_1$ then $v \models \Gamma$, and by induction assumption, $v \models A_1, \Delta$. Because $v \not\models A_1$, $v \models \Delta$. Similar for case $(\neg^R)$.

To show that $(\wedge_1^L)$ preserves the validity, assume that for any assignment v, $v \models \Gamma, A_1 \Rightarrow \Delta$. Then, for any assignment v, if $v \models \Gamma, A_1 \wedge A_2$ then $v \models \Gamma, A_1$ and by induction assumption, $v \models \Delta$. Similar for case $(\wedge_2^L)$.

To show that $(\wedge^R)$ preserves the validity, assume that for any assignment v, $v \models \Gamma \Rightarrow B_1, \Delta$ and $v \models \Gamma \Rightarrow B_2, \Delta$. Then, for any assignment v, if $v \models \Gamma$ then there are two cases: (i) $v \not\models B_1$ or $v \not\models B_2$. By the induction assumption, $v \models \Delta$; (ii) $v \models B_1$ and $v \models B_2$. Then, $v \models A_1 \wedge A_2$, and hence, $v \models A_1 \wedge A_2, \Delta$.

To show that $(\vee^L)$ preserves the validity, assume that for any assignment v, $v \models \Gamma, A_1 \Rightarrow \Delta$ and $v \models \Gamma, A_2 \Rightarrow \Delta$. Then, for any assignment v, if $v \models \Gamma, A_1 \vee A_2$ then $v \models \Gamma$ and $v \models A_1 \vee A_2$. There are two cases: (i) if $v \models A_1$ then by induction assumption, $v \models \Delta$; and (ii) if $v \models A_2$ then by induction assumption, $v \models \Delta$.

To show that $(\vee_1^R)$ preserves the validity, assume that for any assignment v, $v \models \Gamma \Rightarrow B_1, \Delta$. Then, for any assignment v, if $v \models \Gamma$ then by assumption, $v \models B_1, \Delta$. There are two cases: $v \models B_1$, then $v \models B_1 \vee B_2$, and hence, $v \models B_1 \vee B_2, \Delta$; and $v \models \Delta$, and hence, $v \models B_1 \vee B_2, \Delta$. Similar for case $(\vee_2^R)$. □

Theorem 2.1.3 (Completeness theorem) *For any sequent* $\Gamma \Rightarrow \Delta$, *if* $\models_{\mathbf{G}_1} \Gamma \Rightarrow \Delta$ *then* $\vdash_{\mathbf{G}_1} \Gamma \Rightarrow \Delta$.

Proof We will construct a tree T such that either T is a proof tree of $\Gamma \Rightarrow \Delta$ or there is a branch $\gamma \in T$ and an assignment v such that each sequent in γ is not satisfied by v.

The root of T is $\Gamma \Rightarrow \Delta$. A node $\Gamma' \Rightarrow \Delta'$ is a leaf of T if Γ' and Δ' are sets of propositional variables.

Let $\Gamma' \Rightarrow \Delta'$ be a node of T which is not a leaf.

Case $(\neg^L)$. Let $\neg A_1, \ldots, \neg A_n$ be all the formulas in Γ' which begin with $\neg$. Let $\Gamma' = \{\neg A_1, \ldots, \neg A_n\} \cup \Gamma''$. Then, let

$$\Gamma'' \Rightarrow A_1, \ldots, A_n, \Delta'$$

to be a child node of $\Gamma' \Rightarrow \Delta'$.

Case $(\neg^R)$. Let $\neg B_1, \ldots, \neg B_n$ be all the formulas in Δ' which begin with $\neg$. Let $\Delta' = \{\neg B_1, \ldots, \neg B_n\} \cup \Delta''$. Then, let

$$\Gamma', B_1, \ldots, B_n \Rightarrow \Delta''$$

to be a child node of $\Gamma' \Rightarrow \Delta'$.

Case $(\wedge^L)$. Let $A_1^1 \wedge A_1^2, \ldots, A_n^1 \wedge A_n^2$ be all the formulas in Γ' which begin with $\wedge$. Let $\Gamma' = \{A_1^1 \wedge A_1^2, \ldots, A_n^1 \wedge A_n^2\} \cup \Gamma''$. Then, let

$$\Gamma'', A_1^1, A_1^2, \ldots, A_n^1, A_n^2 \Rightarrow \Delta'$$

to be a child node of $\Gamma' \Rightarrow \Delta'$.

Case $(\wedge^R)$. Let $B_1^1 \wedge B_1^2, \ldots, B_n^1 \wedge B_n^2$ be all the formulas in Δ' which begin with $\wedge$. Let $\Delta' = \{B_1^1 \wedge B_1^2, \ldots, B_n^1 \wedge B_n^2\} \cup \Delta''$. Then, $\Gamma' \Rightarrow \Delta'$ has 2^n-many children nodes of the following form:

$$\Gamma' \Rightarrow B_1^{f(1)}, \ldots, B_n^{f(n)}, \Delta'',$$

where f is a function from $\{1, \ldots, n\}$ to $\{1, 2\}$.

Case $(\vee^L)$. Let $A_1^1 \vee A_1^2, \ldots, A_n^1 \vee A_n^2$ be all the formulas in Γ' which begin with $\vee$. Let $\Gamma' = \{A_1^1 \vee A_1^2, \ldots, A_n^1 \vee A_n^2\} \cup \Gamma''$. Then, $\Gamma' \Rightarrow \Delta'$ has 2^n-many children nodes of the following form:

$$\Gamma'', A_1^{f(1)}, \ldots, A_n^{f(n)} \Rightarrow \Delta',$$

where f is a function from $\{1, \ldots, n\}$ to $\{1, 2\}$.

Case $(\vee^R)$. Let $B_1^1 \vee B_1^2, \ldots, B_n^1 \vee B_n^2$ be all the formulas in Δ' which begin with $\vee$. Let $\Delta' = \{B_1^1 \vee B_1^2, \ldots, B_n^1 \vee B_n^2\} \cup \Delta''$. Then, let

$$\Gamma' \Rightarrow B_1^1, B_1^2, \ldots, B_n^1, B_n^2, \Delta''$$

to be a child node of $\Gamma' \Rightarrow \Delta'$.

If each leaf in T is an axiom then it is easy to prove that T is a proof tree of $\Gamma \Rightarrow \Delta$; otherwise, let γ be the branch which leaf is not an axiom, then we define an assignment v by which each sequent on γ is not satisfied.

Let $\Gamma_0 \Rightarrow \Delta_0$ be the leaf of γ which is not an axiom. Define v as follows: for any propositional variable p,

$$v(p) = \begin{cases} 1 \text{ if } p \in \Gamma_0 \\ 0 \text{ if } p \in \Delta_0 \\ 0 \text{ otherwise.} \end{cases}$$

Then, v is well-defined, because $\Gamma_0 \cap \Delta_0 = \emptyset$; and $v \not\models \Gamma_0 \Rightarrow \Delta_0$, because $v \models \Gamma_0$ and $v \not\models \Delta_0$.

For any $\Gamma' \Rightarrow \Delta' \in \gamma$, assume that $v \not\models \Gamma' \Rightarrow \Delta'$. If $\Gamma' \Rightarrow \Delta' = \Gamma \Rightarrow \Delta$ then the proof is complete; otherwise, there is another $\Gamma'' \Rightarrow \Delta'' \in \gamma$ such that $\Gamma' \Rightarrow \Delta'$ is a child node of $\Gamma'' \Rightarrow \Delta''$. There are the following cases for $\Gamma'' \Rightarrow \Delta''$.

Case 1. $\begin{cases} \Gamma' \Rightarrow \Delta' = \Gamma_1, B_1, \ldots, B_n \Rightarrow \Delta_1 \\ \Gamma'' \Rightarrow \Delta'' = \Gamma_1 \Rightarrow \neg B_1, \ldots, \neg B_n, \Delta_1. \end{cases}$ By induction assumption, $v \not\models \Gamma_1, B_1, \ldots, B_n \Rightarrow \Delta_1$, i.e., $v \models \Gamma_1, B_1, \ldots, B_n$ and $v \not\models \Delta_1$. Hence, $v \models \Gamma_1, v \not\models \neg B_1, \ldots, \neg B_n, \Delta_1$.

Case 2. $\begin{cases} \Gamma' \Rightarrow \Delta' = \Gamma_1 \Rightarrow A_1, \ldots, A_n, \Delta_1 \\ \Gamma'' \Rightarrow \Delta'' = \Gamma_1, \neg A_1, \ldots, \neg A_n \Rightarrow \Delta_1. \end{cases}$ By induction assumption, $v \not\models \Gamma_1 \Rightarrow A_1, \ldots, A_n, \Delta_1$, i.e., $v \models \Gamma_1$ and $v \not\models A_1, \ldots, A_n, \Delta_1$. Hence, $v \models \Gamma, v \models \neg A_1, \ldots, \neg A_n$, and $v \not\models \Delta_1$, that is, $v \models \Gamma_1, \neg A_1, \ldots, \neg A_n$ and $v \not\models \Delta_1$.

Case 3. $\begin{cases} \Gamma' \Rightarrow \Delta' = \Gamma_1, A_1^1, A_1^2, \ldots, A_n^1, A_n^2 \Rightarrow \Delta_1 \\ \Gamma'' \Rightarrow \Delta'' = \Gamma_1, A_1^1 \wedge A_1^2, \ldots, A_n^1 \wedge A_n^2 \Rightarrow \Delta_1. \end{cases}$ By induction assumption, $v \not\models \Gamma_1, A_1^1, A_1^2, \ldots, A_n^1, A_n^2 \Rightarrow \Delta_1$, i.e., $v \models \Gamma_1, A_1^1, A_1^2, \ldots, A_n^1, A_n^2$ and $v \not\models \Delta_1$. Hence, $v \models \Gamma_1, A_1^1 \wedge A_1^2, \ldots, A_n^1 \wedge A_n^2$, and $v \not\models \Delta$.

Case 4. There is a function $f : \{1, \ldots, n\} \to \{1, 2\}$ such that

$$\begin{cases} \Gamma' \Rightarrow \Delta' = \Gamma_1 \Rightarrow B_1^{f(1)}, \ldots, B_n^{f(n)}, \Delta_1 \\ \Gamma'' \Rightarrow \Delta'' = \Gamma_1 \Rightarrow B_1^1 \wedge B_1^2, \ldots, B_n^1 \wedge B_n^2, \Delta_1. \end{cases}$$

By induction assumption, $v \not\models \Gamma_1 \Rightarrow B_1^{f(1)}, \ldots, B_n^{f(n)}, \Delta_1$, i.e., $v \models \Gamma_1$ and $v \not\models B_1^{f(1)}, \ldots, B_n^{f(n)}, \Delta_1$. Hence, $v \models \Gamma_1$, and $v \not\models B_1^1 \wedge B_1^2, \ldots, B_n^1 \wedge B_n^2, \Delta$.

Case 5. There is a function $f : \{1, \ldots, n\} \to \{1, 2\}$ such that

$$\begin{cases} \Gamma' \Rightarrow \Delta' = \Gamma_1, A_1^{f(1)}, \ldots, A_n^{f(n)} \Rightarrow \Delta_1 \\ \Gamma'' \Rightarrow \Delta'' = \Gamma_1, A_1^1 \vee A_1^2, \ldots, A_n^1 \vee A_n^2 \Rightarrow \Delta_1. \end{cases}$$

By induction assumption, $v \not\models \Gamma_1, A_1^{f(1)}, \ldots, A_n^{f(n)} \Rightarrow \Delta_1$, i.e., $v \models \Gamma_1, A_1^{f(1)}, \ldots, A_n^{f(n)}$, and $v \not\models \Delta_1$. Hence, $v \models \Gamma_1, A_1^1 \vee A_1^2, \ldots, A_n^1 \vee A_n^2$, and $v \not\models \Delta$.

Case 6. $\begin{cases} \Gamma' \Rightarrow \Delta' = \Gamma_1 \Rightarrow B_1^1, B_1^2, \ldots, B_n^1, B_n^2, \Delta_1 \\ \Gamma'' \Rightarrow \Delta'' = \Gamma_1 \Rightarrow B_1^1 \vee B_1^2, \ldots, B_n^1 \vee B_n^2, \Delta_1. \end{cases}$ By induction assumption, $v \not\models \Gamma_1 \Rightarrow B_1^1, B_1^2, \ldots, B_n^1, B_n^2, \Delta_1$, i.e., $v \models \Gamma_1$, and $v \not\models B_1^1, B_1^2, \ldots, B_n^1, B_n^2, \Delta_1$. Hence, $v \models \Gamma_1$, and $v \not\models B_1^1 \vee B_1^2, \ldots, B_n^1 \vee B_n^2, \Delta$. □

2.2 First-Order Logic

The syntactical objects in first-order logic are terms and formulas, where a term is used to represent an element in a universe, and a formula is to represent a statement about an algebraical structure over the universe.

2.2.1 *Syntax and Semantics*

A logical language of FOL contains the following symbols:

- constant symbols: $c_0, c_1, \ldots$;
- variable symbols: $x_0, x_1, \ldots$;
- function symbols: $f_0, f_1, \ldots$;
- predicate symbols: $p_0, p_1, \ldots$;
- logical connectives and quantifiers: $\neg, \wedge, \vee, \forall$; and
- auxiliary symbols: (,).

Terms are defined as follows:

$$t ::= c|x|f(t_1, \ldots, t_n),$$

where c is a constant symbol; x is a variable (symbol); f is an n-ary function symbol and $t_1, \ldots, t_n$ are terms;

Formulas are strings of the following forms:

$$A ::= p(t_1, \ldots, t_n)|\neg A_1|A_1 \wedge A_2|A_1 \vee A_2|\forall x A_1(x),$$

where p is an n-ary predicate symbol, $t_1, \ldots, t_n$ are terms, and $p(t_1, \ldots, t_n)$ is called atomic.

A model $\mathbf{M}$ is a pair (U, I), where U is a universe and I is an interpretation such that

∘ for any constant symbol c, $I(c) \in U$;
∘ for any n-ary function symbol f, $I(f) : U^n \to U$ is a function; and
∘ for any n-ary predicate symbol p, $I(p) \subseteq U^n$.

An assignment v is a function from variables to U.

The interpretation $t^{I,v}$ of terms t under interpretation I and assignment v is defined as follows:

$$t^{I,v} = \begin{cases} I(c) & \text{if } t = c \\ v(x) & \text{if } t = x \\ I(f)(t_1^{I,v}, \ldots, t_n^{I,v}) & \text{if } t = f(t_1, \ldots, t_n). \end{cases}$$

The interpretation of formulas is defined as follows: a formula A is satisfied in $(\mathbf{M}, v)$, denoted by $\mathbf{M}, v \models A$, if

$$\begin{cases} (t_1^{I,v}, \ldots, t_n^{I,v}) \in p^I & \text{if } A = p(t_1, \ldots, t_n) \\ \mathbf{M}, v \not\models A_1 & \text{if } A = \neg A_1 \\ \mathbf{M}, v \models A_1 \& \mathbf{M}, w \models A_2 & \text{if } A = A_1 \wedge A_2 \\ \mathbf{M}, v \models A_1 \text{ or } \mathbf{M}, w \models A_2 & \text{if } A = A_1 \vee A_2 \\ \mathbf{A} a \in U(\mathbf{M}, v_{x/a} \models A_1(x)) & \text{if } A = \forall x A_1(x), \end{cases}$$

where $v_{x/a}$ is an assignment such that for any variable y,

$$v_{x/a}(y) = \begin{cases} a & \text{if } y = x \\ v(y) & \text{if } y \neq x \end{cases}$$

A formula A is satisfied in $\mathbf{M}$, denoted by $\mathbf{M} \models A$, if for any assignment v,

$$\mathbf{M}, v \models A;$$

and A is valid, denoted by $\models A$, if A is satisfied in any model $\mathbf{M}$.

Let Γ, Δ be sets of formulas. A sequent δ is of form $\Gamma \Rightarrow \Delta$. We say that δ is satisfied in $\mathbf{M}$, denoted by $\mathbf{M} \models_{\mathbf{G}^{\mathrm{FOL}}} \Gamma \Rightarrow \Delta$, if for any assignment v,

$$\mathbf{M}, v \models \Gamma \text{ implies } \mathbf{M}, v \models \Delta,$$

where $\mathbf{M}, v \models \Gamma$ if for each formula $A \in \Gamma$, $\mathbf{M}, v \models A$; and $\mathbf{M}, v \models \Delta$ if for some formula $B \in \Delta$, $\mathbf{M}, v \models B$.

A sequent $\Gamma \Rightarrow \Delta$ is valid, denoted by $\models_{\mathbf{G}^{\mathrm{FOL}}} \Gamma \Rightarrow \Delta$, if for any model $\mathbf{M}$, $\mathbf{M} \models_{\mathbf{G}^{\mathrm{FOL}}} \Gamma \Rightarrow \Delta$.

2.2.2 Gentzen Deduction System

Gentzen deduction system $\mathbf{G}^{\mathrm{FOL}}$ for first-order logic [12, 14] consists of the following axioms and deduction rules:

- **Axioms**:
$$\frac{\Gamma \cap \Delta \neq \emptyset}{\Gamma \Rightarrow \Delta,}$$
where Γ, Δ are sets of atomic formulas.
- **Deduction rules**:

$$(\neg^L)\ \frac{\Gamma \Rightarrow A, \Delta}{\Gamma, \neg A \Rightarrow \Delta} \qquad (\neg^R)\ \frac{\Gamma, A \Rightarrow \Delta}{\Gamma \Rightarrow \neg A, \Delta}$$

$$(\wedge_1^L)\ \frac{\Gamma, A_1 \Rightarrow \Delta}{\Gamma, A_1 \wedge A_2 \Rightarrow \Delta} \qquad (\wedge^R)\ \frac{\Gamma \Rightarrow B_1, \Delta \quad \Gamma \Rightarrow B_2, \Delta}{\Gamma \Rightarrow B_1 \wedge B_2, \Delta}$$

$$(\wedge_2^L)\ \frac{\Gamma, A_2 \Rightarrow \Delta}{\Gamma, A_1 \wedge A_2 \Rightarrow \Delta}$$

$$(\vee^L)\ \frac{\Gamma, A_1 \Rightarrow \Delta \quad \Gamma, A_2 \Rightarrow \Delta}{\Gamma, A_1 \vee A_2 \Rightarrow \Delta} \qquad (\vee_1^R)\ \frac{\Gamma \Rightarrow B_1, \Delta}{\Gamma \Rightarrow B_1 \vee B_2, \Delta}$$

$$(\vee_2^R)\ \frac{\Gamma \Rightarrow B_2, \Delta}{\Gamma \Rightarrow B_1 \vee B_2, \Delta}$$

$$(\forall^L)\ \frac{\Gamma, A_1(t) \Rightarrow \Delta}{\Gamma, \forall x A_1(x) \Rightarrow \Delta} \qquad (\forall^R)\ \frac{\Gamma \Rightarrow B_1(x), \Delta}{\Gamma \Rightarrow \forall x B_1(x), \Delta}$$

where x is a variable not occurring in Γ and Δ; and t is a term.

Definition 2.2.1 A sequent $\Gamma \Rightarrow \Delta$ is provable in $\mathbf{G}^{\mathrm{FOL}}$, denoted by $\vdash_{\mathbf{G}^{\mathrm{FOL}}} \Gamma \Rightarrow \Delta$, if there is a sequence $\Gamma_1 \Rightarrow \Delta_1, \ldots, \Gamma_n \Rightarrow \Delta_n$ of sequents such that $\Gamma_n \Rightarrow \Delta_n = \Gamma \Rightarrow \Delta$, and for each $1 \leq i \leq n$, $\Gamma_i \Rightarrow \Delta_i$ is an axiom or deduced from the previous sequents by one of the deduction rules in $\mathbf{G}^{\mathrm{FOL}}$.

2.2.3 Soundness and Completeness Theorem

Theorem 2.2.2 (Soundness theorem) *For any sequent* $\Gamma \Rightarrow \Delta$, *if* $\vdash_{\mathbf{G}^{\mathrm{FOL}}} \Gamma \Rightarrow \Delta$ *then* $\models_{\mathbf{G}^{\mathrm{FOL}}} \Gamma \Rightarrow \Delta$.

Proof We prove that each axiom is valid and each deduction rule preserves the validity. Fix a model $\mathbf{M}$.

To verify the validity of the axiom, assume that $\Gamma \cap \Delta \neq \emptyset$. For any assignment v, $\mathbf{M}, v \models \Gamma$ implies $\mathbf{M}, v \models \Delta$.

To verify that $(\wedge_1^L)$ preserves the validity, assume that for any assignment v, $\mathbf{M}, v \models \Gamma, A_1$ implies $\mathbf{M}, v \models \Delta$. To show the validity of $\Gamma, A_1 \wedge A_2 \Rightarrow \Delta$, for any assignment v, assume that $\mathbf{M}, v \models \Gamma, A_1 \wedge A_2$. Then, $\mathbf{M}, v \models A_1$, and $\mathbf{M}, v \models \Gamma, A_1$. By induction assumption, $\mathbf{M}, v \models \Delta$.

To verify that $(\wedge^R)$ preserves the validity, assume that for any assignment v, $\mathbf{M}, v \models \Gamma$ implies $\mathbf{M}, v \models \Delta, B_1$; and $\mathbf{M}, v \models \Gamma$ implies $\mathbf{M}, v \models \Delta, B_2$. To show the validity of $\Gamma \Rightarrow B_1 \wedge B_2, \Delta$, for any assignment v, assume that $\mathbf{M}, v \models \Gamma$. Then, by induction assumption, $\mathbf{M}, v \models \Delta, B_1, \mathbf{M}, v \models \Delta, B_2$. If $\mathbf{M}, v \models \Delta$ then $\mathbf{M}, v \models \Delta, B_1 \wedge B_2$; otherwise, $\mathbf{M}, v \models B_1$; $\mathbf{M}, v \models B_2$, and $\mathbf{M}, v \models \Delta, B_1 \wedge B_2$.

To verify that $(\vee^L)$ preserves the validity, assume that for any assignment v, $\mathbf{M}, v \models \Gamma, A_1$ implies $\mathbf{M}, v \models \Delta$; and $\mathbf{M}, v \models \Gamma, A_2$ implies $\mathbf{M}, v \models \Delta$. To show the validity of $\Gamma, A_1 \vee A_2 \Rightarrow \Delta$, for any assignment v, assume that $\mathbf{M}, v \models \Gamma, A_1 \vee A_2$. Then, $\mathbf{M}, v \models \Gamma$, and either $\mathbf{M}, v \models A_1$ or $\mathbf{M}, v \models A_2$. By induction assumption, either implies $\mathbf{M}, v \models \Delta$.

To verify that $(\vee_1^R)$ preserves the validity, assume that for any assignment v, $\mathbf{M}, v \models \Gamma$ implies $\mathbf{M}, v \models \Delta, B_1$. To show the validity of $\Gamma \Rightarrow B_1 \vee B_2, \Delta$, for any assignment v, assume that $\mathbf{M}, v \models \Gamma$. By induction assumption, $\mathbf{M}, v \models \Delta, B_1$. If $\mathbf{M}, v \models \Delta$ then $\mathbf{M}, v \models \Delta, B_1 \vee B_2$; otherwise, $\mathbf{M}, v \models B_1, \mathbf{M}, v \models B_1 \vee B_2$, and $\mathbf{M}, v \models \Delta, B_1 \vee B_2$.

To verify that $(\forall^L)$ preserves the validity, assume that for any assignment v, $\mathbf{M}, v \models \Gamma, A_1(t)$ implies $\mathbf{M}, v \models \Delta$. To show the validity of $\Gamma, \forall x A_1(x) \Rightarrow \Delta$, for any assignment v, assume that $\mathbf{M}, v \models \Gamma, \forall x A_1(x)$. Then, $\mathbf{M}, v \models \Gamma$, and $\mathbf{M}, v \models \forall x A_1(x)$. For any a, $\mathbf{M}, v_{x/a} \models A_1(x)$, and by taking $a = t^{I,v}$, we have $\mathbf{M}, v \models A_1(t)$. By induction assumption, $\mathbf{M}, v \models \Delta$.

To verify that $(\forall^R)$ preserves the validity, assume that for any assignment v, $\mathbf{M}, v \models \Gamma$ implies $\mathbf{M}, v \models \Delta, B_1(x)$. To show the validity of $\Gamma \Rightarrow \forall x B_1(x), \Delta$, for any assignment v, assume that $\mathbf{M}, v \models \Gamma$. By induction assumption, $\mathbf{M}, v \models \Delta, B_1(x)$. If $\mathbf{M}, v \models \Delta$ then $\mathbf{M}, v \models \Delta, \forall x B_1(x)$; otherwise, $\mathbf{M}, v \models B_1(x)$, and for any element a, $\mathbf{M}, v_{x/a} \models \Gamma$, and hence, $\mathbf{M}, v_{x/a} \models B_1(x)$, i.e., $\mathbf{M}, v \models \forall x B_1(x)$. Therefore, $\mathbf{M}, v \models \Delta, \forall x B_1(x)$. □

Theorem 2.2.3 (Completeness theorem) *For any sequent* $\Gamma \Rightarrow \Delta$, *if* $\models_{\mathbf{G}^{\mathrm{FOL}}} \Gamma \Rightarrow \Delta$ *then* $\vdash_{\mathbf{G}^{\mathrm{FOL}}} \Gamma \Rightarrow \Delta$.

Proof Given a sequent $\Gamma \Rightarrow \Delta$, we construct a tree T such that either

(i) for each branch ξ of T, there is a sequent $\Gamma' \Rightarrow \Delta'$ at the leaf of ξ such that $\Gamma' \cap \Delta' \neq \emptyset$; or
(ii) there is a model $\mathbf{M}$ and an assignment v such that $\mathbf{M}, v \not\models \Gamma \Rightarrow \Delta$.

T is constructed as follows:

- the root of T is $\Gamma \Rightarrow \Delta$;
- for a node ξ, if each sequent $\Gamma' \Rightarrow \Delta'$ at ξ is atomic then the node is a leaf;
- otherwise, ξ has the direct child node containing the following sequents:

$$\begin{cases} \Gamma_1 \Rightarrow A, \Delta_1 & \text{if } \Gamma_1, \neg A \Rightarrow \Delta_1 \in \xi \\ \Gamma_1, B \Rightarrow \Delta_1 & \text{if } \Gamma_1 \Rightarrow \neg B, \Delta_1 \in \xi \\ \begin{bmatrix} \Gamma_1, A_1 \Rightarrow \Delta_1 \\ \Gamma_1, A_2 \Rightarrow \Delta_1 \end{bmatrix} & \text{if } \Gamma_1, A_1 \wedge A_2 \Rightarrow \Delta_1 \in \xi \\ \begin{cases} \Gamma_1 \Rightarrow B_1, \Delta_1 \\ \Gamma_1 \Rightarrow B_2, \Delta_1 \end{cases} & \text{if } \Gamma_1 \Rightarrow B_1 \wedge B_2, \Delta_1 \in \xi \\ \begin{cases} \Gamma_1, A_1 \Rightarrow \Delta_1 \\ \Gamma_1, A_2 \Rightarrow \Delta_1 \end{cases} & \text{if } \Gamma_1, A_1 \vee A_2 \Rightarrow \Delta_1 \in \xi \\ \begin{bmatrix} \Gamma_1 \Rightarrow B_1, \Delta_1 \\ \Gamma_1 \Rightarrow B_2, \Delta_1 \end{bmatrix} & \text{if } \Gamma_1 \Rightarrow B_1 \vee B_2, \Delta_1 \in \xi \\ \Gamma_1 \Rightarrow B_1(c), \Delta_1 & \\ c \text{does not occur in} T & \text{if } \Gamma_1 \Rightarrow \forall x B_1(x), \Delta_1 \in \xi, \end{cases}$$

and

- for each $\delta = \Gamma_2, \forall x A'(x) \Rightarrow \Delta_2 \in T$ such that c has not been used for δ, for each child node $\Gamma_3 \Rightarrow \Delta_3$ of δ, let $\Gamma_3 \Rightarrow \Delta_3$ have child node containing sequent $\Gamma_3, A'(c) \Rightarrow \Delta_3$, and we say that c has been used for δ,

where $\begin{bmatrix} \delta_1 \\ \delta_2 \end{bmatrix}$ represents that δ_1, δ_2 are at a same child node; and $\begin{cases} \delta_1 \\ \delta_2 \end{cases}$ represents that δ_1, δ_2 are at different direct child nodes.

Theorem 2.2.4 *If for each branch $\xi \subseteq T$, there is a sequent $\Gamma' \Rightarrow \Delta' \in \xi$ is an axiom in $\mathbf{G}^{\mathrm{FOL}}$ then T is a proof tree of $\Gamma \Rightarrow \Delta$.*

Proof By the definition of T, T is a proof tree of $\Gamma \Rightarrow \Delta$. □

Theorem 2.2.5 *If there is a branch $\xi \subseteq T$ such that each sequent $\Gamma' \Rightarrow \Delta' \in \xi$ is not an axiom in $\mathbf{G}^{\mathrm{FOL}}$ then there is a model $\mathbf{M}$ and an assignment v such that $\mathbf{M}, v \not\models \Gamma \Rightarrow \Delta$.*

Proof Let ξ be a branch of T such that each sequent $\Gamma' \Rightarrow \Delta' \in \xi$ is not an axiom in $\mathbf{G}^{\mathrm{FOL}}$.

Let

$$\Theta^L = \bigcup_{\Gamma' \Rightarrow \Delta' \in \xi} \Gamma',$$
$$\Theta^R = \bigcup_{\Gamma' \Rightarrow \Delta' \in \xi} \Delta'.$$

Define a model $\mathbf{M} = (U, I)$ and an assignment v as follows:

- U is the set of all the constant symbols occurring in T;
- $I(c) = c$ for each constant symbol c, and $I(p) = \{(c_1, \ldots, c_n) : p(c_1, \ldots, c_n) \in \Theta^L\}$;
- $v(x) = x$.

We proved by induction on $\Gamma' \Rightarrow \Delta' \in \xi$ that $v(\Theta^L) = 1$ and $v(\Theta^R) = 0$, that is, for any $\Gamma' \Rightarrow \Delta' \in \xi$, $v(\Gamma') = 1$ and $v(\Delta') = 0$.

We prove by induction on nodes η of ξ that for each sequent $\Gamma' \Rightarrow \Delta'$ at η, $\mathbf{M}, v \models \Gamma'$ and $\mathbf{M}, v \not\models \Delta'$.

Case 1. $\Gamma' \Rightarrow \Delta' = \Gamma_2, \neg A_1 \Rightarrow \Delta_2 \in \eta$. Then, $\Gamma' \Rightarrow \Delta'$ has a direct child node containing $\Gamma_2 \Rightarrow A_1, \Delta_2$. By induction assumption, $\mathbf{M}, v \models \Gamma_2$ and $\mathbf{M}, v \not\models \Delta_2, A_1$. Hence, $\mathbf{M}, v \models \Gamma_2, \neg A_1$ and $\mathbf{M}, v \not\models \Delta_2$.

Case 2. $\Gamma' \Rightarrow \Delta' = \Gamma_2 \Rightarrow \neg B_1, \Delta_2 \in \eta$. Then, $\Gamma' \Rightarrow \Delta'$ has a direct child node containing $\Gamma_2, B_1 \Rightarrow \Delta_2$. By induction assumption, $\mathbf{M}, v \models \Gamma_2, B_1$ and $\mathbf{M}, v \not\models \Delta_2$. Hence, $\mathbf{M}, v \models \Gamma_2$ and $\mathbf{M}, v \not\models \Delta_2, \neg B_1$.

Case 3. $\Gamma' \Rightarrow \Delta' = \Gamma_2, A_1 \wedge A_2 \Rightarrow \Delta_2 \in \eta$. Then, $\Gamma' \Rightarrow \Delta'$ has a direct child node containing $\Gamma_2, A_1 \Rightarrow \Delta_2$ and $\Gamma_2, A_2 \Rightarrow \Delta_2$. By induction assumption, $\mathbf{M}, v \models \Gamma_2, A_1$; $\mathbf{M}, v \models \Gamma_2, A_2$ and $\mathbf{M}, v \not\models \Delta_2$, i.e., $\mathbf{M}, v \models \Gamma_2, A_1 \wedge A_2$ and $\mathbf{M}, v \not\models \Delta_2$.

Case 4. $\Gamma' \Rightarrow \Delta' = \Gamma_2 \Rightarrow B_1 \wedge B_2, \Delta_2 \in \eta$. Then, $\Gamma' \Rightarrow \Delta'$ has a direct child node containing $\Gamma_2 \Rightarrow B_i, \Delta_2$. By induction assumption, $\mathbf{M}, v \models \Gamma_2$ and $\mathbf{M}, v \not\models \Delta_2, B_i$. Hence, $\mathbf{M}, v \models \Gamma_2$ and $\mathbf{M}, v \not\models \Delta_2, B_1 \wedge B_2$.

Case 5. $\Gamma' \Rightarrow \Delta' = \Gamma_2, A_1 \vee A_2 \Rightarrow \Delta_2 \in \eta$. Then, $\Gamma' \Rightarrow \Delta'$ has a direct child node containing $\Gamma_2, A_i \Rightarrow \Delta_2$. By induction assumption, $\mathbf{M}, v \models \Gamma_2, A_i$ and $\mathbf{M}, v \not\models \Delta_2$. Hence, $\mathbf{M}, v \models \Gamma_2, A_1 \vee A_2$ and $\mathbf{M}, v \not\models \Delta_2$.

Case 6. $\Gamma' \Rightarrow \Delta' = \Gamma_2 \Rightarrow B_1 \vee B_2, \Delta_2 \in \eta$. Then, $\Gamma' \Rightarrow \Delta'$ has a direct child node containing $\Gamma_2 \Rightarrow B_1, \Delta_2$ and $\Gamma_2 \Rightarrow B_2, \Delta_2$. By induction assumption, $\mathbf{M}, v \models \Gamma_2$ and $\mathbf{M}, v \not\models \Delta_2, B_1$; $\mathbf{M}, v \not\models \Delta_2, B_2$. Hence, $\mathbf{M}, v \models \Gamma_2$ and $\mathbf{M}, v \not\models \Delta_2, B_1 \vee B_2$.

Case 7. $\Gamma' \Rightarrow \Delta' = \Gamma_2, \forall x A_1(x) \Rightarrow \Delta_2 \in \eta$. Then, $\Gamma' \Rightarrow \Delta'$ has a direct child node containing $\Gamma_2, A_1(d) \Rightarrow \Delta_2$ for each $d \in U$. By induction assumption, $\mathbf{M}, v_{x/d} \models \Gamma_2, A_1(x)$ for each $d \in U$ and $\mathbf{M}, v \not\models \Delta_2$. Hence, $\mathbf{M}, v \models \Gamma_2, \forall x A_1(x)$ and $\mathbf{M}, v \not\models \Delta_2$.

Case 8. $\Gamma' \Rightarrow \Delta' = \Gamma_2 \Rightarrow \forall x B_1(x), \Delta_2 \in \eta$. Then, $\Gamma' \Rightarrow \Delta'$ has a direct child node containing $\Gamma_2 \Rightarrow B_1(c), \Delta_2$. By induction assumption, $\mathbf{M}, v \models \Gamma_2$ and $\mathbf{M}, v \models \Delta_2, B_1(c)$. Hence, $\mathbf{M}, v \models \Gamma_2$ and $\mathbf{M}, v \not\models \Delta_2, \forall x B_1(x)$. □

2.3 Description Logic

Description logic consists of the logical language, the syntax and the semantics defined as follows.

2.3.1 Syntax and Semantics

A logical language for $\mathcal{ALC}$ contains the following symbols:

- (individual) constant symbols: $c_0, c_1, \ldots$;
- atomic concept symbols: $A_0, A_1, \ldots$;
- role symbols: $R_0, R_1, \ldots$;

- concept constructors: $\neg, \sqcap, \sqcup, \exists, \forall$;
- the subsumption relation: $\sqsubseteq$, and
- auxiliary symbols: (,).

Concepts are defined as follows:

$$C ::= A|\neg C|C_1 \sqcap C_2|C_1 \sqcup C_2|\exists R.C|\forall R.C.$$

Primitive statements are:

$$\theta ::= C(c)|R(c,d),$$

where $R(c,d)$ and $A(c)$ are called atomic.

Statements are:

$$\varphi ::= \theta | C \sqsubseteq D.$$

A model **M** is a pair (U, I), where U is a non-empty set (the universe of **M**), and I is an interpretation such that

∘ for any constant c, $I(c) \in U$;
∘ for any atomic concept A, $I(A) \subseteq U$;
∘ for any role R, $I(R) \subseteq U^2$.

The interpretation C^I of a concept C is a subset of U such that

$$C^I = \begin{cases} I(A) & \text{if } C = A \\ U - C_1^I & \text{if } C = \neg C_1 \\ C_1^I \cap C_2^I & \text{if } C = C_1 \sqcap C_2 \\ C_1^I \cup C_2^I & \text{if } C = C_1 \sqcup C_2 \\ \{a \in U : \mathbf{E}b((a,b) \in I(R) \,\&\, b \in C^I)\} & \text{if } C = \exists R.C \\ \{a \in U : \mathbf{A}b((a,b) \in I(R) \Rightarrow b \in C^I)\} & \text{if } C = \forall R.C \end{cases}$$

A statement φ is satisfied in **M**, denoted by $\mathbf{M} \models \varphi$ or $I \models \varphi$, if

$$\begin{cases} I(c) \in C^I & \text{if } \varphi = C(c) \\ (I(c), I(d)) \in I(R) & \text{if } \varphi = R(c,d) \\ C^I \subseteq D^I & \text{if } \varphi = C \sqsubseteq D. \end{cases}$$

A sequent is of form $\Gamma \Rightarrow \Delta$, where Γ, Δ are sets of statements.

Given an interpretation I, we say that I satisfies $\Gamma \Rightarrow \Delta$, denoted by $I \models_{\mathbf{G}^{\mathrm{DL}}} \Gamma \Rightarrow \Delta$, if $I \models \Gamma$ implies $I \models \Delta$, where $I \models \Gamma$ if for each statement $\varphi \in \Gamma$, $I \models \varphi$; and $I \models \Delta$ if for some statement $\psi \in \Delta$, $I \models \psi$.

A sequent $\Gamma \Rightarrow \Delta$ is valid, denoted by $\models_{\mathbf{G}^{\mathrm{DL}}} \Gamma \Rightarrow \Delta$, if for any interpretation I, $I \models_{\mathbf{G}^{\mathrm{DL}}} \Gamma \Rightarrow \Delta$.

In the following subsections, we consider only the primitive statements $C(a)$, $R(a,b)$, and leave $C \sqsubseteq D$ to the last chapter.

2.3.2 Gentzen Deduction System

Gentzen deduction system $\mathbf{G}^{\mathrm{DL}}$ consists of the following axioms and deduction rules. Let Δ, Γ be sets of primitive statements.

- **Axioms:**

$$\Gamma, R(a,b) \Rightarrow R(a,b), \Delta,$$
$$\Gamma, A(a) \Rightarrow A(a), \Delta$$

or equivalently,

$$\frac{\Gamma \cap \Delta \neq \emptyset}{\Gamma \Rightarrow \Delta},$$

where Γ, Δ are sets of atomic statements.

- **Deduction rules for primitive statement parts:**

$$(\neg^L)\ \frac{\Gamma \Rightarrow C(a), \Delta}{\Gamma, \neg C(a) \Rightarrow \Delta} \qquad (\neg^R)\ \frac{\Gamma, C(a) \Rightarrow \Delta}{\Gamma \Rightarrow \neg C(a), \Delta}$$

$$(\sqcap^L_1)\ \frac{\Gamma, C(a) \Rightarrow \Delta}{\Gamma, (C \sqcap D)(a) \Rightarrow \Delta} \qquad (\sqcap^R)\ \frac{\Gamma \Rightarrow C(a), \Delta \ \ \Gamma \Rightarrow D(a), \Delta}{\Gamma \Rightarrow (C \sqcap D)(a), \Delta}$$

$$(\sqcap^L_2)\ \frac{\Gamma, D(a) \Rightarrow \Delta}{\Gamma, (C \sqcap D)(a) \Rightarrow \Delta}$$

$$(\sqcup^L)\ \frac{\Gamma, C(a) \Rightarrow \Delta \ \ \Gamma, D(a) \Rightarrow \Delta}{\Gamma, (C \sqcup D)(a) \Rightarrow \Delta} \qquad (\sqcup^R_1)\ \frac{\Gamma \Rightarrow C(a), \Delta}{\Gamma, \Rightarrow (C \sqcup D)(a), \Delta}$$

$$(\sqcup^R_2)\ \frac{\Gamma \Rightarrow D(a), \Delta}{\Gamma, \Rightarrow (C \sqcup D)(a), \Delta}$$

$$(\forall^L)\ \frac{\Gamma \vdash R(a,d) \ \ \Gamma, C(d) \Rightarrow \Delta}{\Gamma, (\forall R.C)(a) \Rightarrow \Delta} \qquad (\forall^R)\ \frac{R(a,c) \in \Gamma \ \ \Gamma \Rightarrow C(c), \Delta}{\Gamma \Rightarrow (\forall R.C)(a), \Delta}$$

$$(\exists^L)\ \frac{R(a,c) \in \Gamma \ \ \Gamma, C(c) \Rightarrow \Delta}{\Gamma, (\exists R.C)(a) \Rightarrow \Delta} \qquad (\exists^R)\ \frac{\Gamma \vdash R(a,d) \ \ \Gamma \Rightarrow C(d), \Delta}{\Gamma \Rightarrow (\exists R.C)(a), \Delta}$$

where d is a constant and c is a new one, that is, c does not occur in both Γ and Δ. Hence, $R(a,c) \in \Gamma$ means that $R(a,c)$ is enumerated in Γ; and $\Gamma \vdash R(a,d)$ means that $R(a,d)$ is in Γ.

Definition 2.3.1 A sequent $\Gamma \Rightarrow \Delta$ is provable in $\mathbf{G}^{\mathrm{DL}}$, denoted by $\vdash_{\mathbf{G}^{\mathrm{DL}}} \Gamma \Rightarrow \Delta$, if there is a sequence $\Gamma_1 \Rightarrow \Delta_1, \ldots, \Gamma_n \Rightarrow \Delta_n$ of sequents such that $\Gamma_n \Rightarrow \Delta_n = \Gamma \Rightarrow \Delta$, and for each $1 \leq i \leq n$, $\Gamma_i \Rightarrow \Delta_i$ is an axiom or is deduced from the previous sequents by one of the deduction rules in $\mathbf{G}^{\mathrm{DL}}$.

For example, the following sequence of sequents is a deduction in $\mathbf{G}^{\mathrm{DL}}$:

$$\begin{array}{l} C(c) \Rightarrow C(c) \\ \Gamma \vdash R(a,c) \ \ (\forall R.C)(a) \Rightarrow R(a,c) \in \Gamma \ \ C(c) \\ (\forall R.C)(a) \Rightarrow\Rightarrow R(a,c) \in \Gamma \ \ C(c) \\ (\forall R.C)(a) \Rightarrow (\exists R.C)(a). \end{array}$$

Theorem 2.3.2 (Soundness theorem) *For any sequent* $\Gamma \Rightarrow \Delta$, *if* $\vdash_{G^{DL}} \Gamma \Rightarrow \Delta$ *then* $\models_{G^{DL}} \Gamma \Rightarrow \Delta$.

Proof We prove that each axiom is valid and each deduction rule preserves the validity.

Fix an interpretation (U, I).

To verify the validity of the axioms, it is clear that

$$I \models \Gamma, C(a) \text{ implies } I \models C(a), \Delta;$$
$$I \models \Gamma, R(a, b) \text{ implies } I \models R(a, b), \Delta.$$

To verify that $(\neg^L)$ preserves the validity, assume that $I \models \Gamma$ implies $I \models C(a), \Delta$. Assume that $I \models \Gamma, \neg C(a)$. Then by induction assumption, $I \models C(a), \Delta$. Because $I \models \neg C(a), I \models \Delta$.

To verify that $(\sqcap_1^L)$ preserves the validity, assume that $I \models \Gamma, C(a)$ implies $I \models \Delta$. Assume that $I \models \Gamma, (C \sqcap D)(a)$. Then $a^I \in C^I \cap D^I$, and hence, $a^I \in C^I$, i.e., $I \models C(a)$, and hence, $I \models \Gamma, C(a)$, and by induction assumption, $I \models \Delta$.

To verify that $(\sqcup^L)$ preserves the validity, assume that

$$I \models \Gamma, C(a) \text{ implies } I \models \Delta;$$
$$I \models \Gamma, D(a) \text{ implies } I \models \Delta.$$

Assume that $I \models \Gamma, (C \sqcup D)(a)$. Then either $a^I \in C^I$ or $a^I \in D^I$. In either case, by the assumption, we have $I \models \Delta$.

To verify that $(\forall^L)$ preserves the validity, assume that

$$\text{R(a,d)} \in \Delta;$$
$$I \models \text{C(d) implies } I \models \Delta.$$

If $I \models \Gamma, (\forall R.C)(a)$ then for any d, if $R(a, d) \in \Delta$ then $I \models C(d)$. By induction assumption, $I \models \Delta$.

To verify that $(\forall^R)$ preserves the validity, assume that $R(a, c) \in \Gamma$, and $I \models \Gamma$ implies $I \models C(c), \Delta$. Assume that $I \models \Gamma$. Then by induction assumption, either $I \models \Delta$ or $I \models C(c)$. If $I \models \Delta$ then $I \models (\forall R.C)(a), \Delta$; otherwise, assume that $I \not\models (\forall R.C)(a)$. Then, for some constant d, $I \models R(a, d)$ and $I \not\models C(d)$, a contradiction to the assumption $I \models C(d)$.

To verify that $(\exists^L)$ preserves the validity, assume that for any interpretation I, $R(a, c) \in \Gamma$ and $I \models \Gamma, C(c)$ implies $I \models \Delta$. For any interpretation I, if $I \models \Gamma, (\exists R.C)(a)$ then there is a constant symbol d such that $R(a, d) \in \Gamma$ and $I \models C(d)$. By induction assumption, $I \models \Delta$.

To verify that $(\exists^R)$ preserves the validity, assume that for any interpretation I, $\Delta \vdash R(a, d)$ and $I \models \Gamma$ implies $I \models C(d), \Delta$. For any interpretation I, if $I \models \Gamma$ then by induction assumption, $I \models C(d), \Delta$. There are two cases: if $I \models \Delta$ then $I \models (\exists R.C)(a), \Delta$; and if $I \models C(d)$ then $I \models (\exists R.C)(a)$, and hence, $I \models (\exists R.C)(a), \Delta$. □

2.3.3 Completeness Theorem

Theorem 2.3.3 (Completeness theorem) *For any sequent* $\Gamma \Rightarrow \Delta$, *if* $\models_{\mathbf{G}^{\mathrm{DL}}} \Gamma \Rightarrow C$ *then* $\vdash_{\mathbf{G}^{\mathrm{DL}}} \Gamma \Rightarrow C$.

Proof For each sequent δ, we define a finite tree, called the reduction tree for δ, denoted by $T(\delta)$, from which we can obtain either a proof of δ or a show of the invalidity of δ.

This reduction tree $T(\delta)$ for δ contains a sequent at each node, and is constructed in stages as follows.

Stage 0: $T_0(\delta) = \{\delta\}$.

Stage $k(k > 0) : T_k(\delta)$ is defined by cases.

Case 0. If Γ and Δ have any statement in common, write nothing above $\Gamma \Rightarrow \Delta$.

Case 1. Every topmost sequent $\Gamma \Rightarrow \Delta$ in $T_{k-1}(\delta)$ has the same statement in Γ and Δ. Then, stop.

Case 2. Not case 1. $T_k(\delta)$ is defined as follows.

Subcase $(\sqcap^L)$. Let $(C_1 \sqcap D_1)(a_1), \ldots, (C_n \sqcap D_n)(a_n)$ be all the statements in Γ whose outermost logical symbol is $\sqcap$, and to which no reduction has been applied in previous stages by any $(\sqcap^L)$. Then, write down

$$\Gamma, C_1(a_1), D_1(a_1), \ldots, C_n(a_n), D_n(a_n) \Rightarrow \Delta$$

above $\Gamma \Rightarrow \Delta$. We say that a $(\sqcap^L)$ reduction has been applied to $(C_1 \sqcap D_1)(a_1), \ldots,$ $(C_n \sqcap D_n)(a_n)$.

Subcase $(\sqcap^R)$. Let $(C_1 \sqcap D_1)(a_1), \ldots, (C_n \sqcap D_n)(a_n)$ be all the statements in Δ whose outermost logical symbol is $\sqcap$, and to which no reduction has been applied in previous stages by any $(\sqcap^R)$. Then, write down all sequents of the form

$$\Gamma, E_1(a_1), \ldots, E_n(a_n) \Rightarrow \Delta$$

above $\Gamma \Rightarrow \Delta$, where E_i is either C_i or D_i. We say that a $(\sqcap^R)$ reduction has been applied to $(C_1 \sqcap D_1)(a_1), \ldots, (C_n \sqcap D_n)(a_n)$.

Similarly for cases $(\sqcup^R)$ and $(\sqcup^L)$.

Subcase $(\forall^L)$. Let $\forall R_1.C_1(a_1), \ldots, \forall R_n.C_n(a_n)$ be all the formulas in Γ whose outermost logical symbol is $\forall$, to which no reduction has been applied in previous stages. Then, let $b_1, \ldots, b_n$ be n-many new constant symbols having used before the current stage, and write down all the sequents of the form

$$\Gamma, E_1, \ldots, E_n \Rightarrow \Delta$$

above $\Gamma \Rightarrow \Delta$, where E_i is either $\neg R_i(a_i, b_i)$ or $C_i(b_i)$. We say that a $(\forall^L)$ reduction has been applied to $\forall R_1.C_1(a_1), \ldots, \forall R_n.C_n(a_n)$.

Subcase $(\exists^L)$. Let $\exists R_1.C_1(a_1), \ldots, \exists R_n.C_n(a_n)$ be all the formulas in Γ whose outermost logical symbol is $\exists$, let b_i be the first constant at this stage which has not been used for a reduction of $\exists R_i.C_i(a_i)$ and $R_i(a_i, b_i) \in \Delta, i \leq n$. Then write down

$$\Gamma, C_1(b_1), \ldots, C_n(b_n) \Rightarrow \Delta$$

above $\Gamma \Rightarrow \Delta$. We say that a $(\exists^L)$ reduction has been applied to $\exists R_1.C_1(a_1), \ldots, \exists R_n.C_n(a_n)$.

Similarly for cases $(\forall^R)$ and $(\exists^R)$.

Subcase $\neg^L$. Let $\Gamma \Rightarrow \Delta$ be any topmost sequent of the tree which has been defined by stage $k-1$. Let $\neg C_1(a_1), \ldots, \neg C_n(a_n)$ be all the formulas in Γ whose outmost logical symbol is $\neg$, and to which no reduction has been applied in previous stages. Then, write down

$$\Gamma \Rightarrow C_1(a_1), \ldots, C_n(a_n), \Delta$$

above $\Gamma \Rightarrow \Delta$. We say that a $\neg^L$ reduction has been applied to $\neg C_1(a_1), \ldots, \neg C_n(a_n)$.

Subcase $\neg^R$. Let $\neg C_1(a_1), \ldots, \neg C_n(a_n)$ be all the formulas in Δ whose outermost logical symbol is $\neg$, and to which no reduction has been applied in previous stages. Then, write down

$$\Gamma, C_1(a_1), \ldots, C_n(a_n) \Rightarrow \Delta$$

above $\Gamma \Rightarrow \Delta$. We say that a $\neg^R$ reduction has been applied to $\neg C_1(a_1), \ldots, \neg C_n(a_n)$.

The collection of those sequents with the partial order is the reduction tree for δ, denoted by $T(\delta)$.

A sequence $\delta_0, \delta_1, \ldots$ of sequents in $T(\delta)$ is a branch if $\delta_0 = \delta$, δ_{i+1} is immediately above δ_i.

Given a sequent δ, if each branch of $T(\delta)$ is ended with a sequent containing common statements, then it is a routine to construct a proof of δ.

Otherwise, there is a branch $\gamma = \delta_1, \ldots, \delta_n$ of $T(\delta)$ such that there is no rule applicable for δ_n and $\delta_n = \Gamma_n \Rightarrow \Delta_n$ has no common statements in Γ_n and Δ_n.

Let

$$\begin{aligned}\bigcup\Gamma' &= \{C(a), R(a,b), \neg R(a,b) \in \Gamma_i : \Gamma_i \Rightarrow \Delta_i \in \gamma\},\\ \bigcup\Delta' &= \{D(a), R(a,b), \neg R(a,b) \in \Delta_i : \Gamma_i \Rightarrow \Delta_i \in \gamma\}.\end{aligned}$$

We define U to be the set of all the constants occurring in $\cup\Gamma'$ or $\bigcup\Delta'$, and an interpretation I such that for any $a, b \in U$,

$$\begin{aligned}&a \in C^I \text{ iff } C(a) \in \bigcup\Gamma',\\ &(a,b) \in R^I \text{ iff } R(a,b) \in \bigcup\Gamma' \text{ or } \neg R(a,b) \in \bigcup\Delta'.\end{aligned}$$

We prove by induction on δ_n that for any $i \leq n$,

$$I \not\models \delta_i.$$

If $\delta_i = \Gamma' \Rightarrow \Delta'$ is a leaf in $T(\delta)$ then by assumption, there is no literal l common in Γ' and Δ', and for any statement $A(a) \in \Gamma'$, by the definition of I, $A(a) \notin \Delta'$, and $I \models A(a)$, and hence, $I \models \Gamma$; and for any statement $A(a) \in \Delta'$, by the definition of I, $A(a) \notin \Gamma'$, and $I \not\models A(a)$, and hence, $I \not\models \Delta'$;

If $\delta_i = \Gamma' \Rightarrow \Delta'$ is not a leaf and has β as a successor then there are statements $C_1(a), C_2(a)$ such that $\beta = \Gamma'', C_1(a), C_2(a) \Rightarrow \Delta'$ and β is the result of applying $(\sqcap_1^L)$ to δ_i, where $\Gamma' = \Gamma'' \cup \{(C_1 \sqcap C_2)(a)\}$. By induction assumption, $I \not\models \Gamma'', C_1(a), C_2(a) \Rightarrow \Delta'$. Therefore, $I \not\models \Gamma'', (C_1 \sqcap C_2)(a) \Rightarrow \Delta'$.

If $\delta_i = \Gamma' \Rightarrow \Delta'$ is not a leaf and has β_1, β_2 as successors then there are statements $D_1(a), D_2(a)$ such that

$$\beta_1 = \Gamma' \Rightarrow D_1(a), \Delta'',$$
$$\beta_2 = \Gamma' \Rightarrow D_2(a), \Delta'',$$

and β_1, β_2 are the results of applying $(\sqcap^R)$ to δ_i, where $\Delta' = \Delta'' \cup \{(D_1 \sqcap D_2)(a)\}$. There is $j \in \{1, 2\}$ such that $\beta_j \in \delta_{i+1}$, and by induction assumption, $I \not\models \Gamma' \Rightarrow D_j(a), \Delta''$. Hence, $I \not\models \Gamma' \Rightarrow (D_1 \sqcap D_2)(a), \Delta''$.

If $\delta_i = \Gamma' \Rightarrow \Delta'$ is not a leaf and has β_1, β_2 as successors and β_1, β_2 are the results of applying $(\forall^L)$ to δ_i, then there are statements $\neg R(a, d), C(d)$ such that

$$\beta_1 = \Gamma'', \neg R(a, d) \Rightarrow \Delta',$$
$$\beta_2 = \Gamma'', C(d) \Rightarrow \Delta',$$

where $\Gamma' = \Gamma'' \cup \{(\forall R.C)(a)\}$. By the construction of $T(\delta)$, for each constant c, either $\Gamma''', \neg R(a, c) \Rightarrow \Delta'' \in \delta_j (j \geq i)$ or $\Gamma''', C(c) \Rightarrow \Delta'' \in \delta_j (j \geq i)$. By induction assumption, for each constant c, either $I \models \neg R(a, c)$ or $I \models C(c)$, i.e., $I \models (\forall R.C)(a)$. Therefore, $I \not\models \Gamma'', (\forall R.C)(a) \Rightarrow \Delta'$.

If $\delta_i = \Gamma' \Rightarrow \Delta'$ is not a leaf and has β as a successor and β is the result of applying $(\forall^L)$ to δ_i, then there are statements $\neg R(a, c), D(c)$ such that

$$\beta = \Gamma' \Rightarrow \neg R(a, c), D(c), \Delta'',$$

where $\Delta' = \Delta'' \cup \{(\forall R.D)(a)\}$. By induction assumption, $I \not\models \Gamma' \Rightarrow \neg R(a, c), D(c), \Delta''$. Hence, $I \not\models \Gamma' \Rightarrow (\forall R.D)(a), \Delta''$.

Similar for other cases.

This completes the proof.

References

1. A. Avron, I. Lev, Canonical propositional Gentzen-type systems, in *IJCAR 2001*, pp. 529–544
2. F. Baader, D. Calvanese, D.L. McGuinness, D. Nardi, P.F. Patel-Schneider, *The Description Logic Handbook: Theory, Implementation, Applications* (Cambridge University Press, Cambridge, 2003)
3. F. Baader, I. Horrocks, U. Sattler, Chapter 3 Description Logics, in *Handbook of Knowledge Representation*, eds. by F. van Harmelen, V. Lifschitz, B. Porter (Elsevier, Amsterdam, 2007)
4. J. Barwise, An introduction to first-order logic, in *Handbook of Mathematical Logic. Studies in Logic and the Foundations of Mathematics*, ed. by J. Barwise (North-Holland, Amsterdam, 1982)

5. H. Ebbinghaus, J. Flum, W. Thomas, *Mathematical Logic*. Undergraduate Texts in Mathematics, 2nd edn. (Springer, Berlin, 1994)
6. D. Fensel, F. van Harmelen, I. Horrocks, D. McGuinness, P.F. Patel-Schneider, OIL: an ontology infrastructure for the semantic web. IEEE Intell. Syst. **16**(2), 38–45 (2001)
7. E. Franconi, Description logics for natural language processing, AAAI Technical Report FS-94-04, 37-44
8. F. van Harmelen, V. Lifschitz, B. Porter, *Handbook of Knowledge Representation* (Elsevier, Amsterdam, 2008)
9. M. Hofmann, Proof-theoretic approach to description logic, in *Proceedings of LICS 2005*, pp. 229–237
10. I. Horrocks, U. Sattler, Ontology reasoning in the SHOQ(D) description logic, in *Proceedings of the Seventeenth International Joint Conference on Artificial Intelligence* (2001)
11. I. Horrocks, P.F. Patel-Schneider, F. van Harmelen, From SHIQ and RDF to OWL: the making of a web ontology language. J. Web Seman. **1**(1), 7–26 (2003)
12. W. Li, *Mathematical Logic, Foundations for Information Science*. Progress in Computer Science and Applied Logic, vol. 25 (Birkhäuser, Basel, 2010)
13. A. Rademaker, *A Proof Theory for Description Logics*. Briefs in Computer Science (Springer, Berlin, 2012)
14. G. Takeuti, *Proof Theory*, in *Handbook of Mathematical Logic. Studies in Logic and the Foundations of Mathematics*, eds. by J. Barwise (North-Holland, Amsterdam, 1987)

Chapter 3
R-Calculi for Propositional Logic

AGM postulates[1–3] are a set of requirements a revision operator should satisfy, and R-calculus is a revision operator which satisfies the AGM postulates.

R-calculus [5–10] is a Gentzen-typed deduction system to deduce a consistent one $\Gamma' \cup \Delta$ from an inconsistent theory $\Gamma \cup \Delta$ of first-order logic, where $\Gamma' \cup \Delta$ should be a maximal consistent subtheory of $\Gamma \cup \Delta$ (defined in the next section as a minimal change of Γ by Δ), where $\Delta|\Gamma$ is called an R-configuration, Γ is a consistent set of formulas, and Δ is a consistent sets of atomic formulas or the negation of atomic formulas. It was proved that if $\Delta|\Gamma \Rightarrow \Delta|\Gamma'$ is deducible and $\Delta|\Gamma'$ is an R-termination, i.e., there is no R-rule to reduce $\Delta|\Gamma'$ to another R-configuration $\Delta|\Gamma''$, then $\Delta \cup \Gamma'$ is a contraction of Γ by Δ.

AGM-postulates are set for using a formula A to revise a theory K, so that if $K \circ A \Rightarrow K'$ then K' is a maximal consistent subset of $K \cup \{A\}$. Therefore, for AGM postulates, there is an iterated revision and the corresponding axioms are called DP-axioms. R-calculus is to use a theory Δ to revise a formula A, so that if A is consistent with Δ then add A in Δ; otherwise, do nothing.

$\subseteq$-minimal change[4, 11] is with respect to the set-inclusion. That is, if $\Delta|\Gamma \Rightarrow \Delta, \Theta$ be provable in an R-calculus then, Θ is a minimal change of Γ by Δ, that is, (i) $\Theta \cup \Delta$ is consistent, (ii) $\Theta \subseteq \Gamma$, and (iii) for any Θ' with $\Theta \subset \Theta' \subseteq \Gamma$, $\Theta' \cup \Delta$ is inconsistent. Correspondingly, we have the following reduction rule in R-calculus:

$$\frac{\Delta|A_1, \Gamma \Rightarrow \Delta|\Gamma \text{ or } \Delta, A_1|A_2, \Gamma \Rightarrow \Delta, A_1|\Gamma}{\Delta|A_1 \wedge A_2, \Gamma \Rightarrow \Delta|\Gamma},$$

which means that if A_1 is inconsistent with Δ or A_2 is inconsistent with $\Delta \cup \{A_1\}$ then $A_1 \wedge A_2$ is eliminated from $\{A_1 \wedge A_2\} \cup \Gamma$ revised by Δ, even though it may be the case that $\Delta \cup \{A_1\}$ is consistent.

By traditional belief revision, we revise a theory Γ by another theory Δ to eliminate formulas in Γ (or in the theory $\mathrm{Th}(\Gamma)$ of Γ) which negations are deducible by Δ, to make remaining Γ consistent with Δ. To make the result $\Delta|\Gamma$ be a minimal change of

W. Li and Y. Sui, *R-CALCULUS: A Logic of Belief Revision*, Perspectives in Formal Induction, Revision and Evolution,
https://doi.org/10.1007/978-981-16-2944-0_3

Γ by Δ, a (well-ordered) linear ordering of Γ is requited to correspond to a minimal change of Γ by Δ.

A simple way to make $A \circ \Delta$ or $\Delta|A$ consistent is that if Δ is inconsistent with A then set $\Delta|A \Rightarrow \Delta$, i.e., $A \circ \Delta = \Delta$. This condition seems a little strong. For example, to use p to revise $\neg p \wedge q$, a simple result is p. To ensure the minimal change, $\{p, q\}$ is a better result, because q is a minimal change of $\{\neg p, q\}$ by $\{p\}$, or theory $\{q\}$ is a minimal change of theory $\{\neg p, q\}$ by $\{p\}$, such that $\{q\}$ is consistent with $\{p\}$.

Another problem is that AGM postulates are based on deduction of a based logic, and so does R-calculus. To revise Γ by Δ we firstly need decide whether $\Gamma \cup \Delta$ is inconsistent, and if yes, to revise Γ by Δ; otherwise, set $\Delta|\Gamma$ to be $\Delta \cup \Gamma$. We attempt to reduce deciding the inconsistence of $\Gamma \cup \Delta$ to deciding $\Delta \vdash l$ for some literal l (a propositional variable or the negation of a propositional variable).

In traditional belief revision, subset minimal change is considered only. We will give an R-calculus **T** to deduce a configuration $\Delta|\Gamma$ into a consistent theory Δ, Θ by eliminating literals or subformulas in Γ which is inconsistent with Δ, such that (1) Δ, Θ is consistent; (2) Θ is a pseudo-subtheory of Γ (denoted by $\Theta \preceq \Gamma$), that is, each formula in Θ is a pseudo-subformula of some formula A in Γ, denoted by $\Theta \preceq \Gamma$; and (3) Θ is a pseudo-subformula minimal change, i.e., for any Ξ with $\Theta \prec \Xi \preceq \Gamma$, Δ is inconsistent with Ξ. This is $\preceq$-minimal change.

Another minimal change is $\vdash_{\preceq}$-minimal change, and we will given an R-calculus **U** to deduce a configuration $\Delta|\Gamma$ into a consistent theory Δ, Θ by eliminating literals or subformulas in Γ which is inconsistent with Δ, such that (1) Δ, Θ is consistent; (2) Θ is a pseudo-subtheory of Γ, and (3) Θ is a $\vdash_{\preceq}$-minimal change, i.e., for any Ξ with $\Xi \prec \Theta$, either $\Delta, \Xi \vdash \Theta$ and $\Delta, \Theta \vdash \Xi$, or Δ is inconsistent with Ξ.

Because deduction relation $\vdash$ is decidable in propositional logic and undecidable in first-order logic, we consider R-calculus for propositional logic in this chapter for its simplicity.

We should notice the difference between R-calculus and traditional logics in the following point: we take formulas A as the first-class objects, and the deduction relation statements $\Delta \vdash \neg A$ as the second-class objects. In traditional logics, only the first-class objects are considered as the syntactical objects; and the statements of form $\Gamma \vdash \neg A$ as the meta-language objects. In R-calculus, $\Gamma \vdash \neg A$ is taken as a syntactical objects of the second-class, and the deduction rules are about the second-class objects. In another words, R-calculus is based on $\Gamma \vdash \neg A$ to reduce $\Delta|\Gamma$ into a consistent theory Δ, Θ.

This chapter is organized as follows: in the next section we define three kinds of the minimal change: $\subseteq$-minimal change, $\preceq$-minimal change and $\vdash_{\preceq}$-minimal change; the second section gives an R-calculus **S** which is sound and complete with respect to $\subseteq$-minimal change; the third section define an R-calculus **T** which is sound and complete with respect to $\preceq$-minimal change, and the last section gives an R-calculus **U** which is sound and complete with respect to $\vdash_{\preceq}$-minimal change.

3.1 Minimal Changes

A theory Θ (a set of formulas) is maximally consistent if Θ is consistent, and for each formula A, either $\Theta \vdash A$ or $\Theta \vdash \neg A$. Therefore, if Θ is maximally consistent then for any formula A with $\Theta \nvdash A$, $\Theta \cup \{A\}$ is inconsistent.

In propositional logic, set $\{A : v(A) = 1\}$ of true formulas under an assignment v is maximally consistent; and conversely, for any maximally consistent set Θ of formulas, there is an assignment v such that $\{A : v(A) = 1\} = \Theta$. Based on this, completeness theorem of propositional logic is proved.

If Θ is a $\subseteq$-minimal change of Γ by Δ, then for any $A \in \Gamma$, $\Theta \cup \Delta \nvdash A$ implies $\Theta \cup \Delta \vdash \neg A$. Therefore, we call Θ as a $\subseteq$-maximal consistent set of Γ by Δ.

We will define three kinds of the minimal changes: $\subseteq$-minimal change, based on the subset relation $\subseteq$ between theories; $\preceq$-minimal change, based on the pseudo-subformula relation $\preceq$, and $\vdash_{\preceq}$-minimal change, based on the deduction relation $\vdash$ between theories with the pseudo-subformula relation $\preceq$.

3.1.1 Subset-Minimal Change

Definition 3.1.1 Given any consistent theories Γ and Δ, a theory Θ is a subset-minimal ($\subseteq$-minimal) change of Γ by Δ, denoted by $\models_{S} \Delta|\Gamma \Rightarrow \Delta, \Theta$, if (i) $\Theta \subseteq \Gamma$; (ii) Θ is consistent with Δ, and (iii) for any set Ξ with $\Theta \subset \Xi \subseteq \Gamma$, Ξ is inconsistent with Δ.

By definition we have the following

Proposition 3.1.2 *(i) Let Θ be a $\subseteq$-minimal change of Γ by Δ. Then, for any formula $A \in \Gamma$,*

$$\textit{either } \Delta, \Theta \vdash A \textit{ or } \Delta, \Theta \vdash \neg A.$$

(ii) Conversely, if Θ is minimal such that (a) $\Gamma \supseteq \Theta$; (b) Θ is consistent with Δ, and (c) for any formula $A \in \Gamma$, either $\Delta, \Theta \vdash A$ or $\Delta, \Theta \vdash \neg A$ then Θ is a $\subseteq$-minimal change of Γ by Δ. □

The process for computing a $\subseteq$-minimal change Θ of Γ by Δ. Let $\Gamma = \{A_0, A_1, \ldots\}$ be an ordering with a well-ordered ordering $<$, i.e., assume that $A_0 < A_1 < A_2 \cdots$. Define

$$\Theta_0 = \emptyset;$$
$$\Theta_{n+1} = \begin{cases} \Theta_n \cup \{A_n\} & \text{if } \Delta, \Theta_n \nvdash \neg A_n \\ \Theta_n & \text{otherwise;} \end{cases}$$
$$\Theta = \bigcup_n \Theta_n.$$

Then, Θ is a $\subseteq$-minimal change of Γ by Δ, denoted by $\Theta = \sigma(\Gamma^{<}, \Delta)$.

Proposition 3.1.3 $\Theta = \sigma(\Gamma^{<}, \Delta)$ *is a* $\subseteq$*-minimal change of* Γ *by* Δ.

Proof For any formula $A \in \Gamma - \Theta$, by the definition, there is an n such that $A_n = A$, and by the assumption, $\Delta, \Theta_n \vdash \neg A_n$, and by the monotonicity of $\vdash$, $\Delta, \Theta \vdash \neg A_n$, i.e., $\Delta, \Theta \vdash \neg A$. □

Conversely, we have

Proposition 3.1.4 *Given a* $\subseteq$*-minimal change* Θ *of* Γ *by* Δ, *there is an ordering* $<$ *on* Γ *such that*

$$\Theta = \sigma(\Gamma^{<}, \Delta).$$

Proof Given a $\subseteq$-minimal change Θ of Γ by Δ and any ordering $<_1$ on Θ and any ordering $<_2$ on $\Gamma - \Theta$, define an ordering $<$ on Γ such that for any $A, B \in \Gamma$,

$$A < B \text{ iff } \begin{cases} A <_1 B & \text{if } A, B \in \Theta \\ A <_2 B & \text{if } A, B \in \Gamma - \Theta \\ \top & \text{if } A \in \Theta \text{ and } B \in \Gamma - \Theta \end{cases}$$

where $\top$ means *true*. Then, it can be proved that $\Theta = \sigma(\Gamma^{<}, \Delta)$. □

3.1.2 Pseudo-Subformulas-Minimal Change

Subformulas are a basic concept in traditional logics.

Definition 3.1.5 Given a formula A, a formula B is a sub-formula of A, denoted by $B \leq A$, if either $A = B$, or

(i) if $A = \neg A_1$ then $B \leq A_1$;
(ii) if $A = A_1 \vee A_2$ or $A_1 \wedge A_2$ then either $B \leq A_1$ or $B \leq A_2$.

For example, let $A = (p \vee q) \wedge (r \vee s)$. Then,

$$p \vee q, r \vee s \leq A;$$

and

$$p \wedge r, q \wedge r, p \wedge (r \vee s) \nleq A.$$

Definition 3.1.6 Given a formula $A[B_1, \ldots, B_n]$, where $[B_1]$ is an occurrence of B_1 in A, a formula $B = A[\lambda, \ldots, \lambda] = A[B_1/\lambda, \ldots, B_n/\lambda]$, where the occurrence B_i is replaced by the empty formula λ, is called a pseudo-subformula of A, denoted by $B \preceq A$.

For example, let $A = (p \vee q) \wedge (r \vee s)$. Then,

$$p \vee q, r \vee s, p \wedge r, q \wedge r, p \wedge (r \vee s) \preceq A.$$

Proposition 3.1.7 *For any formulas A_1, A_2, B_1 and B_2,*

(i) $B_1 \leq A_1$ *implies* $B_1 \leq A_1 \vee A_2$ *and* $B_1 \leq A_1 \wedge A_2$;
(ii) $B_1 \preceq A_1$ *and* $B_2 \preceq A_2$ *imply* $\neg B_1 \preceq \neg A_1$, $B_1 \vee B_2 \preceq A_1 \vee A_2$ *and* $B_1 \wedge B_2 \preceq A_1 \wedge A_2$. □

Proposition 3.1.8 *For any formulas A and B, if $B \leq A$ then $B \preceq A$.*

Proof By induction on the structure of A. □

Proposition 3.1.9 *$\leq$ and $\preceq$ are partial orderings on the set of all the formulas.* □

Given a formula A, let $P(A)$ be the set of all the pseudo-subformulas of A. Each $B \in P(A)$ is determined by a set $\tau(B) = \{[p_1], \ldots, [p_n]\}$, where each $[p_i]$ is an occurrence of p_i in A, such that

$$B = A[p_1]/\lambda, \ldots, [p_n]/\lambda).$$

Given any $B_1, B_2 \in P(A)$, define

$$B_1 \otimes B_2 = \max\{B : B \preceq B_1, B \preceq B_2\};$$
$$B_1 \oplus B_2 = \min\{B : B \succeq B_1, B \succeq B_2\}.$$

Proposition 3.1.10 *For any pseudo-subformulas $B_1, B_2 \in P(A)$, $B_1 \otimes B_2$ and $B_1 \oplus B_2$ exist.* □

Let $\mathbf{P}(A) = (P(A), \otimes, \oplus, A, \lambda)$ be the lattice with the greatest element A and the least element λ.

Proposition 3.1.11 *For any pseudo-subformulas $B_1, B_2 \in P(A)$, $B_1 \preceq B_2$ if and only if $\tau(B_1) \supseteq \tau(B_2)$. Moreover,*

$$\tau(B_1 \otimes B_2) = \tau(B_1) \cup \tau(B_2);$$
$$\tau(B_1 \oplus B_2) = \tau(B_1) \cap \tau(B_2).$$

□

Definition 3.1.12 A theory Θ is a $\preceq$-minimal change of Γ by Δ, denoted by $\models_{\mathbf{T}} \Delta|\Gamma \Rightarrow \Delta, \Theta$, if

(i) $\Theta \preceq \Gamma$, that is, for each formula $A \in \Theta$, there is a formula $B \in \Gamma$ such that $A \preceq B$,
(ii) $\Theta \cup \Delta$ is consistent, and
(iii) for any theory Ξ with $\Theta \prec \Xi \preceq \Gamma$, $\Xi \cup \Delta$ is inconsistent.

3.1.3 Deduction-Based Minimal Change

Before giving $\vdash_{\preceq}$-minimal change, we give the following

Definition 3.1.13 Given three theories Γ, Δ and Θ, theory Θ is a deduction-minimal ($\vdash$-minimal) change of Γ by Δ, denoted by $\models_{\mathbf{U}'} \Gamma|\Delta \Rightarrow \Delta, \Theta$, if

(i) $\Gamma \vdash \Theta$,
(ii) $\Theta \cup \Delta$ is consistent, and
(iii) for any theory Ξ with $\Gamma \vdash \Xi \vdash \Theta$ and $\Theta \nvdash \Xi$, $\Xi \cup \Delta$ is inconsistent.

The lattice of all the theories under the deduction relation $\vdash$ is dense, that is, for any theories Γ_1, Γ_2, if $\Gamma_1 \vdash \Gamma_2$ and $\Gamma_2 \nvdash \Gamma_1$ then there is a theory Γ such that $\Gamma_1 \vdash \Gamma \vdash \Gamma_2$ and $\Gamma_2 \nvdash \Gamma \nvdash \Gamma_1$. Hence, we cannot find a $\vdash$-minimal change of Γ by Δ.

Therefore, we consider the deduction-based minimal change.

Definition 3.1.14 Given three theories Γ, Δ and Θ, theory Θ is a deduction-based minimal ($\vdash_{\preceq}$-minimal) change of Γ by Δ, denoted by $\models_{\mathbf{U}} \Gamma|\Delta \Rightarrow \Delta, \Theta$, if

(i) $\Theta \cup \Delta$ is consistent;
(ii) $\Theta \preceq \Gamma$, and
(iii) for any theory Ξ with $\Gamma \succeq \Xi \succ \Theta$, either $\Delta, \Xi \vdash \Theta$ and $\Delta, \Theta \vdash \Xi$, or $\Xi \cup \Delta$ is inconsistent.

3.2 R-Calculus for $\subseteq$-Minimal Change

In this section, we will give an R-calculus based on $\subseteq$-minimal change of Γ by Δ, and an Gentzen-typed R-calculus **S** such that for any consistent theories Γ, Δ and Θ, $\Delta|\Gamma \Rightarrow \Theta$ is provable in **S** if and only if Θ is a $\subseteq$-minimal change of Γ by Δ.

3.2.1 R-Calculus **S** *for a Formula*

R-calculus **S** for a formula A consists of the following axioms and deduction rules:

- **Axioms:**

$$(S^{\mathbf{A}})\ \frac{\Delta \nvdash \neg l}{\Delta|l \Rightarrow \Delta, l|} \quad (S_{\mathbf{A}})\ \frac{\Delta \vdash \neg l}{\Delta|l \Rightarrow \Delta|}$$

- **Deduction rules:**

$$(S^{\wedge})\ \frac{\Delta|A_1 \Rightarrow \Delta, A_1| \quad \Delta, A_1|A_2 \Rightarrow \Delta, A_1, A_2|}{\Delta|A_1 \wedge A_2 \Rightarrow \Delta, A_1 \wedge A_2|} \qquad (S^1_{\wedge})\ \frac{\Delta|A_1 \Rightarrow \Delta|}{\Delta|A_1 \wedge A_2 \Rightarrow \Delta|}$$

$$(S^2_{\wedge})\ \frac{\Delta, A_1|A_2 \Rightarrow \Delta, A_1}{\Delta|A_1 \wedge A_2 \Rightarrow \Delta}$$

$$(S^{\vee}_1)\ \frac{\Delta|A_1 \Rightarrow \Delta, A_1|}{\Delta|A_1 \vee A_2 \Rightarrow \Delta, A_1 \vee A_2} \qquad (S_{\vee})\ \frac{\Delta|A_1 \Rightarrow \Delta \quad \Delta|A_2 \Rightarrow \Delta}{\Delta|A_1 \vee A_2 \Rightarrow \Delta}$$

$$(S^{\vee}_2)\ \frac{\Delta|A_2 \Rightarrow \Delta, A_2|}{\Delta|A_1 \vee A_2 \Rightarrow \Delta, A_1 \vee A_2}$$

where the rules (S^*) of the left side are to put a formula of Γ into Θ, and the ones (S_*) of the right side are not to.

Definition 3.2.1 $\Delta|A \Rightarrow \Delta, C$ is provable in **S**, denoted by $\vdash_{\mathbf{S}} \Delta|A \Rightarrow \Delta, C$, if there is a sequence $S_1, \ldots, S_m$ of statements such that

$$\begin{aligned} S_1 &= \Delta|A \Rightarrow \Delta|A_1, \\ S_2 &= \Delta|A_1 \Rightarrow \Delta|A_2, \\ &\ldots \\ S_m &= \Delta|A_{m-1} \Rightarrow \Delta, C; \end{aligned}$$

and for each $i < m$, S_{i+1} is an axiom or is deduced from the previous statements by a deduction rule in **S**.

Theorem 3.2.2 (Soundness theorem) *For any consistent formula set Δ and formula A, if $\Delta|A \Rightarrow \Delta, A$ is provable in* **S** *then $\Delta \cup \{A\}$ is consistent, i.e.,*

$$\vdash_{\mathbf{S}} \Delta|A \Rightarrow \Delta, A \textit{ implies } \models_{\mathbf{S}} \Delta|A \Rightarrow \Delta, A;$$

and if $\Delta|A \Rightarrow \Delta$ is provable in **S** *then $\Delta \cup \{A\}$ is inconsistent, i.e.,*

$$\vdash_{\mathbf{S}} \Delta|A \Rightarrow \Delta \textit{ implies } \models_{\mathbf{S}} \Delta|A \Rightarrow \Delta.$$

Proof We prove the theorem by induction on the structure of formula A.

Assume that $\Delta|A \Rightarrow \Delta, A$ is **S**-provable.

If $A = l$ then $\Delta \nvdash \neg A$, and $\Delta \cup \{A\}$ is consistent;

If $A = A_1 \wedge A_2$ then $\Delta|A_1 \Rightarrow \Delta, A_1$ and $\Delta, A_1|A_2 \Rightarrow \Delta, A_1, A_2$ are **S**-provable. By induction assumption, $\Delta \cup \{A_1\}$ and $\Delta \cup \{A_1, A_2\}$ are consistent, and so is $\Delta \cup \{A_1 \wedge A_2\}$;

If $A = A_1 \vee A_2$ then either $\Delta|A_1 \Rightarrow \Delta, A_1$ or $\Delta|A_2 \Rightarrow \Delta, A_2$ is **S**-provable. By induction assumption, either $\Delta \cup \{A_1\}$ or $\Delta \cup \{A_2\}$ is consistent, and so is $\Delta \cup \{A_1 \vee A_2\}$.

Assume that $\Delta|A \Rightarrow \Delta$ is **S**-provable.

If $A = l$ then $\Delta \vdash \neg A$, and $\Delta \cup \{A\}$ is inconsistent;

If $A = A_1 \wedge A_2$ then either $\Delta|A_1 \Rightarrow \Delta$ or $\Delta, A_1|A_2 \Rightarrow \Delta, A_1$ is **S**-provable, and by induction assumption, either $\Delta \cup \{A_1\}$ or $\Delta \cup \{A_1, A_2\}$ is inconsistent. Hence, $\Delta \cup \{A_1 \wedge A_2\}$ is inconsistent;

If $A = A_1 \vee A_2$ then $\Delta|A_1 \Rightarrow \Delta$ and $\Delta|A_2 \Rightarrow \Delta$ are **S**-provable, and by induction assumption, $\Delta \cup \{A_1\}$ and $\Delta \cup \{A_2\}$ are inconsistent. Hence, $\Delta \cup \{A_1 \vee A_2\}$ is inconsistent. □

Theorem 3.2.3 (Completeness theorem) *For any consistent formula set Δ and formula A, if $\Delta \cup \{A\}$ is consistent then $\Delta|A \Rightarrow \Delta, A$ is provable in* **S**, *i.e.,*

$$\models_{\mathbf{S}} \Delta|A \Rightarrow \Delta, A \textit{ implies } \vdash_{\mathbf{S}} \Delta|A \Rightarrow \Delta, A;$$

and if $\Delta \cup \{A\}$ is inconsistent then $\Delta|A \Rightarrow \Delta$ is provable in **S**, *i.e.,*

$$\models_{\mathbf{S}} \Delta|A \Rightarrow \Delta \textit{ implies } \vdash_{\mathbf{S}} \Delta|A \Rightarrow \Delta.$$

Proof We prove the theorem by induction on the structure of formula A.

Assume that $\Delta \cup \{A\}$ is consistent.

If $A = l$ then $\Delta \nvdash \neg A$, and by $(S^{\mathbf{A}})$, $\Delta|A \Rightarrow \Delta, l$ is **S**-provable;

If $A = A_1 \wedge A_2$ then $\Delta \cup \{A_1\}$ and $\Delta \cup \{A_1, A_2\}$ are consistent, and by induction assumption, $\Delta|A_1 \Rightarrow \Delta, A_1$ is **S**-provable and $\Delta, A_1|A_2 \Rightarrow \Delta, A_1, A_2$ is **S**-provable, and by $(S^{\wedge})$, $\Delta|A_1 \wedge A_2 \Rightarrow \Delta, A_1 \wedge A_2$ is **S**-provable;

If $A = A_1 \vee A_2$ then either $\Delta \cup \{A_1\}$ or $\Delta \cup \{A_2\}$ is consistent, and by induction assumption, either $\Delta|A_1 \Rightarrow \Delta, A_1$ or $\Delta|A_2 \Rightarrow \Delta, A_2$ are **S**-provable, and by $(S_1^{\vee})$ or $(S_2^{\vee})$, $\Delta|A_1 \vee A_2 \Rightarrow \Delta, A_1 \vee A_2$ is **S**-provable.

Assume that $\Delta \cup \{A\}$ is inconsistent.

If $A = l$ then $\Delta \vdash \neg A$, and by $(S_{\mathbf{A}})$, $\Delta|A \Rightarrow \Delta$ is **S**-provable;

If $A = A_1 \wedge A_2$ then either $\Delta \cup \{A_1\}$ or $\Delta \cup \{A_1, A_2\}$ is inconsistent, and by induction assumption, either $\Delta|A_1 \Rightarrow \Delta$ is **S**-provable, or $\Delta, A_1|A_2 \Rightarrow \Delta, A_1|$ is **S**-provable, and by $(S_{\wedge}^1)$ and $(S_{\wedge}^2)$, $\Delta|A_1 \wedge A_2 \Rightarrow \Delta$ is **S**-provable;

If $A = A_1 \vee A_2$ then $\Delta \cup \{A_1\}$ and $\Delta \cup \{A_2\}$ are inconsistent, and by induction assumption, $\Delta|A_1 \Rightarrow \Delta$ and $\Delta|A_2 \Rightarrow \Delta$ are **S**-provable, and by $(S_{\vee})$, $\Delta|A_1 \vee A_2 \Rightarrow \Delta$ is **S**-provable. □

We have the following

Proposition 3.2.4 *There is no formula A such that*

$$\vdash_{\mathbf{S}} \Delta|A \Rightarrow \Delta$$

and

$$\vdash_{\mathbf{S}} \Delta|A \Rightarrow \Delta, A.$$

3.2.2 R-Calculus S for a Theory

Let $\Gamma = (A_1, \ldots, A_n)$, i.e., set $\{A_1, \ldots, A_n\}$ with an ordering $<$ such that $A_1 < A_2 < \cdots < A_n$. Define

$$\Delta|\Gamma = (\cdots((\Delta|A_1)|A_2)\cdots)|A_n.$$

R-calculus **S** for a theory Γ consists of the following axioms and deduction rules:

- **Axioms:**

$$(S^{\mathbf{A}})\ \frac{\Delta \nvdash \neg l}{\Delta|l, \Gamma \Rightarrow \Delta, l|\Gamma} \quad (S_{\mathbf{A}})\ \frac{\Delta \vdash \neg l}{\Delta|l, \Gamma \Rightarrow \Delta|\Gamma}$$

- **Deduction rules:**

$$(S^{\wedge})\ \frac{\begin{array}{c}\Delta|A_1, \Gamma \Rightarrow \Delta, A_1|\Gamma \\ \Delta, A_1|A_2, \Gamma \Rightarrow \Delta, A_1, A_2|\Gamma\end{array}}{\Delta|A_1 \wedge A_2, \Gamma \Rightarrow \Delta, A_1 \wedge A_2|\Gamma} \quad (S^1_{\wedge})\ \frac{\Delta|A_1, \Gamma \Rightarrow \Delta|\Gamma}{\Delta|A_1 \wedge A_2, \Gamma \Rightarrow \Delta|\Gamma}$$

$$(S^2_{\wedge})\ \frac{\Delta, A_1|A_2, \Gamma \Rightarrow \Delta, A_1|\Gamma}{\Delta|A_1 \wedge A_2, \Gamma \Rightarrow \Delta|\Gamma}$$

$$(S^{\vee}_1)\ \frac{\Delta|A_1, \Gamma \Rightarrow \Delta, A_1|\Gamma}{\Delta|A_1 \vee A_2, \Gamma \Rightarrow \Delta, A_1 \vee A_2|\Gamma} \quad (S_{\vee})\ \frac{\begin{array}{c}\Delta|A_1, \Gamma \Rightarrow \Delta|\Gamma \\ \Delta|A_2, \Gamma \Rightarrow \Delta|\Gamma\end{array}}{\Delta|A_1 \vee A_2, \Gamma \Rightarrow \Delta|\Gamma}$$

$$(S^{\vee}_2)\ \frac{\Delta|A_2, \Gamma \Rightarrow \Delta, A_2|\Gamma}{\Delta|A_1 \vee A_2, \Gamma \Rightarrow \Delta, A_1 \vee A_2|\Gamma}$$

Definition 3.2.5 $\Delta|\Gamma \Rightarrow \Delta, \Theta$ is provable in **S**, denoted by $\vdash_{\mathbf{S}} \Delta|\Gamma \Rightarrow \Delta, \Theta$, if there is a sequence $\{S_1, \ldots, S_m\}$ of statements such that

$$\begin{aligned} S_1 &= \Delta|\Gamma \Rightarrow \Delta_1|\Gamma_1, \\ &\cdots \\ S_m &= \Delta_{m-1}|\Gamma_{m-1} \Rightarrow \Delta, \Theta \end{aligned}$$

and for each $i < m$, S_{i+1} is an axiom or is deduced from the previous statements by a deduction rule in **S**.

Theorem 3.2.6 (Soundness theorem) *For any consistent formula sets Θ, Δ and any finite consistent formula set Γ, if $\Delta|\Gamma \Rightarrow \Delta, \Theta$ is provable in **S** then Θ is a ⊆-minimal change of Γ by Δ. That is,*

$$\vdash_{\mathbf{S}} \Delta|\Gamma \Rightarrow \Delta, \Theta \textit{ implies } \models_{\mathbf{S}} \Delta|\Gamma \Rightarrow \Delta, \Theta.$$

Proof We prove the theorem by induction on n.

Assume that $\Delta|\Gamma \Rightarrow \Delta, \Theta$ is provable in **S**.

Let $n = 1$. Then, either $\Theta = A_1$ or $\Theta = \lambda$. If $\Theta = A_1$ then $\Delta \cup \{A_1\}$ is consistent, and Θ is a $\subseteq$-minimal change of A_1 by Δ; otherwise, $\Delta \cup \{A_1\}$ is inconsistent, and $\Theta = \lambda$ is a $\subseteq$-minimal change of A_1 by Δ.

Assume that the theorem holds for n, that is, if $\Delta|\Gamma \Rightarrow \Delta, \Theta$ then Θ is a $\subseteq$-minimal change of Γ by Δ, where $\Gamma = (A_1, \ldots, A_n)$.

Let $\Gamma' = (\Gamma, A_{n+1}) = (A_1, \ldots, A_{n+1})$. Then, if $\Delta|\Gamma' \Rightarrow \Delta, \Theta'$ is provable then $\Delta|\Gamma \Rightarrow \Delta, \Theta$ and $\Delta, \Theta|A_{n+1} \Rightarrow \Delta, \Theta'$ are provable. By case $n = 1$ and induction assumption, Θ' is a $\subseteq$-minimal change of A_{n+1} by $\Delta \cup \Theta$, and Θ is a $\subseteq$-minimal change of Γ by Δ. Therefore, Θ' is a $\subseteq$-minimal change of Γ' by Δ. □

Theorem 3.2.7 (Completeness Theorem) *For any consistent formula sets Θ, Δ and any finite consistent formula set Γ, if Θ is a $\subseteq$-minimal change of Γ by Δ then $\Delta|\Gamma \Rightarrow \Delta, \Theta$ is provable in* **S**. *That is,*

$$\models_{\mathbf{S}} \Delta|\Gamma \Rightarrow \Delta, \Theta \text{ implies } \vdash_{\mathbf{S}} \Delta|\Gamma \Rightarrow \Delta, \Theta.$$

Proof Assume that Θ is a $\subseteq$-minimal change of Γ by Δ. Then, there is an ordering $<$ of Γ such that $\Gamma = (A_1, A_2, \ldots, A_n)$, where $A_1 < A_2 < \cdots < A_n$, and Θ is a maximal subset of Γ such that $\Delta \cup \Theta$ is consistent.

We prove the theorem by induction on n.

Let $n = 1$. If $\Theta = \{A_1\}$ then Δ is consistent with A_1, and $\Delta|A_1 \Rightarrow \Delta, A_1$ is **S**-provable; and if $\Theta = \emptyset$ then Δ is inconsistent with A_1, and $\Delta|A_1 \Rightarrow \Delta, \Delta$ is **S**-provable.

Assume that the theorem holds for n, that is, if Θ is a $\subseteq$-minimal change of Γ by Δ then $\Delta|\Gamma \Rightarrow \Delta, \Theta$ is **S**-provable.

Let $\Gamma' = (\Gamma, A_{n+1}) = (A_1, \ldots, A_{n+1})$ and Θ' is a $\subseteq$-minimal change of Γ' by Δ. Then, Θ' is a $\subseteq$-minimal change of A_{n+1} by $\Delta \cup \Theta$, and $\Delta, \Theta|A_{n+1} \Rightarrow \Delta, \Theta'$ is provable. By induction assumption, $\Delta|\Gamma \Rightarrow \Delta, \Theta$ is **S**-provable and so is $\Delta|\Gamma' \Rightarrow \Delta, \Theta|A_{n+1}$. Hence, $\Delta|\Gamma' \Rightarrow \Delta, \Theta'$ is provable in **S**. □

3.2.3 AGM Postulates $\mathbf{A}^{\subseteq}$ for $\subseteq$-Minimal Change

AGM postulates $\mathbf{A}^{\subseteq}$ for $\subseteq$-minimal change:

- Success: $\Delta \subseteq \Delta|\Gamma$;
- Inclusion: $\Delta|\Gamma \subseteq \Gamma \cup \Delta$;
- Vacuity: $\text{con}(\Delta, \Gamma) \Rightarrow \Delta|\Gamma \vdash\dashv \Delta, \Gamma$;
- Extensionality: $\text{con}(\Delta) \Rightarrow \text{con}(\Delta|\Gamma)$;
- Extensionality: $\Delta \vdash\dashv \Delta' \Rightarrow \Delta|\Gamma = \Delta'|\Gamma$;
- Superexpansion: $(\Delta_1 \cup \Delta_2)||\Gamma \subseteq \Delta_1||\Gamma$, where $\Delta||\Gamma = \Delta|\Gamma - \Delta$;
- Subexpansion: $\text{con}(\Delta_1|\Gamma, \Delta_2) \Rightarrow (\Delta_1|\Gamma), \Delta_2 \vdash\dashv (\Delta_1 \cup \Delta_2)|\Gamma$;
- $\subseteq$-minimal change: $\Delta||\Gamma \subset \Theta \subseteq \Gamma \Rightarrow \text{incon}(\Delta, \Theta)$;
- Closure: $\Delta||\Gamma \subseteq \Theta \subseteq \Gamma \Rightarrow \Delta|\Gamma = \Delta|\Theta$.

Here, $\Delta \vdash\dashv \Delta'$ means that $\Delta \vdash \Delta'$ and $\Delta' \vdash \Delta$.

There is another form of AGM postulates $\mathbf{A}^{\subseteq}$ for ⊆-minimal change: assume that $\Delta|\Gamma \Rightarrow \Delta, \Theta$.

- Success: $\Theta \subseteq \Gamma$;
- Inclusion:
- Vacuity: $\mathrm{con}(\Delta, \Gamma) \Rightarrow \Theta \vdash\dashv \Gamma$;
- Extensionality: $\mathrm{con}(\Delta) \Rightarrow \mathrm{con}(\Delta, \Theta)$;
- Extensionality: $\Delta \vdash\dashv \Delta' \Rightarrow \Theta = \Theta'$;
- Superexpansion: $\Theta_{12} \subseteq \Theta_1$, where $\Delta_1|\Gamma \Rightarrow \Delta_1, \Theta_1$ and $\Delta_1, \Delta_2|\Gamma \Rightarrow \Delta_1, \Delta_2, \Theta_{12}$;
- Subexpansion: $\mathrm{con}(\Delta_1 \cup \Delta, \Delta_2) \Rightarrow \Theta_1 \vdash\dashv \Theta_{12}$;
- ⊆-minimal change: $\Theta \subset \Xi \subseteq \Gamma \Rightarrow \mathrm{incon}(\Delta, \Xi)$;
- Closure: $\Theta \subseteq \Xi \subseteq \Gamma \Rightarrow \Delta|\Xi = \Delta, \Theta$.

Theorem 3.2.8 *Θ satisfies $\mathbf{A}^{\subseteq}$ if and only if Θ is a ⊆-minimal change of Γ by Δ.*

Proof Assume that Θ satisfies $\mathbf{A}^{\subseteq}$. Then,
∘ by Success, $\Theta \subseteq \Gamma$;
∘ by Extensionality, $\Delta \cup \Theta$ is consistent;
∘ if $\Theta \neq \Gamma$ then by Vacuity, $\Gamma \cup \Delta$ is inconsistent;
∘ by ⊆-minimal change, for any $A \in \Gamma - \Delta$, $\Theta \cup \{A\} \cup \Delta$ is inconsistent.
Therefore, Θ is a ⊆-minimal change of Γ by Δ.

Assume that Θ is a ⊆-minimal change of Γ by Δ. Then, there is an ordering $\leq$ on Γ such that $\Theta = \sigma(\Gamma^{<}, \Delta)$. Then, we have

(i) Success: $\sigma(\Gamma^{<}, \Delta) \subseteq \Gamma$,
(ii) Vacuity: $\mathrm{con}(\Delta, \Gamma) \Rightarrow \sigma(\Gamma^{<}, \Delta) \vdash\dashv \Gamma$,
(iii) Extensionality: $\mathrm{con}(\Delta) \Rightarrow \mathrm{con}(\Delta, \sigma(\Gamma^{<}, \Delta))$,
(iv) Extensionality: $\Delta \vdash\dashv \Delta' \Rightarrow \sigma(\Gamma^{<}, \Delta) \vdash\dashv \Theta', \leq$, because the computation $\sigma(\Gamma^{<}, \Delta)$ depends on whether $\Theta_{i-1} \vdash \neg A_i$, and $\Theta_{i-1} \vdash \neg A_i$ if and only if $\Theta'_{i-1} \vdash \neg A_i$, where

$$\Theta'_0 = \Delta';$$
$$\Theta'_i = \begin{cases} \Theta'_{i-1} - \{A_i\} & \text{if } \Delta, \Theta'_{i-1} \vdash \neg A_i \\ \Theta'_{i-1} & \text{otherwise} \end{cases}$$

(v) We have two computations: one is the computation of Θ_1 :

$$\Theta^1_0 = \emptyset;$$
$$\Theta^1_i = \begin{cases} \Theta^1_{i-1} - \{A_i\} & \text{if } \Delta_1, \Theta^1_{i-1} \vdash \neg A_i \\ \Theta^1_{i-1} & \text{otherwise} \end{cases}$$

and $\Theta_1 = \Theta^1_n$, and another is the computation of Θ_{12} :

$$\Theta^{12}_0 = \emptyset;$$
$$\Theta^{12}_i = \begin{cases} \Theta^{12}_{i-1} - \{A_i\} & \text{if } \Delta_1, \Delta_2, \Theta^{12}_{i-1} \vdash \neg A_i \\ \Theta^{12}_{i-1} & \text{otherwise} \end{cases}$$

and $\Theta_{12} = \Theta_n$. By the induction on $i \leq n$, we can prove that $\Theta_i^{12} \cap \Gamma \subseteq \Theta_i^1 \cap \Gamma$, and hence, $\Theta_{12} \subseteq \Theta_1$.

(vi) Subexpansion: assume that $\mathrm{con}(\Theta_1 \cup \Delta_1, \Delta_2)$. Let i_0 be the least such that $\Theta_{i_0}^1 = \Theta_n^1 = \Theta^1 \cup \Delta_1$. Then, $\Theta_{i_0}^1 \nvdash \neg A_{i_0}$, and because $\mathrm{con}(\Theta_1 \cup \Delta_1, \Delta_2)$, $\Theta_{i_0}^1 \cup \Delta_2 \nvdash \neg A_{i_0}$. Hence, if j_0 is the greatest j such that $\Theta_j^{12} \nvdash \neg A_j$ then $j_0 \leq i_0$, that is, $\Theta_{12} \supseteq \Theta_2$. By Superexpansion, $\Theta_{12} \subseteq \Theta_2$, and hence, $\Theta_{12} = \Theta_1$.

(vii) By Theorem 3.2.6, we have $\subseteq$-minimal change.

(viii) By definition of $\subseteq$-minimal change, we have Closure. □

3.3 R-Calculus for $\preceq$-Minimal Change

In this section, we will give an R-calculus **T** which is sound and complete with respect to $\preceq$-minimal changes of Γ by Δ, that is, for any consistent theories Γ, Δ and Θ, $\Delta|\Gamma \Rightarrow \Delta, \Theta$ is provable in **T** if and only if Θ is a $\preceq$-minimal change of Γ by Δ.

The rule $(S_\wedge)$ in **S** may eliminate too much information in Γ. For example,

$$\vdash_{\mathbf{S}} \neg havingarms|havingarms \wedge havinglegs \Rightarrow \neg havingarms.$$

Intuitively we should have

$$\neg\, havingarms|havingarms \wedge havinglegs \Rightarrow \neg havingarms,\ havinglegs.$$

In **T**, we have the above deduction. Formally, we have

$$\begin{aligned} &\vdash_{\mathbf{S}} \neg p|p \wedge q \Rightarrow \neg p; \\ &\vdash_{\mathbf{T}} \neg p|p \wedge q \Rightarrow \neg p, q. \end{aligned}$$

3.3.1 R-Calculus **T** *for a Formula*

R-calculus **T** for a formula A consists of the following axioms and deduction rules:

- **Axioms:**

$$(T^{\mathbf{A}})\ \frac{\Delta \nvdash \neg A}{\Delta|A \Rightarrow \Delta, A} \qquad (T_{\mathbf{A}})\ \frac{\Delta \vdash \neg l}{\Delta|l \Rightarrow \Delta, \lambda}$$

- **Deduction rules:**

$$(T^{\wedge})\ \frac{\Delta|A_1 \Rightarrow \Delta, C_1 \quad \Delta, C_1|A_2 \Rightarrow \Delta, C_1, C_2}{\Delta|A_1 \wedge A_2 \Rightarrow \Delta, C_1 \wedge C_2}$$

$$(T_1^{\vee})\ \frac{\Delta|A_1 \Rightarrow \Delta, C_1 \neq \lambda}{\Delta|A_1 \vee A_2 \Rightarrow \Delta, C_1 \vee A_2}$$

$$(T_2^{\vee})\ \frac{\Delta|A_1 \Rightarrow \Delta, \lambda \quad \Delta|A_2 \Rightarrow \Delta, C_2 \neq \lambda}{\Delta|A_1 \vee A_2 \Rightarrow \Delta, A_1 \vee C_2}$$

$$(T_3^{\vee})\ \frac{\Delta|A_1 \Rightarrow \Delta, \lambda \quad \Delta|A_2 \Rightarrow \Delta, \lambda}{\Delta|A_1 \vee A_2 \Rightarrow \Delta, \lambda}$$

We assume that if C is consistent then

$$\lambda \vee C \equiv C \vee \lambda \equiv C; \quad \lambda \wedge C \equiv C \wedge \lambda \equiv C; \quad \Delta, \lambda \equiv \Delta$$

and if C is inconsistent then

$$\lambda \vee C \equiv C \vee \lambda \equiv \lambda; \quad \lambda \wedge C \equiv C \wedge \lambda \equiv \lambda$$

Rule $(T^{\wedge})$ shows that revision depends on the ordering of subformulas in A_1. That is, if $\Delta|A_1 \wedge A_2 \Rightarrow \Delta, C$ and $\Delta|A_2 \wedge A_1 \Rightarrow \Delta, C'$ are provable in **T** then it is possible that $\Delta, C \nvdash C'$ and $\Delta, C' \nvdash C$. For example, we have both the following deductions:

$$\begin{array}{l}
\neg p \vee \neg q|p \Rightarrow \neg p \vee \neg q, p \\
\neg p \vee \neg q, p|q \Rightarrow \neg p \vee \neg q, p \\
\neg p \vee \neg q|p \wedge q \Rightarrow \neg p \vee \neg q, p \equiv p \wedge \neg q; \\
\neg p \vee \neg q|q \Rightarrow \neg p \vee \neg q, q \\
\neg p \vee \neg q, q|p \Rightarrow \neg p \vee \neg q, q \\
\neg p \vee \neg q|q \wedge p \Rightarrow \neg p \vee \neg q, q \equiv \neg p \wedge q.
\end{array}$$

Therefore, $\neg p \vee \neg q, p \nvdash q$ and $\neg p \vee \neg q, q \nvdash p$.

Definition 3.3.1 $\Delta|A \Rightarrow \Delta, C$ is provable in **T**, denoted by $\vdash_{\mathbf{T}} \Delta|A \Rightarrow \Delta, C$, if there is a sequence $S_1, \ldots, S_m$ of statements such that

$$\begin{aligned}
S_1 &= \Delta|A \Rightarrow \Delta|A_1, \\
&\cdots \\
S_m &= \Delta|A_{m-1} \Rightarrow \Delta, C;
\end{aligned}$$

and for each $i < m$, S_{i+1} is an axiom or is deduced from the previous statements by a deduction rule in **T**.

Theorem 3.3.2 *For any consistent formula set Δ and formula A, there is a formula C such that $C \preceq A$ and $\Delta|A \Rightarrow \Delta, C$ is provable in* **T**.

Proof We prove the theorem by induction on the structure of formula A.

Case $A = l$. Then, if $\Delta \vdash \neg l$ then let $C = \lambda$; if $\Delta \nvdash \neg l$ then let $C = l$. Then, by $(T^{\mathbf{A}})$ and $(T_{\mathbf{A}})$, $\Delta|A \Rightarrow \Delta, C$ is provable in $\mathbf{T}$.

Case $A = A_1 \wedge A_2$. Then, by induction assumption, there are formulas $C_1 \preceq A_1, C_2 \preceq A_2$ such that $\Delta|A_1 \Rightarrow \Delta, C_1$ and $\Delta, C_1|A_2 \Rightarrow \Delta, C_1, C_2$ are provable in $\mathbf{T}$, and by $(T^{\wedge})$, $\Delta|A_1 \wedge A_2 \Rightarrow \Delta, C_1 \wedge C_2$ is provable in $\mathbf{T}$.

Case $A = A_1 \vee A_2$. Then, by induction assumption, there are formulas $C_1 \preceq A_1$ and $C_2 \preceq A_2$ such that $\Delta|A_1 \Rightarrow \Delta, C_1$ and $\Delta|A_2 \Rightarrow \Delta, C_2$ are provable in $\mathbf{T}$. If $C_1 \neq \lambda$ then by $(T_1^{\vee})$, $\vdash_{\mathbf{T}} \Delta|A_1 \vee A_2 \Rightarrow \Delta, C_1 \vee A_2$; if $C_1 = \lambda$ and $C_2 \neq \lambda$ then by $(T_2^{\vee})$, $\vdash_{\mathbf{T}} \Delta|A_1 \vee A_2 \Rightarrow A_1 \vee C_2$; if $C_1 = C_2 = \lambda$ then by $(T_3^{\vee})$, $\vdash_{\mathbf{T}} \Delta|A_1 \vee A_2 \Rightarrow \Delta$. Let

$$C = \begin{cases} C_1 \vee A_2 & \text{if } C_1 \neq \lambda \\ A_1 \vee C_2 & \text{if } C_1 = \lambda \neq C_2 \\ \lambda & \text{otherwise,} \end{cases}$$

and $\vdash_{\mathbf{T}} \Delta|A_1 \vee A_2 \Rightarrow \Delta, C$. □

Lemma 3.3.3 *If Θ is a $\preceq$-minimal change of Γ by Δ and Θ' is a $\preceq$-minimal change of A_{n+1} by $\Delta \cup \Theta$ then Θ' is a $\preceq$-minimal change of $\Gamma \cup \{A_{n+1}\}$ by Δ.*

Proof Assume that Θ is a $\preceq$-minimal change of Γ by Δ and Θ' is a $\preceq$-minimal change of A_{n+1} by $\Delta \cup \Theta$. Then, by definition, Θ' is a $\preceq$-minimal change of $\Gamma \cup \{A_{n+1}\}$ by Δ. □

Lemma 3.3.4 *Assume that Δ is inconsistent with $A_1 \wedge A_2$. If C_1 is a $\preceq$-minimal change of A_1 by Δ and C_2 is a $\preceq$-minimal change of A_2 by $\Delta \cup \{C_1\}$ then $C_1 \wedge C_2$ is a $\preceq$-minimal change of $A_1 \wedge A_2$ by Δ.*

Proof Assume that C_1 is a $\preceq$-minimal changes of A_1 by Δ and C_2 of A_2 by $\Delta \cup \{C_1\}$.

Then, $C_1 \wedge C_2 \preceq A_1 \wedge A_2$, and $\Delta \cup \{C_1 \wedge C_2\}$ is consistent.

For any B with $C_1 \wedge C_2 \prec B \preceq A_1 \wedge A_2$, there are B_1, B_2 such that $B = B_1 \wedge B_2$, and

$$\begin{aligned} C_1 \preceq B_1 \preceq A_1, \\ C_2 \preceq B_2 \preceq A_2. \end{aligned}$$

Then, if $C_1 \prec B_1$ then $\Delta \cup \{B_1\}$ is inconsistent and so is $\Delta \cup \{B_1 \wedge B_2\}$; if $C_2 \prec B_2$ then $\Delta \cup \{C_1, B_2\}$ is inconsistent and so is $\Delta \cup \{B_1 \wedge B_2\}$. □

Lemma 3.3.5 *Assume that Δ is inconsistent with $A_1 \vee A_2$. If C_1, C_2 are $\preceq$-minimal changes of A_1 and A_2 by Δ, respectively, then C' is a $\preceq$-minimal change of $A_1 \vee A_2$ by Δ, where*

$$C' = \begin{cases} C_1 \vee A_2 & \textit{if } C_1 \neq \lambda \\ A_1 \vee C_2 & \textit{if } C_1 = \lambda \textit{ and } C_2 \neq \lambda \\ \lambda & \textit{if } C_1 = C_2 = \lambda \end{cases}$$

Proof It is clear that $C' \preceq A_1 \vee A_2$, and $\Delta \cup \{C'\}$ is consistent.

For any B with $C' \prec B \preceq A_1 \vee A_2$, there are B_1, B_2 such that

$$\begin{array}{l} B = B_1 \vee B_2, \\ C'_1 \preceq B_1 \preceq A_1, \\ C'_2 \preceq B_2 \preceq A_2, \end{array}$$

and either $C'_1 \prec B_1$ or $C'_2 \prec B_2$, where C'_1 is either C_1 or A_1 and C'_2 is either C_2 or A_2.

If $C'_1 \prec B_1$ then $C'_1 = C_1, C'_2 = A_2$, and by induction assumption, $\Delta \cup \{B_1\}$ is inconsistent and so is $\Delta \cup \{B_1 \vee A_2\}$; and if $C'_2 \prec B_2$ then $C'_2 = C_2, C'_1 = A_1$, and by induction assumption, $\Delta \cup \{B_2\}$ is inconsistent and so is $\Delta \cup \{A_1 \vee B_2\}$. □

Theorem 3.3.6 (Soundness theorem) *For any formula set Δ and formulas A, C, if $\Delta|A \Rightarrow \Delta, C$ is provable in* **T** *then C is a $\preceq$-minimal change of A by Δ. That is,*

$$\vdash_{\mathbf{T}} \Delta|A \Rightarrow \Delta, C \textit{ implies } \models_{\mathbf{T}} \Delta|A \Rightarrow \Delta, C.$$

Proof We prove the theorem by induction on the structure of A. Assume that Δ is inconsistent with A.

Case $A = l$, a literal. Either $\Delta|l \Rightarrow \Delta, l$ or $\Delta|l \Rightarrow \Delta$ is provable in **T**. That is, either $C = l$ or $C = \lambda$, which is a $\preceq$-minimal change of A by Δ.

Case $A = A_1 \wedge A_2$. Then, if $\Delta|A_1 \Rightarrow \Delta, C_1$ is provable in **T** then C_1 is a $\preceq$-minimal change of A_1 by Δ, and if $\Delta, C_1|A_2 \Rightarrow \Delta, C_1, C_2$ is provable in **T** then C_2 is a $\preceq$-minimal change of A_2 by $\Delta \cup \{C_1\}$. By Lemma 3.3.4, $C_1 \wedge C_2$ is a $\preceq$-minimal change of $A_1 \wedge A_2$ by Δ.

Case $A = A_1 \vee A_2$. There are C'_1, C'_2 such that $\Delta|A_1 \Rightarrow \Delta, C'_1$ and $\Delta|A_2 \Rightarrow \Delta, C'_2$ are provable in **T**, and

$$B = \begin{cases} C'_1 \vee A_2 \text{ if } C'_1 \neq \lambda \\ A_1 \vee C'_2 \text{ if } C'_1 = \lambda \text{ and} C'_2 \neq \lambda \\ \lambda \qquad\quad \text{if } C'_1 = C'_2 = \lambda \end{cases}$$

By induction assumption, C'_1 and C'_2 are $\preceq$-minimal changes of A_1 and A_2 by Δ, respectively. Then, $\Delta|A_1 \vee A_2 \Rightarrow \Delta, B$ is provable in **T**, and by Lemma 3.3.5, B is a $\preceq$-minimal change of $A_1 \vee A_2$ by Δ. □

We define a relation $\simeq$ on formulas as follows:

(i) for any literal $l, l \simeq l$; and
(ii) if $A_1 \simeq A'_1$ and $A_2 \simeq A'_2$ then $A_1 \vee A_2 \simeq A'_2 \vee A'_1$, and $A_1 \wedge A_2 \simeq A'_2 \wedge A'_1$.

Proposition 3.3.7 *For any formulas A and B, if $A \simeq B$ then $A \vdash\dashv B$.* □

Theorem 3.3.8 (Completeness theorem) *For any formula set* Δ *and formulas* A, C, *if* C *is a* $\preceq$*-minimal change of* A *by* Δ *then there is a formula* A' *such that* $A \simeq A'$ *and* $\Delta|A' \Rightarrow C$ *is* $\mathbf{T}$*-provable. That is,*

$$\models_{\mathbf{T}} \Delta|A \Rightarrow, \Delta, C \textit{ implies } \vdash_{\mathbf{T}} \Delta|A' \Rightarrow, \Delta, C.$$

Proof Let $C \preceq A$ be a $\preceq$-minimal change of A by Δ.

Case $A = l$. Then $C = \lambda$ (if Δ, A is inconsistent) or $C = A$ (if Δ, A is consistent), and $\Delta, A \Rightarrow \Delta, C$ is provable.

Case $A = A_1 \wedge A_2$. Then there are C_1, C_2 such that $C = C_1 \wedge C_2$, and C_1 and C_2 are $\preceq$-minimal changes of A_1 and A_2 by Δ and Δ, C_1, respectively. Hence, $C_1 \wedge C_2$ is a $\preceq$-minimal change of $A_1 \wedge A_2$ by Δ. By induction assumption, there are formulas A_1' and A_2' such that $A_1 \simeq A_1'$, $A_2 \simeq A_2'$; $\Delta|A_1' \Rightarrow \Delta, C_1$ and $\Delta, C_1|A_2' \Rightarrow \Delta, C_1, C_2$ are provable in $\mathbf{T}$, and so is $\Delta|A_1' \wedge A_2' \Rightarrow \Delta, C_1 \wedge C_2$.

Case $A = A_1 \vee A_2$. Then there are C_1 and C_2 such that $C = C'$, and C_1 and C_2 are $\preceq$-minimal changes of A_1 and A_2 by Δ, respectively, where

$$C' = \begin{cases} C_1 \vee A_2 \text{ if } C_1 \neq \lambda \\ A_1 \vee C_2 \text{ if } C_1 = \lambda \text{ and } C_2 \neq \lambda \\ \lambda \qquad\quad \text{if } C_1 = C_2 = \lambda \end{cases}$$

Then, C' is a $\preceq$-minimal change of $A_1 \vee A_2$ by Δ. By induction assumptions, there are formulas A_1, A_2 such that $A_1 \simeq A_1'$, $A_2 \simeq A_2'$; and either (i) $\Delta|A_1' \Rightarrow \Delta, C_1$ or (ii) $\Delta|A_2' \Rightarrow \Delta, C_2$, or (iii) $\Delta|A_1' \Rightarrow \Delta$ and $\Delta|A_2' \Rightarrow \Delta$ are provable in $\mathbf{T}$, and so is $\Delta|A_1' \vee A_2' \Rightarrow \Delta, C'$, where if $C_1 \neq \lambda$ then $\Delta|A_1' \vee A_2' \Rightarrow \Delta, C_1 \vee A_2'$ is provable in $\mathbf{T}$; if $C_1 = \lambda$ and $C_2 \neq \lambda$ then $\Delta|A_1' \vee A_2' \Rightarrow \Delta, A_1' \vee C_2$ is provable in $\mathbf{T}$; and if $C_1 = \lambda$ and $C_2 = \lambda$ then $\Delta|A_1' \vee A_2' \Rightarrow \Delta$ is provable in $\mathbf{T}$. □

3.3.2 *R-Calculus* T *for a Theory*

Let $\Gamma = (A_1, \ldots, A_n)$. Define

$$\Delta|\Gamma = (\cdots((\Delta|A_1)|A_2)\cdots)|A_n.$$

R-calculus $\mathbf{T}$ for a theory Γ consists of the following axioms and deduction rules:

- **Axioms:**

$$(T^{\mathbf{A}}) \frac{\Delta \nvdash \neg A}{\Delta|A, \Gamma \Rightarrow \Delta, A|\Gamma} \qquad (T_{\mathbf{A}}) \frac{\Delta \vdash \neg l}{\Delta|l, \Gamma \Rightarrow \Delta|\Gamma}$$

- **Deduction rules:**

$$(T^{\wedge})\ \frac{\begin{array}{l}\Delta|A_1,\Gamma \Rightarrow \Delta, C_1|\Gamma\\ \Delta, C_1|A_2,\Gamma \Rightarrow \Delta, C_1, C_2, \Gamma\end{array}}{\Delta|A_1 \wedge A_2,\Gamma \Rightarrow \Delta, C_1 \wedge C_2|\Gamma}$$

$$(T_1^{\vee})\ \frac{\Delta|A_1,\Gamma \Rightarrow \Delta, C_1|\Gamma \quad C_1 \neq \lambda}{\Delta|A_1 \vee A_2,\Gamma \Rightarrow \Delta, C_1 \vee A_2|\Gamma}$$

$$(T_2^{\vee})\ \frac{\begin{array}{l}\Delta|A_1,\Gamma \Rightarrow \Delta|\Gamma\\ \Delta|A_2,\Gamma \Rightarrow \Delta, C_2|\Gamma\\ C_2 \neq \lambda\end{array}}{\Delta|A_1 \vee A_2,\Gamma \Rightarrow \Delta, A_1 \vee C_2|\Gamma}$$

$$(T_3^{\vee})\ \frac{\Delta|A_1,\Gamma \Rightarrow \Delta|\Gamma \quad \Delta|A_2,\Gamma \Rightarrow \Delta|\Gamma}{\Delta|A_1 \vee A_2,\Gamma \Rightarrow \Delta|\Gamma}$$

Theorem 3.3.9 (Soundness Theorem) *For any formula sets Θ, Δ and any finite formula set Γ, if $\Delta|\Gamma \Rightarrow \Delta, \Theta$ is provable in* **T** *then Θ is a $\preceq$-minimal change of Γ by Δ. That is,*

$$\vdash_{\mathbf{T}} \Delta|\Gamma \Rightarrow \Delta, \Theta \textit{ implies } \models_{\mathbf{T}} \Delta|\Gamma \Rightarrow \Delta, \Theta.$$

Proof We prove the theorem by induction on n.

Assume that $\Delta|\Gamma \Rightarrow \Delta, \Theta$ is provable in **T**.

Let $n = 1$. By Proposition 3.1.8, $\Theta = C$ for some C is a $\preceq$-minimal change of A by Δ.

Assume that the theorem holds for n, that is, if $\Delta|\Gamma \Rightarrow \Delta, \Theta$ is provable then Θ is a $\preceq$-minimal change of Γ by Δ, where $\Gamma = (A_1, \ldots, A_n)$.

Let $\Gamma' = (\Gamma, A_{n+1}) = (A_1, \ldots, A_{n+1})$. Then, if $\Delta|\Gamma' \Rightarrow \Delta, \Theta'$ is provable then there is a Θ such that $\Delta|\Gamma \Rightarrow \Delta, \Theta$ and $\Delta, \Theta|A_{n+1} \Rightarrow \Delta, \Theta'$ are provable. By the case $n = 1$ and the induction assumption, Θ' is a $\preceq$-minimal change of A_{n+1} by $\Delta \cup \Theta$, and Θ is a $\preceq$-minimal change of Γ by Δ, therefore, Θ' is a $\preceq$-minimal change of Γ' by Δ. □

Theorem 3.3.10 (Completeness Theorem) *For any formula sets Θ, Δ and Γ, if Θ is a $\preceq$-minimal change of Γ by Δ then $\Delta|\Gamma \Rightarrow \Delta, \Theta$ is provable in* **T**. *That is,*

$$\models_{\mathbf{T}} \Delta|\Gamma \Rightarrow \Delta, \Theta \textit{ implies } \vdash_{\mathbf{T}} \Delta|\Gamma \Rightarrow \Delta, \Theta.$$

Proof Assume that Θ is a $\preceq$-minimal change of Γ by Δ. Then, there is an ordering $<$ of Γ such that $\Gamma = (A_1, A_2, \ldots, A_n)$, where $A_1 < A_2 < \cdots < A_n$, and Θ is a $\preceq$-minimal change of Γ by Δ.

We prove the theorem by induction on n.

Let $n = 1$. By Theorem 3.3.8, there is a theory Θ such that $\Delta|A_1 \Rightarrow \Delta, \Theta$ is provable.

Assume that the theorem holds for n, that is, if Θ is a $\preceq$-minimal change of Γ by Δ then $\Delta|\Gamma \Rightarrow \Delta, \Theta$ is provable.

Let $\Gamma' = (\Gamma, A_{n+1}) = (A_1, \ldots, A_{n+1})$ and Θ' is a $\preceq$-minimal change of Γ' by Δ. Then, Θ' is a $\preceq$-minimal change of A_{n+1} by Θ, and $\Theta|A_{n+1} \Rightarrow \Delta, \Theta'$ is provable.

By the induction assumption, $\Delta|\Gamma \Rightarrow \Delta, \Theta$ is provable and $\Delta|\Gamma' \Rightarrow \Delta, \Theta|A_{n+1}$, and hence, $\Delta|\Gamma' \Rightarrow \Delta, \Theta'$ is provable. □

3.3.3 AGM Postulates $\mathbf{A}^{\preceq}$ for $\preceq$-Minimal Change

AGM postulates $\mathbf{A}^{\preceq}$ for $\preceq$-minimal change:

- Success: $\Delta \subseteq \Delta|\Gamma$;
- Inclusion: $(\Delta|\Gamma) - \Delta \preceq \Gamma$;
- Vacuity: $\text{con}(\Delta, \Gamma) \Rightarrow \Delta|\Gamma \vdash\dashv \Delta, \Gamma$;
- Extensionality: $\text{con}(\Delta) \Rightarrow \text{con}(\Delta|\Gamma)$;
- Extensionality: $\Delta \vdash\dashv \Delta' \Rightarrow \Delta|\Gamma \vdash\dashv \Delta'|\Gamma$;
- Superexpansion: $(\Delta_2, \Delta_1|\Gamma) - (\Delta_1 \cup \Delta_2) \preceq \Delta_1|\Gamma$;
- Subexpansion: $\text{con}(\Delta_1|\Gamma, \Delta_2) \Rightarrow (\Delta_1|\Gamma), \Delta_2 \vdash\dashv (\Delta_1 \cup \Delta_2)|\Gamma$;
- The minimal change: $\Theta \prec \Xi \preceq \Gamma \Rightarrow \text{incon}(\Delta, \Xi)$;
- Closure: $\Theta \preceq \Xi \preceq \Gamma \Rightarrow \Theta, \Delta = \Delta|\Xi$.

Remark $Superexpansion$ is implied by $Inclusion$, and $Subexpansion$ is implied by $Sacuity$. □

There is another form of AGM postulates $\mathbf{A}^{\preceq}$ for $\preceq$-minimal change:

- Success: $\Delta \subseteq \Delta, \Theta$;
- Inclusion: $\Theta \preceq \Gamma$;
- Vacuity: $\text{con}(\Delta, \Gamma) \Rightarrow \Delta, \Theta \vdash\dashv \Delta, \Gamma$;
- Extensionality: $\text{con}(\Delta) \Rightarrow \text{con}(\Delta, \Theta)$;
- Extensionality: $\Delta \vdash\dashv \Delta' \Rightarrow \Theta = \Theta'$;
- Minimality: $\Theta \prec \Xi \preceq \Gamma \Rightarrow \text{incon}(\Delta, \Xi)$;
- Closure: $\Theta \preceq \Xi \preceq \Gamma \Rightarrow \Delta|\Xi = \Theta, \Delta$.

Remark Assume that $\Delta_1|\Gamma \Rightarrow \Delta_1, \Theta_1$. It is not true that $\Delta_2, \Delta_1|\Theta_1 \Rightarrow \Delta_2, \Delta_1, \Theta_2$ if and only if $\Delta_2|\Theta_1 \Rightarrow \Delta_2, \Theta_2$.

For example,

$$\begin{aligned} \neg p|p \vee q &\Rightarrow \neg p, p \vee q \\ \neg q, \neg p|p \vee q &\Rightarrow \neg q|\neg p, p \vee q \\ &\Rightarrow \neg q, \neg p|p \vee q \\ &\Rightarrow \neg q, \neg p \end{aligned}$$

where $\Delta_1 = \{\neg p\}, \Delta_2 = \{\neg q\}, \Theta_1 = \{p \vee q\}, \Theta_1 = \lambda$, and

$$\neg q|p \vee q \not\Rightarrow \neg q.$$

□

Definition 3.3.11 Given any set Δ of formulas and formula A, define $\sigma(\Delta, A)$ as follows:

$$\begin{cases} \lambda & \text{if } A = l \text{ and } \Delta \vdash \neg l \\ l & \text{if } A = l \text{ and } \Delta \nvdash \neg l \\ \sigma(\Delta, A_1) & \\ \wedge\sigma(\Delta \cup \sigma(\Delta, A_1), A_2) & \text{if } A = A_1 \wedge A_2 \\ \sigma(\Delta, A_1) \vee A_2 & \text{if } A = A_1 \vee A_2 \text{ and } \sigma(\Delta, A_1) \neq \lambda \\ A_1 \vee \sigma(\Delta, A_2) & \text{if } A = A_1 \vee A_2 \text{ and } \sigma(\Delta, A_2) \neq \lambda \\ \lambda & \text{if } A = A_1 \vee A_2 \text{ and } \sigma(\Delta, A_1) = \sigma(\Delta, A_2) = \lambda \end{cases}$$

Given any set $\Gamma = (A_1, \ldots, A_n)$ of formulas, define

$$\sigma(\Delta, \Gamma) = \sigma(\cdots\sigma(\sigma(\Delta, A_1), A_2), \cdots), A_n).$$

Here, we assume that there is an ordering $\leq$ on Γ and subformulas of each formula A in Γ, because $\sigma(\Delta, A)$ depends on the ordering of subformulas of A.

Proposition 3.3.12 *$\sigma(\Delta, A)$ is a $\preceq$-minimal change of A by Δ, and $\sigma(\Delta, \Gamma)$ is a $\preceq$-minimal change of Γ by Δ.*

Proof By definition and induction on the structure of A. □

Theorem 3.3.13 *For any $\preceq$-minimal change Θ of Γ by Δ, there is an ordering $\leq$ on Γ and subformulas of each formula in Γ such that*

$$\sigma^{\leq}(\Delta, \Gamma) = \Theta.$$

Proof Let C be a $\preceq$-minimal change of A by Δ. Then, if $\Delta \cup \{A\}$ is consistent then $C = A$. By the definition of $\sigma(\Delta, A)$ and induction on the structure of A,

$$\sigma(\Delta, A) = A.$$

Assume that $\Delta \cup \{A\}$ is inconsistent. $C \cup \Delta$ is consistent, and there is a minimal sequence of substitutions $\{B_1/\lambda, \ldots, B_i/\lambda\}$ such that

$$C = A(B_1/\lambda, \ldots, B_i/\lambda).$$

By the definition of $\sigma(\Delta, A)$ and an appropriate ordering of subformulas of A, we have that

$$\sigma(\Delta, A) = A(B_1/\lambda, \ldots, B_i/\lambda).$$

□

Theorem 3.3.14 *A theory Θ satisfies $\mathbf{A}^{\preceq}$ if and only if Θ is a $\preceq$-minimal change of Γ by Δ.*

Proof Assume that Θ satisfies $\mathbf{A}^{\preceq}$.

Then, by Success, $\Theta \subseteq \Gamma$; by Extensionality, $\Delta \cup \Theta$ is consistent; if $\Theta \neq \Gamma$ then by Vacuity, Γ, Δ is inconsistent.

For any Ξ with $\Theta \prec \Xi \preceq \Gamma$, by the $\preceq$-minimality, Δ, Ξ is inconsistent.

Therefore, Θ is a $\preceq$-minimal change of Γ by Δ.

Assume that Θ is a $\preceq$-minimal change of Γ by Δ. Then, there is an ordering $\leq$ such that $\Theta = \sigma^{\leq}(\Delta, \Gamma)$. Then, we have

(i) Success: $\sigma^{\leq}(\Delta, \Gamma) \preceq \Gamma$,
(ii) Vacuity: $\text{con}(\Delta, \Gamma) \Rightarrow \sigma^{\leq}(\Delta, \Gamma) \vdash\dashv \Gamma$,
(iii) Extensionality: $\text{con}(\Delta) \Rightarrow \text{con}(\Delta, \sigma^{\leq}(\Delta, \Gamma))$,
(iv) Extensionality: $\Delta \vdash\dashv \Delta' \Rightarrow \sigma^{\leq}(\Delta, \Gamma) \vdash\dashv \sigma^{\leq}(\Delta, \Gamma)$.
(v) Superexpansion: $(\Delta_2, \Delta_1|\Gamma) - (\Delta_1 \cup \Delta_2) = \Delta_1\Gamma - \Delta_1 \preceq \Delta_1\Gamma$;
(vi) Subexpansion: assume that $\text{con}(\Delta_1 \cup \Gamma_1, \Delta_2)$. Let i_0 be the least such that $\Theta^1_{i_0} = \Theta^1_n = \Theta^1 \cup \Delta_1$. Then, $\Theta^1_{i_0} \nvdash \neg A_{i_0}$, and because $\text{con}(\Theta_1 \cup \Delta_1, \Delta_2)$, $\Theta^1_{i_0} \cup \Delta_2 \nvdash \neg A_{i_0}$. Hence, if j_0 is the greatest j such that $\Theta^{12}_j \nvdash \neg A_j$ then $j_0 \leq i_0$, that is, $\Theta_{12} \supseteq \Theta_2$. By Superexpansion, $\Theta_{12} \subseteq \Theta_2$, and hence, $\Theta_{12} = \Theta_1$.
(vii) Minimal change: $\sigma^{\leq}(\Delta, \Gamma) \prec \Xi \preceq \Gamma$ implies $\text{incon}(\Delta, \Xi)$. By Proposition 3.3.12, we have the $\preceq$-minimality.
(viii) Closure: $\sigma^{\leq}(\Delta, \Gamma) \preceq \Xi \preceq \Gamma$ implies $\sigma^{\leq}(\Delta, \Gamma) = \Delta|\Xi$. □

3.4 R-Calculus for $\vdash_{\preceq}$-Minimal Change

Because the deduction relation $\vdash$ is dense in the set of all the propositional theories and there is no $\vdash$-minimal change, we consider an R-calculus which has a property as close as possible to $\vdash$-minimal change: $\vdash_{\preceq}$-minimal change.

For $\vdash_{\preceq}$-minimal change, we cannot prove that for any theories Γ, Δ and Θ, $\Delta|\Gamma \Rightarrow \Delta, \Theta$ is provable if and only if Θ is a $\vdash_{\preceq}$-minimal change of Γ by Δ. Hence, we require that Γ is a set of formulas in conjunctive normal form. We will give an R-calculus **U** which is proved to be sound and complete with respect to $\vdash_{\preceq}$-minimal change.

In **S**, we have the following deduction:

$$\vdash_{\mathbf{T}} \neg p, \neg r|(p \vee q) \wedge (r \vee s) \Rightarrow \neg p, \neg r, (p \vee q) \wedge (r \vee s),$$

which is equivalent to $\neg p, \neg r, q \wedge s$. In **U**, we will have the following deduction:

$$\vdash_{\mathbf{U}} \neg p, \neg r|(p \vee q) \wedge (r \vee s) \Rightarrow \neg p, \neg r, q \wedge s.$$

3.4.1 R-Calculus U for a Formula

Assume that A is in conjunctive normal form.

R-calculus **U** for a formula A consisting of the following axioms and deduction rules:

- **Axioms:**

$$(U^{\mathbf{A}})\ \frac{\Delta \nvdash \neg l}{\Delta|l \Rightarrow \Delta, l} \qquad (U_{\mathbf{A}})\ \frac{\Delta \vdash \neg l}{\Delta|l \Rightarrow \Delta, \lambda}$$

- **Deduction rules:**

$$(U^{\wedge})\ \frac{\Delta|A_1 \Rightarrow \Delta, C_1}{\Delta|A_1 \wedge A_2 \Rightarrow \Delta, C_1|A_2}$$

$$(U^{\vee})\ \frac{\Delta|A_1 \Rightarrow \Delta, C_1 \quad \Delta|A_2 \Rightarrow \Delta, C_2}{\Delta|A_1 \vee A_2 \Rightarrow \Delta, C_1 \vee C_2}$$

where if C is consistent then

$$\begin{aligned} &\lambda \vee C \equiv C \vee \lambda \equiv C \\ &\lambda \wedge C \equiv C \wedge \lambda \equiv C; \\ &\Delta, \lambda \equiv \Delta \end{aligned}$$

and if C is inconsistent then

$$\begin{aligned} &\lambda \vee C \equiv C \vee \lambda \equiv \lambda \\ &\lambda \wedge C \equiv C \wedge \lambda \equiv \lambda. \end{aligned}$$

Example 3.4.1 For example,

$$\begin{aligned} \neg p|p \wedge r &\Rightarrow (\neg p|p)|r \\ &\Rightarrow \neg p|r \Rightarrow \neg p, r \\ \neg p|s \wedge p &\Rightarrow (\neg p|s)|p \\ &\Rightarrow \neg p, s|p \Rightarrow \neg p, s \\ \neg p|(p \wedge r) \vee (s \wedge p) &\Rightarrow \neg p, r \vee s; \end{aligned}$$

and

$$\begin{aligned} \neg p|p &\Rightarrow \neg p, \lambda \\ \neg p|r &\Rightarrow \neg p, r \\ \neg p|p \vee r &\Rightarrow \neg p, r \vee \lambda \equiv \neg p, r \\ \neg p, r|p &\Rightarrow \neg p, r, \lambda \\ \neg p, r|s &\Rightarrow \neg p, r, s \\ \neg p, r|p \vee s &\Rightarrow \neg p, r, \lambda \vee s \equiv \neg p, r, s \\ \neg p|(p \vee r) \wedge (p \vee s) &\Rightarrow \neg p, r|p \vee s \\ &\Rightarrow \neg p, r, s \equiv \neg p, r \wedge s. \end{aligned}$$

is a deduction. Equivalently, we have the following deductions:

$$\begin{aligned}\neg p|r &\Rightarrow \neg p, r\\ \neg p|s &\Rightarrow \neg p, s\\ \neg p|r \vee s &\Rightarrow \neg p, r \vee s\\ \neg p|(p \wedge r) \vee (s \wedge p) &\Rightarrow \neg p|p \wedge (r \vee s)\\ &\Rightarrow (\neg p|p)|r \vee s\\ &\Rightarrow p|r \vee s \equiv p, r \vee s;\end{aligned}$$

and

$$\begin{aligned}\neg p|p &\Rightarrow \neg p, \lambda\\ \neg p|r \wedge s &\Rightarrow (\neg p|r)|s\\ &\Rightarrow (\neg p, r)|s\\ &\Rightarrow \neg p, r, s \equiv \neg p, r \wedge s\\ \neg p|(p \vee r) \wedge (p \vee s) &\Rightarrow \neg p|p \vee (r \wedge s)\\ &\Rightarrow \neg p, \lambda \vee (r \wedge s) \equiv \neg p, r \wedge s.\end{aligned}$$

Theorem 3.4.2 *For any consistent set Δ of formulas and formula A in conjunctive normal form, there is a formula C such that*

(1) $\Delta|A \Rightarrow \Delta, C$ *is provable in* $\mathbf{U}$*;*
(2) $C \preceq A$, *and*
(3) $\Delta \cup \{C\}$ *is consistent, and for any D with $C \prec D \preceq A$, either $\Delta, C \vdash D$ and $\Delta, D \vdash C$, or $\Delta \cup \{D\}$ is inconsistent.*

Proof We prove the theorem by induction on the structure of A.

Case $A = l$, a literal. Then by assumption, if Δ, l is consistent then $\vdash_{\mathbf{U}} \Delta|l \Rightarrow \Delta, l$ and let $C = l$; and if Δ, l is inconsistent, then $\Delta \vdash \neg l$, and by $(U_{\mathbf{A}})$, $\vdash_{\mathbf{U}} \Delta|l \Rightarrow \Delta$, and let $C = \lambda$. C satisfies (2) and (3).

Case $A = A_1 \wedge A_2$. Then by induction assumption, there are C_1, C_2 such that

$$\begin{aligned}&\vdash_{\mathbf{U}} \Delta|A_1 \Rightarrow \Delta, C_1;\\ &\vdash_{\mathbf{U}} \Delta, C_1|A_2 \Rightarrow \Delta, C_1, C_2.\end{aligned}$$

By $(U^{\wedge})$, $\vdash_{\mathbf{U}} \Delta|A_1 \wedge A_2 \Rightarrow \Delta, C_1 \wedge C_2$, and let $C = C_1 \wedge C_2$. C satisfies (3), because for any D with $C \prec D \preceq A_1 \wedge A_2$, there are formulas D_1, D_2 such that

$$C_1 \preceq D_1; C_2 \preceq D_2.$$

By the induction assumption, either

$$\begin{aligned}&\Delta, C_1 \vdash D_1; \Delta, D_1 \vdash C_1;\\ &\Delta, C_2 \vdash D_2; \Delta, D_2 \vdash C_2;\end{aligned}$$

or $\Delta \cup \{D_1\}$ or $\Delta \cup \{D_2\}$ are inconsistent. Therefore, we have

$$\Delta, C_1 \wedge C_2 \vdash D_1 \wedge D_2;$$
$$\Delta, D_1 \wedge D_2 \vdash C_1 \wedge C_2,$$

or $\Delta \cup \{D_1 \wedge D_2\}$ is inconsistent.

Case $A = A_1 \vee A_2$. By assumption, A_1, A_2 are disjunctions of sets of literals, say A_1', A_2', respectively. By induction assumption, there are C_1, C_2 such that $C_1' \subseteq A_1', C_2' \subseteq A_2'$, and

$$\vdash_{\mathrm{U}} \Delta|A_1 \Rightarrow \Delta, C_1$$
$$\vdash_{\mathrm{U}} \Delta|A_2 \Rightarrow \Delta, C_2.$$

By $(U^{\vee})$, $\vdash_{\mathrm{U}} \Delta|A_1 \vee A_2 \Rightarrow \Delta, C_1 \vee C_2$. To prove that $C_1 \vee C_2$ satisfies (3), for any D with $C_1 \vee C_2 \prec D \preceq A_1 \vee A_2$, there are formulas D_1, D_2 such that

$$C_1 \preceq D_1 \preceq A_1$$
$$C_2 \preceq D_2 \preceq A_2,$$

and

$$C_1' \subseteq D_1' \subseteq A_1'$$
$$C_2' \subseteq D_2' \subseteq A_2',$$

and either

$$\Delta, C_1 \vdash D_1; \Delta, D_1 \vdash C_1;$$
$$\Delta, C_2 \vdash D_2; \Delta, D_2 \vdash C_2;$$

or $\Delta \cup \{D_1\}$ is inconsistent and $\Delta \cup \{D_2\}$ is inconsistent. Because

$$\Delta \vdash \neg(\textstyle\bigwedge_{l \in D_1' - C_1'} l$$
$$\Delta \vdash \neg(\textstyle\bigwedge_{l \in D_2' - C_2'} l$$

we have

$$\Delta, C_1 \vee C_2 \vdash D_1 \vee D_2;$$
$$\Delta, D_1 \vee D_2 \vdash C_1 \vee C_2.$$

□

Remark We have the following facts about the consistence and inconsistence:

$$\mathrm{incon}(A_1 \wedge A_2, \Delta) \equiv \mathrm{incon}(A_1, \Delta) \text{ or } \mathrm{incon}(A_2, \Delta \cup \{A_1\})$$
$$\mathrm{incon}(A_1 \vee A_2, \Delta) \equiv \mathrm{incon}(A_1, \Delta)\&\mathrm{incon}(A_2, \Delta),$$

and

$$\mathrm{con}(A_1 \wedge A_2, \Delta) \equiv \mathrm{con}(A_1, \Delta)\&\mathrm{con}(A_2, \Delta \cup \{A_1\})$$
$$\mathrm{con}(A_1 \vee A_2, \Delta) \equiv \mathrm{con}(A_1, \Delta) \text{ or } \mathrm{con}(A_2, \Delta),$$

□

Theorem 3.4.3 *Assume that $\Delta|A \Rightarrow \Delta, C$ is provable in* **U**. *If A is consistent with Δ then $\Delta, A \vdash C$ and $\Delta, C \vdash A$.*

Proof Assume that $\Delta|A \Rightarrow \Delta, C$ is provable in **U**. We prove the theorem by induction on the structure of A.

Case $A = l$. By $(U^{\mathbf{A}})$, $C = l$, and

$$\Delta, l \vdash l.$$

Case $A = A_1 \wedge A_2$. By the induction assumption, there are formulas C_1 and C_2 such that

$$\begin{array}{l} \vdash_{\mathbf{U}} \Delta|A_1 \Rightarrow \Delta, C_1, \\ \vdash_{\mathbf{U}} \Delta|A_2 \Rightarrow \Delta, C_2, \end{array}$$

and $C = C_1 \wedge C_2$. Because A is consistent with Δ iff A_1 and A_2 are consistent with Δ and $\Delta \cup \{C_1\}$, respectively, we have

$$\begin{array}{l} \Delta, A_1 \vdash C_1; \Delta, C_1 \vdash A_1; \\ \Delta, C_1, A_2 \vdash C_2; \Delta, C_1, C_2 \vdash A_2; \end{array}$$

and hence,

$$\begin{array}{l} \Delta, A_1 \wedge A_2 \vdash C_1 \wedge C_2; \\ \Delta, C_1 \wedge C_2 \vdash A_1 \wedge A_2. \end{array}$$

That is,

$$\Delta, A \vdash C; \Delta, C \vdash A.$$

Case $A = A_1 \vee A_2$. By the induction assumption, there are formulas C_1 and C_2 such that

$$\begin{array}{l} \vdash_{\mathbf{U}} \Delta|A_1 \Rightarrow \Delta, C_1, \\ \vdash_{\mathbf{U}} \Delta|A_2 \Rightarrow \Delta, C_2, \end{array}$$

and $C = C_1 \vee C_2$. Because A is consistent with Δ iff either A_1 is consistent with Δ or A_2 is consistent with Δ, we have

$$\begin{array}{l} \Delta, A_1 \vdash C_1; \Delta, C_1 \vdash A_1; \\ \Delta, A_2 \vdash C_2; \Delta, C_2 \vdash A_2; \end{array}$$

and hence,

$$\begin{array}{l} \Delta, A_1 \vee A_2 \vdash C_1 \vee C_2; \\ \Delta, C_1 \vee C_2 \vdash A_1 \vee A_2. \end{array}$$

That is,

$$\Delta, A \vdash C; \Delta, C \vdash A.$$

□

Theorem 3.4.4 *Assume that* $\Delta|A \Rightarrow \Delta, C$ *is provable in* $\mathbf{U}$. *Then,* C *is a* $\vdash_{\preceq}$-*minimal change of* A *by* Δ.

Proof Assume that $\Delta|A \Rightarrow \Delta, C$ is provable in $\mathbf{U}$. We prove the theorem by induction on the structure of A.

Case $A = l$. If $C = l$ then By $(U^{\mathbf{A}})$, $\Delta \cup \{l\}$ is consistent, and l is a $\vdash_{\preceq}$-minimal change of l by Δ; if $C = \lambda$ then By $(U_{\mathbf{A}})$, $\Delta \cup \{l\}$ is inconsistent, and λ is a $\vdash_{\preceq}$-minimal change of l by Δ.

Case $A = A_1 \wedge A_2$. There are formulas C_1 and C_2 such that

$$\vdash_{\mathbf{U}} \Delta|A_1 \Rightarrow \Delta, C_1,$$
$$\vdash_{\mathbf{U}} \Delta, C_1|A_2 \Rightarrow \Delta, C_1, C_2,$$

and $C = C_1 \wedge C_2$. By the induction assumption, C_1 is a $\vdash_{\preceq}$-minimal change of A_1 by Δ, and C_2 is a $\vdash_{\preceq}$-minimal change of A_2 by $\Delta \cup \{C_1\}$. Therefore, C is a $\vdash_{\preceq}$-minimal change of $A_1 \wedge A_2$. For any formula D with $C \prec D \preceq A_1 \wedge A_2$, there are formulas D_1, D_2 such that $C_1 \preceq D_1 \preceq A_1, C_2 \preceq D_2 \preceq A_2$, and either

$$\Delta, C_1 \vdash D_1; \Delta, D_1 \vdash C_1;$$
$$\Delta, C_2 \vdash D_2; \Delta, D_2 \vdash C_2;$$

or $\Delta \cup \{D_1\}$ or $\Delta \cup \{D_2\}$ is inconsistent. Hence, either

$$\Delta, C_1 \wedge C_2 \vdash D_1 \wedge D_2; \Delta, D_1 \wedge D_2 \vdash C_1 \wedge C_2;$$

or $\Delta \cup \{D_1 \wedge D_2\}$ is inconsistent.

Case $A = A_1 \vee A_2$. There are formulas C_1 and C_2 such that

$$\vdash_{\mathbf{U}} \Delta|A_1 \Rightarrow \Delta, C_1,$$
$$\vdash_{\mathbf{U}} \Delta|A_2 \Rightarrow \Delta, C_2,$$

and $C = C_1 \vee C_2$. By the induction assumption, C_1 is a $\vdash_{\preceq}$-minimal change of A_1 by Δ, and C_2 is a $\vdash_{\preceq}$-minimal change of A_2 by Δ. Therefore, C is a $\vdash_{\preceq}$-minimal change of $A_1 \vee A_2$.

For any formula D with $C \prec D \preceq A_1 \vee A_2$, there are formulas D_1, D_2 such that $C_1 \preceq D_1 \preceq A_1, C_2 \preceq D_2 \preceq A_2$, and either

$$\Delta, C_1 \vdash D_1; \Delta, D_1 \vdash C_1;$$
$$\Delta, C_2 \vdash D_2; \Delta, D_2 \vdash C_2;$$

or $\Delta \cup \{D_1\}$ or $\Delta \cup \{D_2\}$ is inconsistent. Hence, either

$$\Delta, C_1 \vee C_2 \vdash D_1 \vee D_2; \Delta, D_1 \vee D_2 \vdash C_1 \vee C_2;$$

or $\Delta \cup \{D_1 \wedge D_2\}$ is inconsistent. □

Therefore, we have the following corollary.

Theorem 3.4.5 **U** *is sound and complete with respect to* $\vdash_{\preceq}$*-minimal change.* □

Notice that C depends on the ordering of conjunctures of A. For example,

$$\begin{aligned} p|(\neg p \vee q) \wedge (\neg q \vee r) \wedge \neg r &\Rightarrow p, q|(\neg q \vee r) \wedge \neg r \\ &\Rightarrow p, q, r|\neg r \\ &\Rightarrow p, q, r; \\ p|\neg r \wedge (\neg q \vee r) \wedge (\neg p \vee q) &\Rightarrow p, \neg r|(\neg q \vee r) \wedge (\neg p \vee q) \\ &\Rightarrow p, \neg r, \neg q|\neg p \vee q \\ &\Rightarrow p, \neg r, \neg q \end{aligned}$$

are two deductions in **T**.

3.4.2 *R-Calculus* **U** *for a Theory*

Assume that Γ is a set of formulas in conjunctive normal form. Let $\Gamma = (A_1, \ldots, A_n)$. Define

$$\Delta|\Gamma = (\cdots((\Delta|A_1)|A_2)\cdots)|A_n.$$

R-calculus **U** for $\vdash_{\preceq}$-minimal change consists of the following axioms and deduction rules:

- **Axioms:**

$$(U^{\mathbf{A}})\ \frac{\Delta \nvdash \neg l}{\Delta|l, \Gamma \Rightarrow \Delta, l|\Gamma} \qquad (U_{\mathbf{A}})\ \frac{\Delta \vdash \neg l}{\Delta|l, \Gamma \Rightarrow \Delta|\Gamma}$$

- **Deduction rules:**

$$(U^{\wedge})\ \frac{\Delta|A_1, \Gamma \Rightarrow \Delta, C_1|\Gamma}{\Delta|A_1 \wedge A_2, \Gamma \Rightarrow \Delta, C_1|A_2, \Gamma}$$

$$(U^{\vee})\ \frac{\begin{array}{c}\Delta|A_1, \Gamma \Rightarrow \Delta, C_1|\Gamma \\ \Delta|A_2, \Gamma \Rightarrow \Delta, C_2|\Gamma\end{array}}{\Delta|A_1 \vee A_2, \Gamma \Rightarrow \Delta, C_1 \vee C_2|\Gamma}$$

Definition 3.4.6 $\Delta|\Gamma \Rightarrow \Delta, \Theta$ is provable in **U**, denoted by $\vdash_{\mathbf{U}} \Delta|\Gamma \Rightarrow \Delta, \Theta$, if there is a sequence $\{S_1, \ldots, S_m\}$ of statements such that

$$\begin{aligned} S_1 &= \Delta|\Gamma \Rightarrow \Delta_1|\Gamma_1, \\ &\cdots \\ S_m &= \Delta_{m-1}|\Gamma_{m-1} \Rightarrow \Delta, \Theta \end{aligned}$$

and for each $i < m$, S_{i+1} is an axiom or is deduced from the previous statements by a deduction rule in **U**.

Theorem 3.4.7 (Soundness Theorem) *For any consistent formula sets* Θ, Δ *and any finite consistent formula set* Γ, *if* $\Delta|\Gamma \Rightarrow \Delta, \Theta$ *is provable in* **U** *then* Θ *is a* $\vdash_{\preceq}$*-minimal change of* Γ *by* Δ. *That is,*

$$\vdash_{\mathbf{U}} \Delta|\Gamma \Rightarrow \Delta, \Theta \textit{ implies } \models_{\mathbf{U}} \Delta|\Gamma \Rightarrow \Delta, \Theta.$$

□

Theorem 3.4.8 (Completeness Theorem) *For any consistent formula sets* Θ, Δ *and any finite consistent formula set* Γ, *if* Θ *is a* $\vdash_{\preceq}$*-minimal change of* Γ *by* Δ *then* $\Delta|\Gamma \Rightarrow \Delta, \Theta$ *is provable in* **U**. *That is,*

$$\models_{\mathbf{U}} \Delta|\Gamma \Rightarrow \Delta, \Theta \textit{ implies } \vdash_{\mathbf{U}} \Delta|\Gamma \Rightarrow \Delta, \Theta.$$

□

References

1. C.E. Alchourrón, P. Gärdenfors, D. Makinson, On the logic of theory change: partial meet contraction and revision functions. J. Symb. Logic **50**, 510–530 (1985)
2. S.O. Hansson, *A Textbook of Belief Dynamics, Theory Change and Database Updating* (Kluwer, Dordrecht, 1999)
3. S.O. Hansson, Ten philosophical problems in belief revision. J. Logic Comput. **13**, 37–49 (2003)
4. H. Katsuno, A.O. Mendelzon, Propositional knowledge base revision and minimal change. Artif. Intell. **52**, 263–294 (1991)
5. W. Li, Y. Sui, The R-calculus based-on addition instead of cancelation. To appear in Frontier in Computer Science
6. W. Li, Y. Sui, The sound and complete R-calculi with respect to pseudo-revision and pre-revision. Int. J. Intell. Sci. **3**, 110–117 (2013)
7. W. Li, Y. Sui, The set-based and inference-based R-calculus
8. W. Li, Y. Sui, The sound and complete R-calculi with pseudo-subtheory minimal change property. To appear in International Journal of Software and Informatics
9. W. Li, Y. Sui, M. Sun, The sound and complete R-calculus for revising propositional theories. Sci. China: Inf. Sci. **58**, 092101:1–092101:12 (2015)
10. H. Rott, M.-A. Williams (eds.), *Frontiers of Belief Revision* (Kluwer, Dordrecht, 2001)
11. K. Satoh, Nonmonotonic reasoning by minimal belief revision, in *Proceedings of the Int'l Conference on Fifth Generation Computer Systems*, Tokyo (1988), pp. 455–462

Chapter 4
R-Calculi for Description Logics

A concept α is a pair (X, Y), where X is the intent of α and Y is the extent of α, satisfying the Galois' connection:

$$X \textit{ is the set of the properties which each element in } Y \textit{ satisfies},$$

and

$$Y \textit{ is the set of the elements satisfying each property in } X.$$

Traditional logics are to formalize deduction; and description logics [3–5, 7] are to formalize concepts [2], where deduction and concepts are two main ingredients in classical informal logic.

Description logics [2–5] are a class of logics in which

- the logical symbols contain the concept constructors, such as $\neg, \sqcup, \sqcap, \forall$; and the subsumption relation $\sqsubseteq$ between concepts; and
- the nonlogical symbols contain atomic concept symbols, atomic role symbols, and individual constant symbols.

Here, we consider one description logic which logical language given as in Chap. 2.

R-calculus for description logics is between the one for propositional logic and the one for first-order logic, because the complexity of the deduction relation of description logics is between the ones of the deduction relation of the former and of the latter. Even though the quantifier $\forall$ occurs in the concept constructor $\forall R$, its equivalent form in first-order logic is a guarded first-order formula, and the deduction relation in the guarded first-order logic is decidable [1, 6]. Therefore, for R-calculi of description logics, we can decompose a statement $C(a)$ into an atomic statement $A(a)$, where A is an atomic concept, and two sets of deduction rules: one for $C(a)$ consistent with Δ, and another for $C(a)$ inconsistent with Δ.

There are also $\subseteq$-minimal change, $\preceq$-minimal change and $\vdash_{\preceq}$-minimal change in description logics and the corresponding R-calculi $\mathbf{S}^{\mathrm{DL}}$, $\mathbf{T}^{\mathrm{DL}}$ and $\mathbf{U}^{\mathrm{DL}}$ which are

W. Li and Y. Sui, *R-CALCULUS: A Logic of Belief Revision*, Perspectives in Formal Induction, Revision and Evolution,
https://doi.org/10.1007/978-981-16-2944-0_4

sound and complete with respect to $\subseteq$-minimal change, $\preceq$-minimal change and $\vdash_{\preceq}$-minimal change, respectively, where $\preceq$ is the pseudo-subconcept relation between concepts, which is neither the subconcept relation nor the para-subconcept relation between concepts.

4.1 R-Calculus for $\subseteq$-Minimal Change

Being parallel to R-calculus $\mathbf{S}$ for propositional logic, R-calculus $\mathbf{S}^{\mathrm{DL}}$ in description logic is sound and complete with respect to $\subseteq$-minimal change.

Here, if $C(a)$ is consistent with Δ we denote by $\models_{\mathbf{S}^{\mathrm{DL}}} \Delta|C(a) \Rightarrow \Delta, C(a)$; and if $C(a)$ is inconsistent with Δ we denote by $\models_{\mathbf{S}^{\mathrm{DL}}} \Delta|C(a) \Rightarrow \Delta$. For a set Γ of statements, if a theory Θ is a $\subseteq$-minimal change of Γ by Δ then we denote

$$\models_{\mathbf{S}^{\mathrm{DL}}} \Delta|\Gamma \Rightarrow \Delta, \Theta.$$

4.1.1 R-Calculus $\mathbf{S}^{\mathrm{DL}}$ for a Statement

The deduction rules in R-calculus $\mathbf{S}^{\mathrm{DL}}$ are the composing rules, which compose sub-statements (e.g., $C_1(a), C_2(a)$) in the precondition of a rule into a complex statement (e.g., $(C_1 \sqcap C_2)(a)$) in the postcondition of the rule.

R-calculus $\mathbf{S}^{\mathrm{DL}}$ for a statement $C(a)$ consists of the following axioms and deduction rules:

- **Axioms:**

$$(S^{\mathbf{A}})\ \frac{\Delta \nvdash \neg A(c)}{\Delta|A(c) \Rightarrow \Delta, A(c)} \quad (S_{\mathbf{A}})\ \frac{\Delta \vdash \neg A(c)}{\Delta|A(a) \Rightarrow \Delta}$$

- **Deduction rules:**

$$(S^{\neg})\ \frac{\Delta \nvdash A(c)}{\Delta|\neg A(c) \Rightarrow \Delta, \neg A(c)} \qquad (S_{\neg})\ \frac{\Delta \vdash A(c)}{\Delta|\neg A(a) \Rightarrow \Delta}$$

$$(S^{\sqcap})\ \frac{\begin{array}{c}\Delta|C_1(a) \Rightarrow \Delta, C(a)\\ \Delta, C_1(a)|C_2(a) \Rightarrow \Delta, C_1(a), C_2(a)\end{array}}{\Delta|(C_1 \sqcap C_2)(a) \Rightarrow \Delta, (C_1 \sqcap C_2)(a)} \qquad (S^1_{\sqcap})\ \frac{\Delta|C_1(a) \Rightarrow \Delta}{\Delta|(C_1 \sqcap C_2)(a) \Rightarrow \Delta}$$

$$(S^2_{\sqcap})\ \frac{\Delta, C_1(a)|C_2(a) \Rightarrow \Delta, C_1(a)}{\Delta|(C_1 \sqcap C_2)(a) \Rightarrow \Delta}$$

$$(S^{\sqcup}_1)\ \frac{\Delta|C_1(a) \Rightarrow \Delta, C_1(a)}{\Delta|(C_1 \sqcup C_2)(a) \Rightarrow \Delta, (C_1 \sqcup C_2)(a)} \qquad (S_{\sqcup})\ \frac{\begin{array}{c}\Delta|C_1(a) \Rightarrow \Delta\\ \Delta|C_2(a) \Rightarrow \Delta\end{array}}{\Delta|(C_1 \sqcup C_2)(a) \Rightarrow \Delta}$$

$$(S^{\sqcup}_2)\ \frac{\Delta|C_2(a) \Rightarrow \Delta, C_2(a)}{\Delta|(C_1 \sqcup C_2)(a) \Rightarrow \Delta, (C_1 \sqcup C_2)(a)}$$

$$(S^{\forall})\ \frac{R(a,c) \in \Delta \quad \Delta|C(c) \Rightarrow \Delta, C(c)}{\Delta|(\forall R.C)(a) \Rightarrow \Delta, (\forall R.C)(a)} \qquad (S_{\forall})\ \frac{\Delta \vdash R(a,d) \in \Delta \quad \Delta|C(d) \Rightarrow \Delta}{\Delta|(\forall R.C)(a) \Rightarrow \Delta}$$

$$(S^{\exists})\ \frac{\Delta \vdash R(a,d) \quad \Delta|C(d) \Rightarrow \Delta, C(d)}{\Delta|(\exists R.C)(a) \Rightarrow \Delta, (\exists R.C)(a)} \qquad (S_{\exists})\ \frac{R(a,c) \in \Delta \quad \Delta|C(c) \Rightarrow \Delta}{\Delta|(\exists R.C)(a) \Rightarrow \Delta}$$

where d is a constant and c is a new constant.

Remark Because Δ is not contradictory to $R(a, b)$, whether Δ contradicts with $(\forall R.C)(a)$ depends on both whether Δ has $R(a, d)$ and whether Δ contradicts with $C(d)$; and whether Δ contradicts with $(\exists R.C)(a)$ only depends on whether Δ contradicts with $C(c)$. □

Definition 4.1.1 $\Delta|C(a) \Rightarrow \Delta, C^i(a)$ is provable in $\mathbf{S}^{\mathrm{DL}}$, denoted by $\vdash_{\mathbf{S}^{\mathrm{DL}}} \Delta|C(a) \Rightarrow \Delta, C^i(a)$, if there is a sequence $\theta_1, \ldots, \theta_m$ of statements such that

$$\begin{aligned} \theta_1 &= \Delta|C(a) \Rightarrow \Delta|C_1(a), \\ &\ldots \\ \theta_m &= \Delta|C_{m-1}(a) \Rightarrow \Delta, C^i(a); \end{aligned}$$

and for each $j < m$, $\Delta|C_j(a) \Rightarrow \Delta|C_{j+1}(a)$ is an axiom or is deduced from the previous statements by a deduction rule in $\mathbf{S}^{\mathrm{DL}}$, where $i \in \{0, 1\}$, $C^1(a) = C(a)$ and $C^0(a) = \lambda$, the empty string.

Intuitively, we firstly decompose $C(a)$ into literals according to the structure of $C(a)$, and then delete/add literals by rules $(S^{\mathbf{A}})$, $(S_{\mathbf{A}})/(S^{\neg})$, $(S_{\neg})$.

Example 4.1.2 Let

$$\begin{aligned} \Delta &= \{(\exists R.(A_1 \sqcap A_2))(a), (\exists R.(A_3 \sqcap A_4)(a)\} \\ C(a) &= (\forall R.((\neg A_1 \sqcap \neg A_3) \sqcup (\neg A_2 \sqcap \neg A_4)))(a). \end{aligned}$$

We have the following deduction:

$$\begin{aligned} &\Delta \vdash R(a, d_1), (A_1 \sqcap A_2)(d_1) \\ &\Delta|(\neg A_1 \sqcap \neg A_3)(d_1) \Rightarrow \Delta \\ &\Delta|(\neg A_2 \sqcap \neg A_4)(d_1) \Rightarrow \Delta \\ &\Delta|((\neg A_1 \sqcap \neg A_3) \sqcup (\neg A_2 \sqcap \neg A_4))(d_1) \Rightarrow \Delta \\ &\Delta|C(a) \Rightarrow \Delta. \end{aligned}$$

Theorem 4.1.3 (Soundness theorem) *For any consistent theory Δ and statement $C(a)$, if $\vdash_{\mathbf{S}^{\mathrm{DL}}} \Delta|C(a) \Rightarrow \Delta, C^i(a)$ then if $i = 0$ then $\Delta \cup \{C(a)\}$ is inconsistent, i.e.,*

$$\vdash_{\mathbf{S}^{\mathrm{DL}}} \Delta|C(a) \Rightarrow \Delta \text{ implies } \models_{\mathbf{S}^{\mathrm{DL}}} \Delta|C(a) \Rightarrow \Delta;$$

otherwise, $\Delta \cup \{C(a)\}$ is consistent, i.e.,

$$\vdash_{\mathbf{S}^{\mathrm{DL}}} \Delta|C(a) \Rightarrow \Delta, C(a) \text{ implies } \models_{\mathbf{S}^{\mathrm{DL}}} \Delta|C(a) \Rightarrow \Delta, C(a).$$

Proof Assume that $\Delta|C(a) \Rightarrow \Delta, C^i(a)$ is provable. We prove the theorem by induction on the structure of C.

Case $C(a) = B(a)$, where $B ::= A|\neg A$. Then $\vdash_{\mathbf{S}^{\mathrm{DL}}} \Delta|B(a) \Rightarrow \Delta, B^i(a)$ only if

$$\begin{cases} \Delta \vdash \neg B(a) \text{ if } i = 0 \\ \Delta \nvdash \neg B(a) \text{ if } i = 1; \end{cases}$$

that is,

$$\begin{cases} \text{incon}(\Delta, B(a)) \text{ if } i = 0 \\ \text{con}(\Delta, B(a)) \quad \text{if } i = 1. \end{cases}$$

Case $C(a) = (C_1 \sqcap C_2)(a)$. If $\vdash_{\mathbf{S}^{\mathrm{DL}}} \Delta|(C_1 \sqcap C_2)(a) \Rightarrow \Delta, (C_1 \sqcap C_2)(a)$ then

$$\vdash_{\mathbf{S}^{\mathrm{DL}}} \Delta|C_1(a) \Rightarrow \Delta, C_1(a);$$
$$\vdash_{\mathbf{S}^{\mathrm{DL}}} \Delta, C_1(a)|C_2(a) \Rightarrow \Delta, C_1(a), C_2(a).$$

By induction assumption, $\Delta \cup \{C_1(a)\}$ and $\Delta \cup \{C_1(a), C_2(a)\}$ are consistent, and so is $\Delta \cup \{(C_1 \sqcap C_2)(a)\}$;

If $\vdash_{\mathbf{S}^{\mathrm{DL}}} \Delta|(C_1 \sqcap C_2)(a) \Rightarrow \Delta$ then either

$$\vdash_{\mathbf{S}^{\mathrm{DL}}} \Delta|C_1(a) \Rightarrow \Delta$$

or

$$\vdash_{\mathbf{S}^{\mathrm{DL}}} \Delta, C_1(a)|C_2(a) \Rightarrow \Delta, C_1(a).$$

By induction assumption, either $\Delta \cup \{C_1(a)\}$ is inconsistent or $\Delta \cup \{C_1(a), C_2(a)\}$ is inconsistent, which implies $\Delta \cup \{(C_1 \sqcap C_2)(a)\}$ is inconsistent.

Case $C(a) = (C_1 \sqcup C_2)(a)$. If $\vdash_{\mathbf{S}^{\mathrm{DL}}} \Delta|(C_1 \sqcup C_2)(a) \Rightarrow \Delta, (C_1 \sqcup C_2)(a)$ then either $\vdash_{\mathbf{S}^{\mathrm{DL}}} \Delta|C_1(a) \Rightarrow \Delta, C_1(a)$ or $\vdash_{\mathbf{S}^{\mathrm{DL}}} \Delta|C_2(a) \Rightarrow \Delta, C_2(a)$. By induction assumption, either $\Delta \cup \{C_1(a)\}$ is consistent or $\Delta \cup \{C_2(a)\}$ is consistent, either of which implies $\Delta \cup \{(C_1 \sqcup C_2)(a)\}$ is consistent;

If $\vdash_{\mathbf{S}^{\mathrm{DL}}} \Delta|(C_1 \sqcup C_2)(a) \Rightarrow \Delta$ then $\vdash_{\mathbf{S}^{\mathrm{DL}}} \Delta|C_1(a) \Rightarrow \Delta$ and $\vdash_{\mathbf{S}^{\mathrm{DL}}} \Delta|C_2(a) \Rightarrow \Delta$. By induction assumption, $\Delta \cup \{C_1(a)\}$ is inconsistent and $\Delta \cup \{C_2(a)\}$ is inconsistent, which imply $\Delta \cup \{(C_1 \sqcup C_2)(a)\}$ is inconsistent.

Case $C(a) = (\forall R.C_1)(a)$. If $\vdash_{\mathbf{S}^{\mathrm{DL}}} \Delta|(\forall R.C_1)(a) \Rightarrow \Delta, (\forall R.C_1)(a)$ then for any constant c with $R(a, c)$ being enumerated in Δ, $\vdash_{\mathbf{S}^{\mathrm{DL}}} \Delta|C_1(c) \Rightarrow \Delta, C_1(c)$. By induction assumption, $\Delta \cup \{C_1(c)\}$ is consistent, and so is $\Delta \cup \{(\forall R.C_1)(a)\}$.

If $\vdash_{\mathbf{S}^{\mathrm{DL}}} \Delta|(\forall R.C_1)(a) \Rightarrow \Delta$ then for some constant d with $\Delta \vdash R(a, d)$, $\vdash_{\mathbf{S}^{\mathrm{DL}}} \Delta|C_1(d) \Rightarrow \Delta$. By induction assumption, $\Delta \cup \{C_1(d)\}$ is inconsistent, and so is $\Delta \cup \{(\forall R.C_1)(a)\}$.

Case $C(a) = (\exists R.C_1)(a)$. If $\vdash_{\mathbf{S}^{\mathrm{DL}}} \Delta|(\exists R.C_1)(a) \Rightarrow \Delta, (\exists R.C_1)(a)$ then for some constant d with $\Delta \vdash R(a, d)$, $\vdash_{\mathbf{S}^{\mathrm{DL}}} \Delta|C_1(d) \Rightarrow \Delta, C_1(d)$. By induction assumption, $\Delta \cup \{C_1(d)\}$ is consistent, and so $\Delta \cup \{(\exists R.C_1)(a)\}$;

If $\vdash_{\mathbf{S}^{\mathrm{DL}}} \Delta|(\exists R.C_1)(a) \Rightarrow \Delta$ then for any constant c with $R(a, c) \in \Delta$, $\vdash_{\mathbf{S}^{\mathrm{DL}}} \Delta|C_1(c) \Rightarrow \Delta$. By induction assumption, $\Delta \cup \{C_1(c)\}$ is inconsistent, and so is $\Delta \cup \{(\forall R.C_1)(a)\}$. □

Theorem 4.1.4 (Completeness theorem) *For any consistent theory* Δ *and statement* $C(a)$, *if* $\Delta \cup \{C(a)\}$ *is consistent then* $\Delta|C(a) \Rightarrow \Delta, C(a)$ *is provable in* $\mathbf{S}^{\mathrm{DL}}$, *i.e.,*

$$\models_{\mathbf{S}^{\mathrm{DL}}} \Delta|C(a) \Rightarrow \Delta, C(a) \text{ implies } \vdash_{\mathbf{S}^{\mathrm{DL}}} \Delta|C(a) \Rightarrow \Delta, C(a);$$

and if $\Delta \cup \{C(a)\}$ *is inconsistent then* $\Delta|C(a) \Rightarrow \Delta$ *is provable in* $\mathbf{S}^{\mathrm{DL}}$, *i.e.,*

$$\models_{\mathbf{S}^{\mathrm{DL}}} \Delta|C(a) \Rightarrow \Delta \text{ implies } \vdash_{\mathbf{S}^{\mathrm{DL}}} \Delta|C(a) \Rightarrow \Delta.$$

Proof We prove the theorem by induction on the structure of C.

Case $C(a) = B(a)$, where $B ::= A|\neg A$. If $\Delta \cup \{B(a)\}$ is consistent then $\Delta \nvdash \neg B(a)$, by $(S^{\mathbf{A}})$ and $(S^{\neg})$,

$$\vdash_{\mathbf{S}^{\mathrm{DL}}} \Delta|B(a) \Rightarrow \Delta, B(a);$$

and if $\Delta \cup \{B(a)\}$ is inconsistent then $\Delta \vdash \neg B(a)$, by $(S_{\mathbf{A}})$ and $(S_{\neg})$,

$$\vdash_{\mathbf{S}^{\mathrm{DL}}} \Delta|B(a) \Rightarrow \Delta.$$

Case $C(a) = (C_1 \sqcap C_2)(a)$. If $\Delta \cup \{(C_1 \sqcap C_2)(a)\}$ is consistent then $\Delta \cup \{C_1(a)\}$ and $\Delta \cup \{C_1(a), C_2(a)\}$ are consistent, and by induction assumption,

$$\begin{array}{l} \vdash_{\mathbf{S}^{\mathrm{DL}}} \Delta|C_1(a) \Rightarrow \Delta, C_1(a) \\ \vdash_{\mathbf{S}^{\mathrm{DL}}} \Delta, C_1(a)|C_2(a) \Rightarrow \Delta, C_1(a), C_2(a). \end{array}$$

By $(S^{\sqcap})$, we have that $\vdash_{\mathbf{S}^{\mathrm{DL}}} \Delta|(C_1 \sqcap C_2)(a) \Rightarrow \Delta, (C_1 \sqcap C_2)(a)$;

If $\Delta \cup \{(C_1 \sqcap C_2)(a)\}$ is inconsistent then either $\Delta \cup \{C_1(a)\}$ is inconsistent or $\Delta \cup \{C_1(a)\} \cup \{C_2(a)\}$ is inconsistent. By induction assumption, either $\vdash_{\mathbf{S}^{\mathrm{DL}}} \Delta|C_1(a) \Rightarrow \Delta$, or $\vdash_{\mathbf{S}^{\mathrm{DL}}} \Delta, C_1(a)|C_2(a) \Rightarrow \Delta, C_1(a)$. By $(S^1_{\sqcap})$ or $(S^2_{\sqcap})$, we have that $\vdash_{\mathbf{S}^{\mathrm{DL}}} \Delta|(C_1 \sqcap C_2)(a) \Rightarrow \Delta$.

Case $C(a) = (C_1 \sqcup C_2)(a)$. If $\Delta \cup \{(C_1 \sqcup C_2)(a)\}$ is consistent then either $\Delta \cup \{C_1(a)\}$ or $\Delta \cup \{C_2(a)\}$ are consistent, and by induction assumption, either

$$\vdash_{\mathbf{S}^{\mathrm{DL}}} \Delta|C_1(a) \Rightarrow \Delta, C_1(a),$$

or

$$\vdash_{\mathbf{S}^{\mathrm{DL}}} \Delta|C_2(a) \Rightarrow \Delta, C_2(a).$$

By $(S^{\sqcap}_1)$ or $(S^{\sqcap}_2)$, we have that $\vdash_{\mathbf{S}^{\mathrm{DL}}} \Delta|(C_1 \sqcup C_2)(a) \Rightarrow \Delta, (C_1 \sqcup C_2)(a)$;

If $\Delta \cup \{(C_1 \sqcup C_2)(a)\}$ is inconsistent then $\Delta \cup \{C_1(a)\}$ and $\Delta \cup \{C_2(a)\}$ are inconsistent. By induction assumption,

$$\begin{array}{l} \vdash_{\mathbf{S}^{\mathrm{DL}}} \Delta|C_1(a) \Rightarrow \Delta, \\ \vdash_{\mathbf{S}^{\mathrm{DL}}} \Delta|C_2(a) \Rightarrow \Delta. \end{array}$$

By $(S_{\sqcup})$, we have that $\vdash_{\mathbf{S}^{\mathrm{DL}}} \Delta|(C_1 \sqcup C_2)(a) \Rightarrow \Delta$.

Case $C(a) = (\forall R.C_1)(a)$. If $\Delta \cup \{(\forall R.C_1)(a)\}$ is consistent then for any constant c with $R(a, c) \in \Delta$, $\Delta \cup \{C_1(c)\}$ is consistent, and by induction assumption,

$$\vdash_{\mathbf{S}^{\mathrm{DL}}} \Delta|C_1(c) \Rightarrow \Delta, C_1(c).$$

By $(S^{\forall})$, we have that $\vdash_{\mathbf{S}^{\mathrm{DL}}} \Delta|(\forall R.C_1)(a) \Rightarrow \Delta, (\forall R.C_1)(a)$;

If $\Delta \cup \{(\forall R.C_1)(a)\}$ is inconsistent then there is a constant d such that $\Delta \vdash R(a, d)$ and $\Delta \cup \{C_1(d)\}$ is inconsistent. By induction assumption, $\vdash_{\mathbf{S}^{\mathrm{DL}}} \Delta|C_1(d) \Rightarrow \Delta$. By $(S_{\forall})$, we have that $\vdash_{\mathbf{S}^{\mathrm{DL}}} \Delta|(\forall R.C_1)(a) \Rightarrow \Delta$.

Case $C(a) = (\exists R.C_1)(a)$. If $\Delta \cup \{(\exists R.C_1)(a)\}$ is consistent then there is a constant d such that $\Delta \vdash R(a, d)$ and $\Delta \cup \{C_1(d)\}$ is consistent, and by induction assumption,

$$\vdash_{\mathbf{S}^{\mathrm{DL}}} \Delta|C_1(d) \Rightarrow \Delta, C_1(d).$$

By $(S^{\exists})$, we have that $\vdash_{\mathbf{S}^{\mathrm{DL}}} \Delta|(\exists R.C_1)(a) \Rightarrow \Delta, (\exists R.C_1)(a)$;

If $\Delta \cup \{(\exists R.C_1)(a)\}$ is inconsistent then for any constant c with $R(a, c) \in \Delta$, $\Delta \cup \{C_1(c)\}$ is inconsistent. By induction assumption, $\vdash_{\mathbf{S}^{\mathrm{DL}}} \Delta|C_1(c) \Rightarrow \Delta$. By $(S_{\exists})$, we have that $\vdash_{\mathbf{S}^{\mathrm{DL}}} \Delta|(\exists R.C_1)(c) \Rightarrow \Delta$. □

4.1.2 R-Calculus $\mathbf{S}^{\mathrm{DL}}$ for a Set of Statements

Let Γ be a finite consistent set of statements such that $\Gamma = \{C_1(a), \ldots, C_n(a)\}$. Define

$$\Delta|\Gamma = (\cdots((\Delta|C_1(a))|C_2(a))|\cdots)|C_n(a).$$

Correspondingly, we have the following

Theorem 4.1.5 *For any consistent theories Δ, Γ and Θ, if $\Delta|\Gamma \Rightarrow \Delta, \Theta$ is provable in $\mathbf{S}^{\mathrm{DL}}$ then Θ is a $\subseteq$-minimal change of Γ by Δ. That is,*

$$\vdash_{\mathbf{S}^{\mathrm{DL}}} \Delta|\Gamma \Rightarrow \Delta, \Theta \textit{ implies } \models_{\mathbf{S}^{\mathrm{DL}}} \Delta|\Gamma \Rightarrow \Delta, \Theta.$$

Proof We prove the theorem by induction on n.

Assume that $\Delta|\Gamma \Rightarrow \Delta, \Theta$ is provable in $\mathbf{S}^{\mathrm{DL}}$.

Let $n = 1$. Then, either $\Theta = C_1(a_1)$ or $\Theta = \lambda$. If $\Theta = C_1(a_1)$ then $\Delta \cup \{C_1(a_1)\}$ is consistent, and Θ is a $\subseteq$-minimal change of $C_1(a_1)$ by Δ; otherwise, $\Delta \cup \{C_1(a_1)\}$ is inconsistent, and $\Theta = \lambda$ is a $\subseteq$-minimal change of $C_1(a_1)$ by Δ.

Assume that the theorem holds for n, that is, if $\Delta|\Gamma \Rightarrow \Delta, \Theta$ then Θ is a $\subseteq$-minimal change of Γ by Δ, where $\Gamma = (C_1(a_1), \ldots, C_n(a_n))$.

Let $\Gamma' = (\Gamma, C_{n+1}(a_{n+1})) = (C_1(a_1), \ldots, C_{n+1}(a_{n+1}))$. Then, if $\Delta|\Gamma' \Rightarrow \Delta, \Theta'$ is provable then $\Delta|\Gamma \Rightarrow \Delta, \Theta$ and $\Delta, \Theta|C_{n+1}(a_{n+1}) \Rightarrow \Delta, \Theta'$ are provable. By the case $n = 1$ and induction assumption, Θ' is a $\subseteq$-minimal change of $C_{n+1}(a_{n+1})$ by $\Delta \cup \Theta$, and Θ is a $\subseteq$-minimal change of Γ by Δ, therefore, Θ' is a $\subseteq$-minimal change of Γ' by Δ. □

Theorem 4.1.6 *For any consistent theories Δ and Γ and any $\subseteq$-minimal change Θ of Γ by Δ, $\Delta|\Gamma \Rightarrow \Theta$ is provable in $\mathbf{S}^{\mathrm{DL}}$. That is,*

$$\models_{\mathbf{S}^{\mathrm{DL}}} \Delta|\Gamma \Rightarrow \Theta \text{ implies } \vdash_{\mathbf{S}^{\mathrm{DL}}} \Delta|\Gamma \Rightarrow \Theta.$$

Proof Assume that Θ is a $\subseteq$-minimal change of Γ by Δ. Then, there is an ordering $<$ of Γ such that $\Gamma = (C_1(a_1), C_2(a_2), \ldots, C_n(a_n))$, where $C_1(a_1) < C_2(a_2) < \cdots < C_n(a_n)$, and Θ is a maximal subset of Γ such that $\Delta \cup \Theta$ is consistent.

We prove the theorem by induction on n.

Let $n = 1$. By the last theorem, if $\Theta = \{C_1(a_1)\}$ then $\Delta|C_1(a_1) \Rightarrow \Delta, C_1(a_1)$ is provable; and if $\Theta = \emptyset$ then $\Delta|C_1(a_1) \Rightarrow \Delta, \Delta$ is provable.

Assume that the theorem holds for n, that is, if Θ is a $\subseteq$-minimal change of Γ by Δ then $\Delta|\Gamma \Rightarrow \Delta, \Theta$ is provable.

Let $\Gamma' = (\Gamma, C_{n+1}(a_{n+1})) = (C_1(a_1), \ldots, C_{n+1}(a_{n+1}))$ and Θ' is a $\subseteq$-minimal change of Γ' by Δ. Then, Θ' is a $\subseteq$-minimal change of $C_{n+1}(a_{n+1})$ by $\Delta \cup \Theta$, and $\Delta, \Theta|C_{n+1}(a_{n+1}) \Rightarrow \Delta, \Theta'$ is provable. By induction assumption, $\Delta|\Gamma \Rightarrow \Delta, \Theta$ is provable and so is $\Delta|\Gamma' \Rightarrow \Delta, \Theta|C_{n+1}(a_{n+1})$, and hence, $\Delta|\Gamma' \Rightarrow \Delta, \Theta'$ is provable in **S**. □

4.2 R-Calculus for ⪯-Minimal Change

There are concepts *para-subconcepts* and *pseudo-subconcepts* in description logics, corresponding to the subformulas and pseudo-subformulas, and based on which there is an R-calculus $\mathbf{T}^{\mathrm{DL}}$ sound and complete with respect to $\preceq$-minimal change.

4.2.1 Pseudo-Subconcept-Minimal Change

In description logics, a concept C is a subconcept of D, denoted by $C \sqsubseteq D$, which is true in an interpretation I if $C^I \subseteq D^I$.

The para-subconcepts correspond to subformulas in traditional logics.

Definition 4.2.1 Given any concept C, a concept D is a para-subconcept of C, denoted by $D \leq C$, if either $C = D$, or

(i) if $C = \neg C_1$ then $D \leq C_1$;

(ii) if $C = C_1 \sqcap C_2$ or $C_1 \sqcup C_2$ then either $D \leq C_1$ or $D \leq C_2$.

For example, let $C = (A_1 \sqcup A_2) \sqcap (A_3 \sqcup A_4)$. Then,

$$A_1 \sqcup A_2, A_3 \sqcup A_4 \leq C;$$

and

$$A_1 \sqcap A_3, A_2 \sqcap A_4, A_1 \sqcap (A_3 \sqcup A_4) \not\leq C.$$

Definition 4.2.2 Given a concept $C[D_1, \ldots, D_n]$, where $[D_1]$ is an occurrence of D_1 in C, a concept $D = C[\lambda, \ldots, \lambda] = C[D_1/\lambda, \ldots, D_n/\lambda]$, where the occurrence

D_i is replaced by the empty concept λ, is called a pseudo-subconcept of C, denoted by $D \preceq C$.

For example let $C = (A_1 \sqcup A_2) \sqcap (A_3 \sqcup A_4)$. Then,

$$A_1 \sqcup A_2, A_3 \sqcup A_4, A_1 \sqcap A_3, A_2 \sqcap A_3, A_1 \sqcap (A_3 \sqcup A_4) \preceq C.$$

Proposition 4.2.3 *For any concepts C_1, C_2, D_1 and D_2,*
(i) $D_1 \leq C_1$ implies $D_1 \leq C_1 \sqcup C_2$ and $D_1 \leq C_1 \sqcap C_2$;
(ii) $D_1 \preceq C_1$ and $D_2 \preceq C_2$ imply $\neg D_1 \preceq \neg C_1$, $D_1 \sqcup D_2 \preceq C_1 \sqcup C_2$ and $D_1 \sqcap D_2 \preceq C_1 \sqcap C_2$. □

Proposition 4.2.4 *For any concepts C and D, if $D \leq C$ then $D \preceq C$.* □

Proof By induction on the structure of C. □

Proposition 4.2.5 *$\leq$ and $\preceq$ are partial orderings on the set of all the concepts.* □

Given a concept C, let $P(C)$ be the set of all the pseudo-subconcepts of C. Each $D \in P(C)$ is determined by a set $\tau(D) = \{[A_1], \ldots, [A_n]\}$, where each $[A_i]$ is an occurrence of A_i in C, such that

$$D = C[A_1]/\lambda, \ldots, [A_n]/\lambda).$$

Given any $D_1, D_2 \in P(C)$, define

$$\begin{aligned} D_1 \curlywedge D_2 &= \max\{D : D \preceq D_1, D \preceq D_2\}; \\ D_1 \curlyvee D_2 &= \min\{D : D \succeq D_1, D \succeq D_2\}. \end{aligned}$$

Proposition 4.2.6 *For any pseudo-subconcepts $D_1, D_2 \in P(C)$, $D_1 \curlywedge D_2$ and $D_1 \curlyvee D_2$ exist.*

Let $\mathbf{P}(C) = (P(C), \curlywedge, \curlyvee, C, \lambda)$ be the lattice with the greatest element C and the least element λ.

Proposition 4.2.7 *For any pseudo-subconcepts $D_1, D_2 \in P(C)$, $D_1 \preceq D_2$ if and only if $\tau(D_1) \supseteq \tau(D_2)$. Moreover,*

$$\begin{aligned} \tau(D_1 \curlywedge D_2) &= \tau(D_1) \cup \tau(D_2); \\ \tau(D_1 \curlyvee D_2) &= \tau(D_1) \cap \tau(D_2). \end{aligned}$$

Definition 4.2.8 A theory Θ is a $\preceq^{\mathrm{DL}}$-minimal change of Γ by Δ, denoted by $\models_{\mathrm{T^{DL}}} \Delta|\Gamma \Rightarrow \Delta, \Theta$, if

(i) $\Theta \preceq \Gamma$, that is, for each statement $C(a) \in \Theta$, there is a concept D such that $C \preceq D$ and $D(a) \in \Gamma$,

(ii) $\Theta \cup \Delta$ is consistent, and

(iii) for any theory Ξ with $\Theta \prec \Xi \preceq \Gamma$, $\Xi \cup \Delta$ is inconsistent.

4.2.2 R-Calculus $\mathbf{T}^{\mathrm{DL}}$ for a Statement

R-calculus $\mathbf{T}^{\mathrm{DL}}$ for a statement $C(a)$ consists of the following axioms and deduction rules:

- **Axioms:**

$$(T^{\mathbf{A}})\ \frac{\Delta \nvdash \neg C(a)}{\Delta|C(a) \Rightarrow \Delta, C(a)} \qquad (T_{\mathbf{A}})\ \frac{\Delta \vdash \neg B(a)}{\Delta|B(a) \Rightarrow \Delta}$$

- **Deduction rules:**

$$(T^{\wedge})\ \frac{\Delta|C_1(a) \Rightarrow \Delta, D_1(a)}{\Delta|(C_1 \sqcap C_2)(a) \Rightarrow \Delta, D_1(a)|C_2(a)}$$

$$(T_1^{\vee})\ \frac{\Delta|C_1(a) \Rightarrow \Delta, D_1(a)\ \ D_1 \neq \perp}{\Delta|(C_1 \sqcup C_2)(a) \Rightarrow \Delta, (D_1 \sqcup C_2)(a)}$$

$$(T_2^{\vee})\ \frac{\Delta|C_1(a) \Rightarrow \Delta\ \ \Delta|C_2(a) \Rightarrow \Delta, D_2(a)\ \ D_2 \neq \perp}{\Delta|(C_1 \sqcup C_2)(a) \Rightarrow \Delta, (C_1 \sqcup D_2)(a)}$$

$$(T_3^{\vee})\ \frac{\Delta|C_1(a) \Rightarrow \Delta\ \ \Delta|C_2(a) \Rightarrow \Delta}{\Delta|(C_1 \sqcup C_2)(a) \Rightarrow \Delta}$$

$$(T^{\forall})\ \frac{\Delta \vdash R(a,d)\ \ \Delta|C_1(d) \Rightarrow \Delta, D_1(d)}{\Delta|(\forall R.C_1)(a) \Rightarrow \Delta|(\forall R.D_1)(a)}$$

$$(T^{\exists})\ \frac{R(a,c) \in \Delta\ \ \Delta|C_1(c) \Rightarrow \Delta, D_1(c)}{\Delta|(\exists R.C_1)(a) \Rightarrow \Delta, (\exists R.D_1)(a)}$$

where $B(a) ::= A(a)|\neg A(a)$ for some atomic concept A; and d is a constant, and c is a new constant not occurring in Δ.

To understand rule $(T^{\forall})$, assume that there are constants $d_1, \ldots, d_n$, and pseudo-subconcepts $D_1, \ldots, D_n$ of C such that $R(a, a_1), \ldots, R(a, a_n) \in \Delta$, and

$$\begin{aligned} &\vdash_{\mathbf{T}^{\mathrm{DL}}} \Delta|C(d_1) \Rightarrow \Delta, D_1(d_1), \\ &\vdash_{\mathbf{T}^{\mathrm{DL}}} \Delta|D_1(d_2) \Rightarrow \Delta, D_2(d_2), \\ &\cdots \\ &\vdash_{\mathbf{T}^{\mathrm{DL}}} \Delta|D_{n-1}(d_n) \Rightarrow \Delta, D_n(d_n). \end{aligned}$$

Let $D = D_n$. Then,

$$\vdash_{\mathbf{T}^{\mathrm{DL}}} \Delta|(\forall R.C)(a) \Rightarrow \Delta, (\forall R.D)(a).$$

Definition 4.2.9 $\Delta|C(a) \Rightarrow \Delta, D(a)$ is provable in $\mathbf{T}^{\mathrm{DL}}$, denoted by $\vdash_{\mathbf{T}^{\mathrm{DL}}} \Delta|C(a) \Rightarrow \Delta, D(a)$, if there is a sequence $\theta_1, \ldots, \theta_m$ of statements such that

$$\begin{aligned} \theta_1 &= \Delta|C(a) \Rightarrow \Delta|C_2(a), \\ &\cdots \\ \theta_m &= \Delta|C_m(a) \Rightarrow \Delta, D(a); \end{aligned}$$

and for each $j \leq m$, $\Delta|C_j(a) \Rightarrow \Delta|C_{j+1}(a)$ is an axiom or is deduced from the previous statements by a deduction rule, where $C_1 = C$, $C_{m+1} = D$.

Example 4.2.10 Let

$$\begin{array}{l}\Delta = \{(\exists R.(A_1 \sqcap A_2))(a), (\exists R.A_3)(a), (\exists R.A_4)(a)\} \\ C(a) = (\forall R.((\neg A_1 \sqcap \neg A_3) \sqcup (\neg A_2 \sqcap \neg A_4)))(a).\end{array}$$

Then,

$$\begin{array}{l}\Delta \vdash (A_1 \sqcap A_2)(d), R(a,d) \\ \Delta|(\neg A_1 \sqcap \neg A_3)(d) \Rightarrow \Delta|(\neg A_3)(d), \\ \Delta|((\neg A_1 \sqcap \neg A_3) \sqcup (\neg A_2 \sqcap \neg A_4))(d) \Rightarrow \Delta, (\neg A_3 \sqcup (\neg A_2 \sqcap \neg A_4))(d) \\ \Delta|C(a) \Rightarrow \Delta, (\forall R.(\neg A_3 \sqcup (\neg A_2 \sqcap \neg A_4)))(a).\end{array}$$

Example 4.2.11 Let

$$\begin{array}{l}\Delta = \{(\exists R.(A_1 \sqcap A_2))(a), (\exists R.(A_3 \sqcap A_4)(a)\} \\ C(a) = (\forall R.((\neg A_1 \sqcap \neg A_3) \sqcup (\neg A_2 \sqcap \neg A_4)))(a).\end{array}$$

We have the following deduction:

$$\begin{array}{l}\Delta \vdash R(a,d_1), (A_1 \sqcap A_2)(d_1) \\ \Delta|((\neg A_1 \sqcap \neg A_3) \sqcup (\neg A_2 \sqcap \neg A_4))(d_1) \Rightarrow \Delta, (\neg A_3 \sqcup (\neg A_2 \sqcap \neg A_4))(d_1) \\ \Delta \vdash R(a,d_2), (A_3 \sqcap A_4)(d_2) \\ \Delta|(\neg A_3 \sqcup (\neg A_2 \sqcap \neg A_4))(d_2) \Rightarrow \Delta, (\neg A_3 \sqcup \neg A_2)(d_2); \\ \Delta|C(a) \Rightarrow \Delta, (\forall R.(\neg A_3 \sqcup \neg A_2))(a).\end{array}$$

Theorem 4.2.12 (Soundness theorem) *For any statement set* Δ *and statements* $C(a), D(a)$,

$$\vdash_{\mathbf{T}^{\mathrm{DL}}} \Delta|C(a) \Rightarrow \Delta, D(a) \text{ implies } \models_{\mathbf{T}^{\mathrm{DL}}} \Delta|C(a) \Rightarrow \Delta, D(a).$$

Proof We prove the theorem by induction on the structure of C. Assume that $\vdash_{\mathbf{T}^{\mathrm{DL}}} \Delta|C(a) \Rightarrow \Delta, D(a)$.

Case $C = B$, where $B ::= A|\neg A$. Then, by $(T^{\mathbf{A}})$ and $(T_{\mathbf{A}})$,

$$D(a) = \begin{cases} A(a) & \text{if } C = A \text{ and } \Delta \nvdash \neg A(a) \\ \neg A(a) & \text{if } C = \neg A \text{ and } \Delta \nvdash A(a) \\ \lambda & \text{otherwise,} \end{cases}$$

and it is clear that $D(a)$ is a $\preceq^{\mathrm{DL}}$-minimal change of $C(a)$ by Δ.

Case $C = C_1 \sqcap C_2$. Then, by $(T^{\wedge})$,

$$\begin{array}{l}\vdash_{\mathbf{T}^{\mathrm{DL}}} \Delta|C_1(a) \Rightarrow \Delta, D_1(a), \\ \vdash_{\mathbf{T}^{\mathrm{DL}}} \Delta, C_1(a)|C_2(a) \Rightarrow \Delta, D_1(a), D_2(a),\end{array}$$

and by induction assumption, $D_1(a)$ is a $\preceq^{\rm DL}$-minimal change of $C_1(a)$ by Δ, and $D_2(a)$ is a $\preceq^{\rm DL}$-minimal change of $C_2(a)$ by $\Delta \cup \{C_1(a)\}$. Therefore, $(D_1 \sqcap D_2)(a)$ is a $\preceq^{\rm DL}$-minimal change of $(C_1 \sqcap C_2)(a)$ by Δ.

Case $C = C_1 \sqcup C_2$. Then, there are the following three cases:

(i) if $\vdash_{\mathbf{T}^{\rm DL}} \Delta|C_1(a) \Rightarrow \Delta, D_1(a)$ and $D_1 \neq \perp$ then by $(\vee_1^T)$, $\vdash_{\mathbf{T}^{\rm DL}} \Delta|(C_1 \sqcup C_2)(a) \Rightarrow \Delta, (D_1 \sqcup C_2)(a)$, and by induction assumption, $D_1(a)$ is a $\preceq^{\rm DL}$-minimal change of $C_1(a)$ by Δ, and $(D_1 \sqcup C_2)(a)$ is a $\preceq^{\rm DL}$-minimal change of $(C_1 \sqcup C_2)(a)$ by Δ;

(ii) if

$$\vdash_{\mathbf{T}^{\rm DL}} \Delta|C_1(a) \Rightarrow \Delta, D_1(a), D_1 = \perp$$
$$\vdash_{\mathbf{T}^{\rm DL}} \Delta|C_2(a) \Rightarrow \Delta, D_2(a), D_2 \neq \perp,$$

then by $(\vee_2^T)$, $\vdash_{\mathbf{T}^{\rm DL}} \Delta|(C_1 \sqcup C_2)(a) \Rightarrow \Delta, (C_1 \sqcup D_2)(a)$, and by induction assumption, $D_2(a)$ is a $\preceq^{\rm DL}$-minimal change of $C_2(a)$ by Δ, and $(C_1 \sqcup D_2)(a)$ is a $\preceq^{\rm DL}$-minimal change of $(C_1 \sqcup C_2)(a)$ by Δ.

(iii) if

$$\vdash_{\mathbf{T}^{\rm DL}} \Delta|C_1(a) \Rightarrow \Delta, D_1(a),$$
$$\vdash_{\mathbf{T}^{\rm DL}} \Delta|C_2(a) \Rightarrow \Delta, D_2(a),$$
$$D_1 = D_2 = \perp,$$

then by $(\vee_3^T)$, $\vdash_{\mathbf{T}^{\rm DL}} \Delta|(C_1 \sqcup C_2)(a) \Rightarrow \Delta, \lambda$, and by induction assumption, λ is a $\preceq^{\rm DL}$-minimal change of $C_1(a)$ by Δ, and a $\preceq^{\rm DL}$-minimal change of $C_2(a)$ by Δ. Therefore, λ is a $\preceq^{\rm DL}$-minimal change of $(C_1 \sqcup C_2)(a)$ by Δ.

Case $C = \forall R.C_1$. Then, by $(T^\forall)$, there are constants $d_1, \ldots, d_n$ such that $R(a, d_1), \ldots, R(a, d_n) \in \Delta$, and

$$\frac{\begin{array}{l}\Delta|C_1(d_1) \Rightarrow \Delta|C_2(d_1)\\ \Delta|C_2(d_2) \Rightarrow \Delta|C_3(d_2)\\ \ldots\\ \Delta|C_n(d_n) \Rightarrow \Delta, C(d_n)\end{array}}{\Delta|(\forall R.C_1)(a) \Rightarrow \Delta, (\forall R.C)(a).}$$

By induction assumption, for any $i \leq n+1$, $C_i(d_i)$ is a $\preceq^{\Delta}$-minimal change of $C_{i-1}(d_i)$, where $C_{n+1} = C$. Therefore, $(\forall R.C)(a)$ is a $\preceq^{\Delta}$-minimal change of $(\forall R.C_1)(a)$, because for any concept E with $\forall R.C \prec E \preceq \forall R.C_1$, there is a concept E' and $i \leq n-1$ such that $E = \forall R.E'$ and $C_i(d_{i+1}) \prec E'(d_{i+1}) \preceq C_{i-1}(d_{i+1})$. By induction assumption, Δ is inconsistent with $E'(d_{i+1})$ and $(\forall R.E')(a)$.

Case $C = \exists R.C_1$. Then, by $(T^\exists)$, there is a constant c and concept $D_1 \preceq C_1$ such that $R(a, c) \in \Delta$, and

$$\frac{\Delta|C_1(c) \Rightarrow \Delta|D_1(c)}{\Delta|(\exists R.C_1)(a) \Rightarrow \Delta, (\exists R.D_1)(a).}$$

By induction assumption, $D_1(c)$ is a $\preceq^{\Delta}$-minimal change of $C_1(c)$. Therefore, $(\exists R.D_1)(a)$ is a $\preceq^{\Delta}$-minimal change of $(\exists R.C_1)(a)$, because for any concept E with $\forall R.D_1 \prec E \preceq \forall R.C_1$, there is a concept E' such that $E = \exists R.E'$ and $D_1(c) \prec E'(c) \preceq C_1(c)$. By induction assumption, Δ is inconsistent with $E'(c)$ and $(\exists R.E')(a)$. □

Theorem 4.2.13 (Completeness theorem) *For any statement set* Δ *and statements* $C(a), D(a)$,

$$\models_{\mathbf{T}^{\mathrm{DL}}} \Delta|C(a) \Rightarrow \Delta, D(a) \textit{ implies } \vdash_{\mathbf{T}^{\mathrm{DL}}} \Delta|C(a) \Rightarrow \Delta, D(a).$$

Proof We prove the theorem by induction on the structure of C. Assume that $\models_{\mathbf{T}^{\mathrm{DL}}} \Delta|C(a) \Rightarrow \Delta, D(a)$.

Case $C = B$, where $B ::= A|\neg A$. Then, by the definition of $\preceq^{\mathrm{DL}}$-minimal change,

$$D(a) = \begin{cases} A(a) & \text{if } C = A \text{ and } \Delta \nvdash \neg A(a) \\ \neg A(a) & \text{if } C = \neg A \text{ and } \Delta \nvdash A(a) \\ \lambda & \text{otherwise,} \end{cases}$$

and by $(T^{\mathbf{A}})$ and $(T_{\mathbf{A}})$, we have

$$\vdash_{\mathbf{T}^{\mathrm{DL}}} \Delta|B(a) \Rightarrow \Delta, D(a).$$

Case $C = C_1 \sqcap C_2$. Then, there are statements $D_1(a)$ and $D_2(a)$ such that $D_1(a)$ is a $\preceq^{\mathrm{DL}}$-minimal change of $C_1(a)$ by Δ, and $D_2(a)$ is a $\preceq^{\mathrm{DL}}$-minimal change of $C_2(a)$ by $\Delta \cup \{D_1(a)\}$. By induction assumption, we have

$$\begin{array}{l} \vdash_{\mathbf{T}^{\mathrm{DL}}} \Delta|C_1(a) \Rightarrow \Delta, D_1(a), \\ \vdash_{\mathbf{T}^{\mathrm{DL}}} \Delta, C_1(a)|C_2(a) \Rightarrow \Delta, C_1(a), C_2(a), \end{array}$$

and by $(T^{\wedge})$, we have

$$\vdash_{\mathbf{T}^{\mathrm{DL}}} \Delta|(C_1 \sqcap C_2)(a) \Rightarrow \Delta, (D_1 \sqcap D_2)(a).$$

Case $C = C_1 \sqcup C_2$. Then, either

(i) there is a statement $D_1(a) \neq \lambda$ such that $D_1(a)$ is a $\preceq^{\mathrm{DL}}$-minimal change of $C_1(a)$ by Δ, so that $(D_1 \sqcup C_2)$ is a $\preceq^{\mathrm{DL}}$-minimal change of $(C_1 \sqcup C_2)(a)$ by Δ, or

(ii) there is a statement $D_2(a) \neq \lambda$ such that $D_2(a)$ is a $\preceq^{\mathrm{DL}}$-minimal change of $C_2(a)$ by Δ, so that $(C_1 \sqcup D_2)$ is a $\preceq^{\mathrm{DL}}$-minimal change of $(C_1 \sqcup C_2)(a)$ by Δ, or

(iii) λ is a $\preceq^{\mathrm{DL}}$-minimal change of $(C_1 \sqcup C_2)(a)$ by Δ.

By induction assumption, we have one of the following

$$\begin{array}{l}\vdash_{\mathbf{T}^{\mathrm{DL}}} \Delta|C_1(a) \Rightarrow \Delta, D_1(a), D_1 \neq \perp;\\ \vdash_{\mathbf{T}^{\mathrm{DL}}} \Delta|C_2(a) \Rightarrow \Delta, D_2(a), D_1 = \perp, D_2 \neq \perp;\\ \vdash_{\mathbf{T}^{\mathrm{DL}}} \Delta|C_1(a) \Rightarrow \Delta, \lambda;\ \vdash_{\mathbf{T}^{\mathrm{DL}}} \Delta|C_2(a) \Rightarrow \Delta, \lambda,\end{array}$$

and by one of $(T_1^{\vee})$, $(T_2^{\vee})$ and $(T_3^{\vee})$, we have

$$\vdash_{\mathbf{T}^{\mathrm{DL}}} \Delta|(C_1 \sqcup C_2)(a) \Rightarrow \Delta, (D_1 \sqcup D_2)(a).$$

Case $C = \forall R.C_1$. Then, there are concepts $D_1, \ldots, D_n, D$ and constants $d_1 = a, \ldots, d_n$ such that $D(d_n)$ is a $\preceq^{\mathrm{DL}}$-minimal change of $D_n(d_n)$ by Δ, and for any $i \leq n$, $D_i(d_i)$ is a $\preceq^{\mathrm{DL}}$-minimal change of $D_{i-1}(d_i)$. By induction assumption, we have

$$\begin{array}{l}\vdash_{\mathbf{T}^{\mathrm{DL}}} \Delta|D_{i-1}(d_i) \Rightarrow \Delta, D_i(d_i),\\ \vdash_{\mathbf{T}^{\mathrm{DL}}} \Delta|(\forall R.D_{i-1})(a) \Rightarrow \Delta|(\forall R.D_i)(a),\end{array}$$

and by $(T^{\forall})$, we have

$$\vdash_{\mathbf{T}^{\mathrm{DL}}} \Delta|(\forall R.C_1)(a) \Rightarrow \Delta, (\forall R.D)(a).$$

Case $C = \exists R.C_1$. Then, there is a concept D_1 and constant c such that $D_1(c)$ is a $\preceq^{\mathrm{DL}}$-minimal change of $C_1(c)$ by Δ. By induction assumption, we have

$$\vdash_{\mathbf{T}^{\mathrm{DL}}} \Delta|C_1(c) \Rightarrow \Delta, D_1(c),$$

and by $(T^{\exists})$, we have

$$\vdash_{\mathbf{T}^{\mathrm{DL}}} \Delta|(\exists R.C_1)(a) \Rightarrow \Delta, (\exists R.D_1)(a).$$

□

4.2.3 R-Calculus $\mathbf{T}^{\mathrm{DL}}$ for a Set of Statements

R-calculus $\mathbf{T}^{\mathrm{DL}}$ for a set Γ of statements consists of the following axioms and deduction rules:

- **Axioms:**

$$(T^{\mathrm{A}})\ \frac{\Delta \nvdash \neg C(a)}{\Delta|C(a), \Gamma \Rightarrow \Delta, C(a)|\Gamma} \qquad (T_{\mathrm{A}})\ \frac{\Delta \vdash \neg B(a)}{\Delta|B(a), \Gamma \Rightarrow \Delta|\Gamma}$$

- **Deduction rules:**

$$(T^{\wedge})\ \frac{\Delta|C_1(a),\Gamma\Rightarrow\Delta,D_1(a)|\Gamma}{\Delta|(C_1\sqcap C_2)(a),\Gamma\Rightarrow\Delta,D_1(a)|C_2(a),\Gamma}$$

$$(T_1^{\vee})\ \frac{\Delta|C_1(a),\Gamma\Rightarrow\Delta,D_1(a)|\Gamma\quad D_1\neq\bot}{\Delta|(C_1\sqcup C_2)(a),\Gamma\Rightarrow\Delta,(D_1\sqcup C_2)(a)|\Gamma}$$

$$(T_2^{\vee})\ \frac{\Delta|C_1(a),\Gamma\Rightarrow\Delta|\Gamma\quad \Delta|C_2(a),\Gamma\Rightarrow\Delta,D_2(a)|\Gamma\quad D_2\neq\bot}{\Delta|(C_1\sqcup C_2)(a),\Gamma\Rightarrow\Delta,(C_1\sqcup D_2)(a)|\Gamma}$$

$$(T_3^{\vee})\ \frac{\Delta|C_1(a),\Gamma\Rightarrow\Delta|\Gamma\quad \Delta|C_2(a),\Gamma\Rightarrow\Delta|\Gamma}{\Delta|(C_1\sqcup C_2)(a),\Gamma\Rightarrow\Delta|\Gamma}$$

$$(T^{\forall})\ \frac{\Delta\vdash R(a,d)\quad \Delta|C_1(d),\Gamma\Rightarrow\Delta,D_1(d)|\Gamma}{\Delta|(\forall R.C_1)(a),\Gamma\Rightarrow\Delta|(\forall R.D_1)(a),\Gamma}$$

$$(T^{\exists})\ \frac{R(a,c)\in\Delta\quad \Delta|C_1(c),\Gamma\Rightarrow\Delta,D_1(c)|\Gamma}{\Delta|(\exists R.C_1)(a),\Gamma\Rightarrow\Delta,(\exists R.D_1)(a)|\Gamma}$$

where d is a constant and c is a new constant.

Definition 4.2.14 $\Delta|\Gamma\Rightarrow\Delta,\Theta$ is provable in $\mathbf{T}^{\mathrm{DL}}$, denoted by $\vdash_{\mathbf{T}^{\mathrm{DL}}}\Delta|\Gamma\Rightarrow\Delta,\Theta$, if there is a sequence $S_1,\ldots,S_m$ of statements such that

$$\begin{aligned}S_1&=\Delta|\Gamma\Rightarrow\Delta_1|\Gamma_1,\\ &\cdots\\ S_m&=\Delta_{m-1}|\Gamma_{m-1}\Rightarrow\Delta,\Theta;\end{aligned}$$

and for each $i<m$, S_{i+1} is an axiom or is deduced from the previous statements by a deduction rule in $\mathbf{T}^{\mathrm{DL}}$.

Theorem 4.2.15 (Soundness Theorem) *For any statement sets Θ,Δ and any finite statement set Γ, if $\Delta|\Gamma\Rightarrow\Delta,\Theta$ is provable in $\mathbf{T}^{\mathrm{DL}}$ then Θ is a $\preceq$-minimal change of Γ by Δ. That is,*

$$\vdash_{\mathbf{T}^{\mathrm{DL}}}\Delta|\Gamma\Rightarrow\Delta,\Theta \textit{ implies } \models_{\mathbf{T}^{\mathrm{DL}}}\Delta|\Gamma\Rightarrow\Delta,\Theta.$$

□

Theorem 4.2.16 (Completeness Theorem) *For any statement sets Θ,Δ and finite statement set Γ, if Θ is a $\preceq^{\mathrm{DL}}$-minimal change of Γ by Δ then $\Delta|\Gamma\Rightarrow\Delta,\Theta$ is provable in $\mathbf{T}^{\mathrm{DL}}$. That is,*

$$\models_{\mathbf{T}^{\mathrm{DL}}}\Delta|\Gamma\Rightarrow\Delta,\Theta \textit{ implies } \vdash_{\mathbf{T}^{\mathrm{DL}}}\Delta|\Gamma\Rightarrow\Delta,\Theta.$$

□

4.3 Discussion on R-Calculus for $\vdash_{\preceq}$-Minimal Change

Similar to R-calculus **U** in the last chapter, we define R-calculus $\mathbf{U}^{\mathrm{DL}}$ for $\vdash_{\preceq}$-minimal change as follows.

Definition 4.3.1 A statement $D(a)$ is a $\vdash_{\preceq}^{\mathrm{DL}}$-minimal change of $C(a)$ by Δ, denoted by $\models_{\mathbf{U}^{\mathrm{DL}}} \Delta|C(a) \Rightarrow \Delta, D(a)$, if

(i) $D(a) \preceq C(a)$, that is, $D \preceq C$;
(ii) $D(a)$ is consistent with Δ, and
(iii) for any statement $E(a)$ with $D \preceq E \preceq C$, either $\Delta, D(a) \vdash E(a)$ and $\Delta, E(a) \vdash D(a)$, or $\Delta \cup \{E(a)\}$ is inconsistent.

Definition 4.3.2 A theory Θ is a $\vdash_{\preceq}^{\mathrm{DL}}$-minimal change of Γ by Δ, denoted by $\models_{\mathbf{U}^{\mathrm{DL}}} \Delta|\Gamma \Rightarrow \Delta, \Theta$, if

(i) $\Theta \preceq \Gamma$, that is, for each statement $C(a) \in \Theta$, there is a statement $D(a) \in \Gamma$ such that $C \preceq D$,
(ii) $\Theta \cup \Delta$ is consistent, and
(iii) for any theory Ξ with $\Theta \preceq \Xi \preceq \Gamma$, either $\Delta, \Theta \vdash \Xi$ and $\Delta, \Xi \vdash \Theta$, or $\Xi \cup \Delta$ is inconsistent.

For $\vdash_{\preceq}^{\mathrm{DL}}$-minimal change, we cannot prove that for any theories Γ, Δ and Θ, $\Delta|\Gamma \Rightarrow \Delta, \Theta$ is provable if and only if Θ is a $\vdash_{\preceq}^{\mathrm{DL}}$-minimal change of Γ by Δ. Hence, we require that Γ is a set of statements in which the concepts are in quantifer-prefix normal form (simply called in the normal form), where the concepts without quantifiers are in conjunctive normal form.

R-calculus $\mathbf{U}^{\mathrm{DL}}$ for a statement $C(a)$ should consist of the following axioms and deduction rules:

- **Axioms:**

$$(U^{\mathbf{A}})\ \frac{\Delta \nvdash \neg C(a)}{\Delta|C(a) \Rightarrow \Delta, C(a)} \qquad (U_{\mathbf{A}})\ \frac{\Delta \vdash \neg B(a)}{\Delta|B(a) \Rightarrow \Delta}$$

- **Deduction rules:**

$$(U^{\wedge})\ \frac{\Delta|C_1(a) \Rightarrow \Delta, D_1(a)}{\Delta|(C_1 \sqcap C_2)(a) \Rightarrow \Delta, D_1(a)|C_2(a)}$$

$$(U^{\vee})\ \frac{\Delta|C_1(a) \Rightarrow \Delta, D_1(a) \quad \Delta|C_2(a) \Rightarrow \Delta, D_2(a)}{\Delta|(C_1 \sqcup C_2)(a) \Rightarrow \Delta, (D_1 \sqcup D_2)(a)}$$

$$(U^{\forall})\ \frac{\Delta \vdash R(a,d) \quad \Delta|C_1(d) \Rightarrow \Delta, D_1(d)}{\Delta|(\forall R.C_1)(a) \Rightarrow \Delta|(\forall R.D_1)(a)}$$

$$(U^{\exists})\ \frac{R(a,c) \in \Delta \quad \Delta|C_1(c) \Rightarrow \Delta, D_1(c)}{\Delta|(\exists R.C_1)(a) \Rightarrow \Delta, (\exists R.D_1)(a)}$$

where d is a constant and c is a new constant.

But in fact, R-calculus $\mathbf{U}^{\mathrm{DL}}$ is not sound, shown in the following example.

Example 4.3.3 For example, let $\Delta = \{R(a, d_1), R(a, d_2), \neg A_1(d_1), \neg A_2(d_2)\}$. We have the following deduction:

$$\begin{array}{l}\Delta|\forall R.(A \sqcup A_1 \sqcup A_2)(a) \Rightarrow \Delta|(A \sqcup A_1 \sqcup A_2)(d_1)\\ \Delta|(A \sqcup A_1 \sqcup A_2)(d_1) \Rightarrow \Delta, (A \sqcup A_2)(d_1)\\ \Delta|\forall R.(A \sqcup A_1 \sqcup A_2)(a) \Rightarrow \Delta|\forall R.(A \sqcup A_2)(a)\\ \Delta|(A \sqcup A_2)(d_2) \Rightarrow \Delta, A(d_2)\\ \Delta|\forall R.(A \sqcup A_2)(a) \Rightarrow \Delta, (\forall R.A)(a)\\ \Delta|\forall R.(A \sqcup A_1 \sqcup A_2)(a) \Rightarrow \Delta, (\forall R.A)(a)\end{array}$$

and

$$\Delta, \forall R.(A \sqcup A_2)(a) \nvdash (\forall R.A)(a),$$

even though $\Delta, \forall R.(A \sqcup A_2)(a)$ is consistent and $(\forall R.A)(a) \prec \forall R.(A \sqcup A_2)(a) \preceq \forall R.(A \sqcup A_1 \sqcup A_2)(a)$.

We do not know whether is a sound and complete R-calculus $\mathbf{U}^{\mathrm{DL}}$ for $\vdash_{\preceq}$-minimal change in description logic.

References

1. H. Andréka, J. van Benthem, I. Németi, Modal languages and bounded fragments of predicate logic. ILLC Research Report ML-96-03 (1996)
2. A. Artale, E. Franconi, Temporal description logics, in *Handbook of Temporal Reasoning in Artificial Intelligence* (2005)
3. F. Baader, D. Calvanese, D.L. McGuinness, D. Nardi, P.F. Patel-Schneider, *The Description Logic Handbook: Theory, Implementation, Applications* (Cambridge University Press, Cambridge, 2003)
4. F. Baader, I. Horrocks, U. Sattler, Chapter 3 Description Logics, in *Handbook of Knowledge Representation*, eds. by F. van Harmelen, V. Lifschitz, B. Porter (Elsevier, Amsterdam, 2007)
5. D. Fensel, F. van Harmelen, I. Horrocks, D. McGuinness, P.F. Patel-Schneider, OIL: an ontology infrastructure for the semantic web. IEEE Intell. Syst. **16**, 38–45 (2001)
6. E. Grädel, P. Kolaitis, M. Vardi, On the decision problem for two-variable first-order logic. Bull. Symb. Logic **3**, 53–69 (1997)
7. I. Horrocks, Ontologies and the semantic web. Commun. ACM **51**, 58–67 (2008)
8. I. Horrocks, U. Sattler, Ontology reasoning in the SHOQ(D) description logic, in *Proceedings of the Seventeenth International Joint Conference on Artificial Intelligence* (2001)

Chapter 5
R-Calculi for Modal Logic

We consider R-calculus for modal logic [3, 6] and give R-calculi $\mathbf{S^M}$, $\mathbf{T^M}$ and $\mathbf{U^M}$, which are sound and complete with respect to $\subseteq^{\mathbf{M}}$-minimal change, $\preceq^{\mathbf{M}}$-minimal change and $\vdash^{\mathbf{M}}_{\preceq}$-minimal change, respectively.

Propositional modal logic is translated into the guarded first-order logic and decidable.

In propositional modal logic, consistence has the following properties:

- $\Delta \cup \{a : \Box A\}$ is consistent if and only if for any constant symbol c such that $\Delta \vdash R(a, c)$, $\Delta \cup \{c : A\}$ is consistent; correspondingly, $\Delta \cup \{a : \Box A\}$ is inconsistent if and only if for some constant symbol d, $\Delta \vdash R(a, d)$ and $\Delta \cup \{d : A\}$ is inconsistent;
- $\Delta \cup \{a : \Diamond A\}$ is consistent if and only if for some constant symbol d, $\Delta \vdash R(a, d)$ and $\Delta \cup \{d : A\}$ is consistent; correspondingly, $\Delta \cup \{a : \Diamond A\}$ is inconsistent if and only if for any constant symbol c such that $\Delta \vdash R(a, c)$, $\Delta \cup \{c : A\}$ is inconsistent.

Just as in Hoare logic [1, 2, 4–6], where a program is taken as a modality which changes the properties from a state to another, R-calculus can be represented as a modal logic, in which the revising theory Δ is taken as a modality and Γ as a precondition, so that $\Delta|\Gamma \Rightarrow \Theta, \Delta$ is represented as a formula $\Gamma[\Delta]\Theta$ in modal logic.

W. Li and Y. Sui, *R-CALCULUS: A Logic of Belief Revision*, Perspectives in Formal Induction, Revision and Evolution,
https://doi.org/10.1007/978-981-16-2944-0_5

5.1 Propositional Modal Logic

A logical language of PML contains the following symbols:

- propositional variables: $p_0, p_1, \ldots$;
- logical connectives: $\neg, \wedge, \vee$, and
- modalities: $\Box, \Diamond$.

Formulas are defined as follows:

$$A ::= p|\neg A_1|A_1 \wedge A_2|A_1 \vee A_2|\Box A_1|\Diamond A_1.$$

A frame F is a pair (W, R), where W is a set of possible worlds, and $R \subseteq W^2$ is a binary relation on W.

A model $\mathbf{M}$ is a pair (F, v), where F is a frame and v is an assignment such that for any propositional symbol p and possible world $w \in W$, $v(p, w) \in \{0, 1\}$.

A formula A is satisfied at w, denoted by $\mathbf{M}, w \models A$, if

$$\begin{cases} v(p, w) = 1 & \text{if } A = p \\ \mathbf{M}, w \not\models A_1 & \text{if } A = \neg A_1 \\ \mathbf{M}, w \models A_1 \& \mathbf{M}, w \models A_2 & \text{if } A = A_1 \wedge A_2 \\ \mathbf{M}, w \models A_1 \text{ or } \mathbf{M}, w \models A_2 & \text{if } A = A_1 \vee A_2 \\ \mathbf{A}w' \in W((w, w') \in R \Rightarrow \mathbf{M}, w' \models A_1) & \text{if } A = \Box A_1 \\ \mathbf{E}w' \in W((w, w') \in R \& \mathbf{M}, w' \models A_1) & \text{if } A = \Diamond A_1. \end{cases}$$

A formula A is satisfied in $\mathbf{M}$, denoted by $\mathbf{M} \models A$, if for any possible world w, $\mathbf{M}, w \models A$; and A is valid, denoted by $\models A$, if A is satisfied in any model $\mathbf{M}$.

Let a be a label interpreted to be a possible world. Given a formula A, $a : A$ means that A is true in the possible world labeled by a. Given a model $\mathbf{M}$, let f be a mapping from labels to W. Then, we define that $\mathbf{M}, w \models a : A$ if $\mathbf{M}, f(a) \models A$.

Let Γ, Δ be sets of labeled formulas. A sequent $\Gamma \Rightarrow \Delta$ is satisfied in $\mathbf{M}$, denoted by $\mathbf{M} \models \Gamma \Rightarrow \Delta$, if for any possible world w, $\mathbf{M}, w \models \Gamma$ implies $\mathbf{M}, w \models \Delta$, where $\mathbf{M}, w \models \Gamma$ if for each formula $A \in \Gamma$, $\mathbf{M}, w \models A$; and $\mathbf{M}, w \models \Delta$ if for some formula $B \in \Delta$, $\mathbf{M}, w \models B$.

A sequent $\Gamma \Rightarrow \Delta$ is valid, denoted by $\models_{\mathbf{M}} \Gamma \Rightarrow \Delta$, if for any model $\mathbf{M}$, $\mathbf{M} \models \Gamma \Rightarrow \Delta$.

Gentzen deduction system $\mathbf{G_M}$ consists of the following axioms and deduction rules:

- **Axiom**:

$$(\mathbf{A}) \ \frac{\Gamma \cap \Delta \neq \emptyset}{\Gamma \Rightarrow \Delta}$$

where Γ, Δ are sets of atomic labeled formulas of form $a : p$; p is a propositional variable and a is a label.

• **Deduction rules:**

$$(\neg^L)\ \frac{\Gamma \Rightarrow a : A, \Delta}{\Gamma, a : \neg A \Rightarrow \Delta} \qquad (\neg^R)\ \frac{\Gamma, a : A \Rightarrow \Delta}{\Gamma \Rightarrow a : \neg A, \Delta}$$

$$(\wedge_1^L)\ \frac{\Gamma, a : A \Rightarrow \Delta}{\Gamma, a : A \wedge B \Rightarrow \Delta} \qquad (\wedge^R)\ \frac{\Gamma \Rightarrow a : A, \Delta \quad \Gamma \Rightarrow a : B, \Delta}{\Gamma \Rightarrow a : A \wedge B, \Delta}$$

$$(\wedge_2^L)\ \frac{\Gamma, a : B \Rightarrow \Delta}{\Gamma, a : A \wedge B \Rightarrow \Delta}$$

$$(\vee^L)\ \frac{\Gamma, a : A \Rightarrow \Delta \quad \Gamma, a : B \Rightarrow \Delta}{\Gamma, a : A \vee B \Rightarrow \Delta} \qquad (\vee_1^R)\ \frac{\Gamma \Rightarrow a : A, \Delta}{\Gamma \Rightarrow a : A \vee B, \Delta}$$

$$(\vee_2^R)\ \frac{\Gamma \Rightarrow a : B, \Delta}{\Gamma \Rightarrow a : A \vee B, \Delta}$$

$$(\Box^L)\ \frac{\begin{array}{l}\Gamma \vdash R(a, d)\\ \Gamma, R(a, d) \Rightarrow \Delta\end{array}}{\Gamma, a : \Box A \Rightarrow \Delta} \qquad (\Box^R)\ \frac{\begin{array}{l}R(a, c) \in \Gamma\\ \Gamma \Rightarrow c : A, \Delta\end{array}}{\Gamma \Rightarrow a : \Box A, \Delta}$$

$$(\Diamond^L)\ \frac{\begin{array}{l}R(a, c) \in \Gamma\\ \Gamma, c : A \Rightarrow \Delta\end{array}}{\Gamma, a : \Diamond A \Rightarrow \Delta} \qquad (\Diamond^R)\ \frac{\begin{array}{l}\Gamma \vdash R(a, d)\\ \Gamma \Rightarrow d : A, \Delta\end{array}}{\Gamma \Rightarrow a : \Diamond A, \Delta}$$

where c is a new label, and d a label. Here, $R(a, c) \in \Gamma$ means that $R(a, c)$ is enumerated in Γ; $\Gamma \vdash R(a, c)$ means that $R(a, c)$ is in Γ.

Then we have the following

Theorem 5.1.1 (Soundness theorem) *For any sequent* $\Gamma \Rightarrow \Delta$,

$$\vdash_{\mathbf{M}} \Gamma \Rightarrow \Delta \textit{ implies } \models_{\mathbf{M}} \Gamma \Rightarrow \Delta.$$

Proof We prove that each axiom is valid and each deduction rule preserves the validity.

Fix a models $\mathbf{M} = (F, v)$ and a possible world w. Assume that there is a surjective function f from labels to possible worlds such that $R(a, b)$ holds if and only if $(f(a), f(b)) \in R$.

For axiom (**A**), assume that $\mathbf{M}, w \models \Gamma, a : p$. That is, $\mathbf{M}, f(a) \models p$, and hence, $\mathbf{M}, w \models a : p, \Delta$.

For $(\neg^L)$, assume that $\mathbf{M}, w \models \Gamma \Rightarrow a : A, \Delta$. To show that $\mathbf{M}, w \models \Gamma, a : \neg A \Rightarrow \Delta$, assume that $\mathbf{M}, w \models \Gamma, a : \neg A$. Then, $\mathbf{M}, w \models \Gamma$ and by the assumption that $\mathbf{M}, w \models \Gamma \Rightarrow a : A, \Delta$, we have $\mathbf{M}, w \models a : A, \Delta$. Because $\mathbf{M}, w \models a : \neg A, \mathbf{M}, w \not\models a : A$, hence, $\mathbf{M}, w \models \Delta$. Similar for case $(\neg^R)$.

For $(\wedge)$ and $(\vee)$, the proofs are similar to the ones in propositional logic and omitted here.

For $(\Box^L)$, assume that either $\Gamma \nvdash R(a, d)$ or $\mathbf{M}, w \models \Gamma, d : A \Rightarrow \Delta$. To show that $\mathbf{M}, w \models \Gamma, a : \Box A \Rightarrow \Delta$, assume that $\mathbf{M}, w \models \Gamma, a : \Box A$. Then, for any label b with $(f(a), f(b)) \in R, \mathbf{M}, f(b) \models A$. Hence, for any possible world w' with $(w, w') \in R$,

$$\mathbf{M}, w' \models A,$$

i.e., $\mathbf{M}, w \models b : A$. By induction assumption, we have $\mathbf{M}, w \models \Delta$.

For $(\Box^R)$, assume that $R(a, c) \in \Gamma$ and $\mathbf{M}, w \models \Gamma \Rightarrow c : A, \Delta$. To show that $\mathbf{M}, w \models \Gamma \Rightarrow a : \Box A, \Delta$, assume that $\mathbf{M}, w \models \Gamma$. By assumption, $\mathbf{M}, w \models c : A, \Delta$. If $\mathbf{M}, w \models \Delta$ then $\mathbf{M}, w \models a : \Box A, \Delta$. If $\mathbf{M}, w \models c : A$ then because c is a new label, for any w' with $\mathbf{M}, w' \vdash \Gamma$ and $(a, w') \in R, \mathbf{M}, w' \models A$, and hence, $\mathbf{M}, w \models \Box A$, i.e., $\mathbf{M}, w \models a : \Box A, \Delta$.

For $(\Diamond^L)$, assume that $R(a, c) \in \Gamma$ and $\mathbf{M}, w \models \Gamma, c : A \Rightarrow \Delta$. To show that $\mathbf{M}, w \models \Gamma, a : \Diamond A \Rightarrow \Delta$, assume that $\mathbf{M}, w \models \Gamma, a : \Diamond A$. Then, there is a label c with $(f(a), f(c)) \in R, \mathbf{M}, f(c) \models A$. Hence, for some possible world w' with $(w, w') \in R$,

$$\mathbf{M}, w' \models A,$$

i.e., $\mathbf{M}, w \models c : A$. By induction assumption, we have $\mathbf{M}, w \models \Delta$.

For $(\Diamond^R)$, assume that $\Gamma \vdash R(a, d)$ and $\mathbf{M}, w \models \Gamma \Rightarrow d : A, \Delta$. To show that $\mathbf{M}, w \models \Gamma \Rightarrow a : \Diamond A, \Delta$, assume that $\mathbf{M}, w \models \Gamma$. By assumption, $\Gamma \vdash R(a, d)$ and $\mathbf{M}, w \models d : A, \Delta$. If $\mathbf{M}, w \models \Delta$ then $\mathbf{M}, w \models a : \Diamond A, \Delta$. If $\mathbf{M}, w \models d : A$ then $\mathbf{M}, w \models a : \Diamond A$, and hence, $\mathbf{M}, w \models a : \Box A, \Delta$. □

Theorem 5.1.2 (Completeness theorem) *For any sequent* $\Gamma \Rightarrow \Delta$,

$$\models_{\mathbf{M}} \Gamma \Rightarrow \Delta \textit{ implies } \vdash_{\mathbf{M}} \Gamma \Rightarrow \Delta.$$

Proof For each sequent S, define a (possibly infinite) tree, called the reduction tree for S, denoted by $T(S)$, from which we can obtain either a cut-free proof of S or an interpretation not satisfying S.

This reduction tree $T(S)$ for S contains a sequent at each node, and is constructed in stages as follows.

Stage 0: $T_0(S) = \{S\}$.

Stage $k(k > 0) : T_k(S)$ is defined by cases.

Case 1. Every topmost sequent in $T_{k-1}(S)$ has a formula common to its antecedent and succedent. Then, stop.

Case 2. Not case 1. $T_k(S)$ is defined as follows. Let $\Gamma \Rightarrow \Delta$ be any topmost sequent of the tree which has been defined by stage $k - 1$.

Subcase $\neg^L$. Let $a_1 : \neg A_1, \ldots, a_n : \neg A_n$ be all the formulas in Γ whose outmost logical symbol is $\neg$, and to which no reduction has been applied in previous stages. Then, write down

$$\Gamma \Rightarrow \Delta, a_1 : A_1, \ldots, a_n : A_n$$

above $\Gamma \Rightarrow \Delta$. We say that a $\neg^L$-reduction has been applied to $a_1 : \neg A_1, \ldots, a_n : \neg A_n$.

Subcase $\neg^R$. Let $a_1 : \neg A_1, \ldots, a_n : \neg A_n$ be all the formulas in Δ whose outermost logical symbol is $\neg$, and to which no reduction has been applied in previous stages. Then, write down

$$a_1 : A_1, \ldots, a_n : A_n, \Gamma \Rightarrow \Delta$$

above $\Gamma \Rightarrow \Delta$. We say that a $\neg^R$-reduction has been applied to $a_1 : \neg A_1, \dots, a_n : \neg A_n$.

Subcase $\wedge^L$. Let $a_1 : A_1 \wedge B_1, \dots, a_n : A_n \wedge B_n$ be all the formulas in Γ whose outermost logical symbol is $\wedge$, and to which no reduction has been applied in previous stages. Then, write down

$$a_1 : A_1, a_1 : B_1, a_2 : A_2, a_2 : B_2, \dots, a_n : A_n, a_n : B_n, \Gamma \Rightarrow \Delta$$

above $\Gamma \to \Delta$. We say that a $\wedge^L$-reduction has been applied to $a_1 : A_1 \wedge B_1, \dots, a_n : A_n \wedge B_n$.

Subcase $\wedge^R$. Let $a_1 : A_1 \wedge B_1, \dots, a_n : A_n \wedge B_n$ be all the formulas in Δ whose outermost logical symbol is $\wedge$, and to which no reduction has been applied in previous stages. Then, write down all sequents of the form

$$\Gamma \Rightarrow \Delta, a_1 : C_1, \dots, a_n : C_n,$$

above $\Gamma \Rightarrow \Delta$, where C_i is either A_i or B_i. We say that a $\wedge^R$-reduction has been applied to $a_1 : A_1 \wedge B_1, \dots, a_n : A_n \wedge B_n$.

Subcase $\vee^L$. Let $a_1 : A_1 \vee B_1, \dots, a_n : A_n \vee B_n$ be all the formulas in Γ whose outermost logical symbol is $\vee$, and to which no reduction has been applied in previous stages. Then, write down all sequents of the form

$$\Gamma, a_1 : C_1, \dots, a_n : C_n \Rightarrow \Delta,$$

above $\Gamma \Rightarrow \Delta$, where C_i is either A_i or B_i. We say that a $\vee^L$-reduction has been applied to $a_1 : A_1 \vee B_1, \dots, a_n : A_n \vee B_n$.

Subcase $\vee^R$. Let $a_1 : A_1 \vee B_1, \dots, a_n : A_n \vee B_n$ be all the formulas in Γ whose outermost logical symbol is $\vee$, and to which no reduction has been applied in previous stages. Then, rite down

$$\Gamma \Rightarrow a_1 : A_1, a_1 : B_1, a_2 : A_2, a_2 : B_2, \dots, a_n : A_n, a_n : B_n, \Delta$$

above $\Gamma \vee \Delta$. We say that a $\vee^R$-reduction has been applied to $a_1 : A_1 \vee B_1, \dots, a_n : A_n \vee B_n$.

Subcase $\Box^L$. Let $a_1 : \Box A_1, \dots, a_n : \Box A_n$ be all the formulas in Γ whose outermost logical symbol is $\Box$. Let d_i be the first label at this stage which has not been used for a reduction of $a_i : \Box A_i$ for $1 \leq i \leq n$. Then, write down

$$\Gamma, R(a_1, d_1), d_1 : A_1, \dots, R(a_n, d_n), d_n : A_n \Rightarrow \Delta$$

above $\Gamma \Rightarrow \Delta$. We say that a $\Box^L$-reduction has been applied to $a_1 : \Box A_1, \dots, a_n : \Box A_n$.

Subcase $\Box^R$. Let $a_1 : \Box A_1, \dots, a_n : \Box A_n$ be all the formulas in Δ whose outermost logical symbol is $\Box$, and to which no reduction has been applied in previous stages. Then, let $c_1, \dots, c_n$ be the first n labels (in the list of variables) which are *not* available at this stage. Then write down

$$\Gamma \Rightarrow \Delta, \neg R(a_1, c_1), c_1 : A_1, \ldots, \neg R(a_n, c_n), c_n : A_n, \Delta$$

above $\Gamma \Rightarrow \Delta$. We say that a $\Box^R$-reduction has been applied to $a_1 : \Box A_1, \ldots, a_n : \Box A_n$. Notice that $c_1, \ldots, c_n$ are new available labels.

Subcase $\Diamond^L$. Let $a_1 : \Diamond A_1, \ldots, a_n : \Diamond A_n$ be all the formulas in Γ whose outermost logical symbol is $\Diamond$, and to which no reduction has been applied in previous stages. Then, let $c_1, \ldots, c_n$ be the first n labels (in the list of variables) which are *not* available at this stage. Then write down

$$\Gamma, R(a_1, c_1), c_1 : A_1, \ldots, R(a_n, c_n), c_n : A_n \Rightarrow \Delta$$

above $\Gamma \Rightarrow \Delta$. We say that a $\Diamond^L$-reduction has been applied to $a_1 : \Diamond A_1, \ldots, a_n : \Diamond A_n$. Notice that $c_1, \ldots, c_n$ are new available labels.

Subcase $\Diamond^R$. Let $a_1 : \Diamond A_1, \ldots, a_n : \Diamond A_n$ be all the formulas in Δ whose outermost logical symbol is $\Diamond$. Let d_i be the first label at this stage which has not been used for a reduction of $a_i : \Diamond A_i$ for $1 \leq i \leq n$. Then, write down

$$\Gamma \Rightarrow C_1, \ldots, C_n, \Delta$$

above $\Gamma \Rightarrow \Delta$, where C_i is $R(a_i, d_i)$ or $d_i : A_i$. We say that a $\Diamond^R$-reduction has been applied to $a_1 : \Diamond A_1, \ldots, a_n : \Diamond A_n$.

Case 0. If Γ and Δ have any formula in common, or Γ and Δ have no formula in common and the reductions described above are not applicable, then $\Gamma \Rightarrow \Delta$ is a leaf. This is the reduction tree for S, denoted by $T(S)$.

A sequence $S_0, \ldots$ of sequents in $T(S)$ is a branch if $S_0 = S$, S_{i+1} is immediately above S_i.

For a sequent S, if each branch of $T(S)$ ends with a sequent whose antecedent and succedent contain a formula in common, then $T(s)$ is a proof tree of S.

Otherwise, there is a branch, say $S_0, \ldots, S_n$, such that $\Gamma_n \cap \Delta_n = \emptyset$, where $S_i = \Gamma_i \Rightarrow \Delta_i$. Define $\cup\Gamma$ be the set of all formulas occurring in Γ_i for some i, and let $\cup\Delta$ be the set of all the formulas occurring in Δ_j for some j.

We define an interpretation in which every formula in $\cup\Gamma$ holds and no formula in $\cup\Delta$ holds. Thus, S does not hold in the interpretation.

First notice that by the choice of the branch, $\cup\Gamma$ and $\cup\Delta$ have no atomic formula in common. Let W be the set of all the labels, and R consist of pairs (a, c) such that $\neg R(a, c) \in \cup\Delta$, where c is produced by ($\Box^R$) with $a : \Box A$. Define interpretation v such that for any propositional symbol p and label $a \in W$,

$$v(p, a) = 1 \text{ iff } a : p \in \cup\Gamma.$$

Then, v satisfies every formula in $\cup\Gamma$, and no formula in $\cup\Delta$.

The proof is by induction on the structure of formula A. Other cases are similar to the proofs in Chap. 2 and we consider the following case:

Case $A = \Box A_1$.

Subcase 1. $a : A \in \bigcup\Gamma$. Let i be the least number such that $A \in \Gamma_i$. Hence, $A \in \Gamma_j$ for all $j > i$. It is sufficient to show that all substitution instances $d : A(a)$, for $d \in W$ with $\Gamma \vdash R(a, d)$, are satisfied by $f(d)$, i.e., all these substitution instances are in $\bigcup\Gamma$. This is evident for the construction.

Subcase 2. $A \in \bigcup\Delta$. Consider the step at which $a : A$ was used to define an upper sequent from $\Gamma_i \Rightarrow \Delta$ (or $\Gamma_i \to \Delta_i^1, a : A, \Delta_i^2$). It looks like this:

$$\frac{\begin{array}{l}R(a,c) \in \Gamma \\ \Gamma_{i+1} \Rightarrow c : A_1, \Delta_{i+1}\end{array}}{\Gamma_i \Rightarrow a : A, \Delta_i}$$

By induction assumption, $c : A_1$ is not satisfied by v; by definition of R, $R(a, c)$ is satisfied. Hence, $a : A$ is not satisfied by v either.

Case $A = \Diamond A_1$.

Subcase 1. $a : A \in \bigcup\Gamma$. Let i be the least number such that $A \in \Gamma_i$. Hence, $A \in \Gamma_j$ for all $j > i$. It is sufficient to show that some substitution instance $c : A(a)$, for $c \in W$ with $R(a, c) \in \Gamma$, are satisfied by v, i.e., some substitution instance is in $\bigcup\Gamma$. This is evident for the construction.

Subcase 2. $A \in \bigcup\Delta$. Consider the step at which $a : A$ was used to define an upper sequent from $\Gamma_i \Rightarrow \Delta$ (or $\Gamma_i \to \Delta_i^1, a : A, \Delta_i^2$). It looks like this:

$$\frac{\begin{array}{l}\Gamma \vdash R(a,d) \\ \Gamma_{i+1} \Rightarrow d : A, \Delta_{i+1}\end{array}}{\Gamma_i \Rightarrow a : A, \Delta_i}$$

By induction assumption, A_1 is not satisfied by v; by definition of R, $R(a, c)$ is satisfied. Hence, $a : A$ is not satisfied by v either.

This completes the proof. □

5.2 R-Calculus $\mathbf{S}^{\mathbf{M}}$ for $\subseteq$-Minimal Change

Definition 5.2.1 For two consistent theories Δ and Γ of modal logic, a theory Θ is a $\subseteq$-minimal change of Γ by Δ, denoted by $\models_{\mathbf{M}} \Delta|\Gamma \Rightarrow \Delta, \Theta$, if Θ is consistent with Δ; $\Theta \subseteq \Gamma$, and for any theory Ξ with $\Theta \subset \Xi \subseteq \Gamma$, Ξ is inconsistent with Δ.

R-calculus $\mathbf{S}^{\mathbf{M}}$ consists of the following deduction rules:

- **Axioms:**

$$(M^+)\ \frac{\Delta \nvdash \neg l}{\Delta|a : l, \Gamma \Rightarrow \Delta, a : l|\Gamma} \qquad (M^-)\ \frac{\Delta \vdash \neg l}{\Delta|a : l, \Gamma \Rightarrow \Delta|\Gamma}$$

- **Deduction rules:**

$$(M^{\wedge})\ \frac{\begin{array}{l}\Delta|a:A_1,\Gamma\Rightarrow\Delta,a:A_1|\Gamma\\ \Delta,a:A_1|a:A_2,\Gamma\Rightarrow\Delta,a:A_1,a:A_2|\Gamma\end{array}}{\Delta|a:A_1\wedge A_2,\Gamma\Rightarrow\Delta,a:A_1\wedge A_2|\Gamma}$$

$$(M_{\wedge}^1)\ \frac{\Delta|a:A_1,\Gamma\Rightarrow\Delta|\Gamma}{\Delta|a:A_1\wedge A_2,\Gamma\Rightarrow\Delta|\Gamma}$$

$$(M_{\wedge}^2)\ \frac{\Delta,a:A_1|a:A_2,\Gamma\Rightarrow\Delta,a:A_1|\Gamma}{\Delta|a:A_1\wedge A_2,\Gamma\Rightarrow\Delta|\Gamma}$$

$$(M_1^{\vee})\ \frac{\Delta|a:A_1,\Gamma\Rightarrow\Delta,a:A_1|\Gamma}{\Delta|a:A_1\vee A_2,\Gamma\Rightarrow\Delta,a:A_1\vee A_2|\Gamma}$$

$$(M_2^{\vee})\ \frac{\Delta|a:A_2,\Gamma\Rightarrow\Delta,a:A_2|\Gamma}{\Delta|a:A_1\vee A_2,\Gamma\Rightarrow\Delta,a:A_1\vee A_2|\Gamma}$$

$$(M_{\vee})\ \frac{\begin{array}{l}\Delta|a:A_1,\Gamma\Rightarrow\Delta|\Gamma\\ \Delta|a:A_2,\Gamma\Rightarrow\Delta|\Gamma\end{array}}{\Delta|a:A_1\vee A_2,\Gamma\Rightarrow\Delta|\Gamma}$$

$$(M^{\Box})\ \frac{\begin{array}{l}R(a,c)\in\Delta\\ \Delta|c:A,\Gamma\Rightarrow\Delta,c:A|\Gamma\end{array}}{\Delta|a:\Box A,\Gamma\Rightarrow\Delta,a:\Box A|\Gamma}$$

$$(M_{\Box})\ \frac{\begin{array}{l}\Delta\vdash R(a,d)\\ \Delta|d:A,\Gamma\Rightarrow\Delta|\Gamma\end{array}}{\Delta|a:\Box A,\Gamma\Rightarrow\Delta|\Gamma}$$

$$(M^{\Diamond})\ \frac{\begin{array}{l}\Delta\vdash R(a,d)\\ \Delta|d:A,\Gamma\Rightarrow\Delta,d:A|\Gamma\end{array}}{\Delta|a:\Diamond A,\Gamma\Rightarrow\Delta,a:\Diamond A|\Gamma}$$

$$(M_{\Diamond})\ \frac{\begin{array}{l}R(a,c)\in\Delta\\ \Delta|c:A,\Gamma\Rightarrow\Delta|\Gamma\end{array}}{\Delta|a:\Diamond A,\Gamma\Rightarrow\Delta|\Gamma}$$

where c is a new label, and d a label.

Definition 5.2.2 $\Delta|a:A\Rightarrow\Delta,a:C$ is provable in $\mathbf{S^M}$, denoted by $\vdash_{\mathbf{S^M}}\Delta|a:A\Rightarrow\Delta,a:C$, if there is a sequence $S_1,\ldots,S_m$ of statements such that

$$\begin{array}{l}S_1=\Delta|a:A_1\Rightarrow\Delta|a:A_1',\\ \ldots\\ S_m=\Delta|a:A_m\Rightarrow\Delta,a:A_m';\\ A_1=A,\\ A_m'=C;\end{array}$$

and for each $i<m$, S_{i+1} is an axiom or is deduced from the previous statements by a deduction rule in $\mathbf{S^M}$.

Theorem 5.2.3 (Completeness theorem) *For any consistent formula set Δ and formula A, if $\Delta\cup\{A\}$ is consistent then for any label a, $\Delta|a:A\Rightarrow\Delta,a:A$ is provable in $\mathbf{S^M}$, i.e.,*

$$\models_{\mathbf{S^M}} \Delta|a : A \Rightarrow \Delta, a : A \textit{ implies } \vdash_{\mathbf{S^M}} \Delta|a : A \Rightarrow \Delta, a : A;$$

and if $\Delta \cup \{A\}$ *is inconsistent then* $\Delta|a : A \Rightarrow \Delta$ *is provable, i.e.,*

$$\models_{\mathbf{S^M}} \Delta|a : A \Rightarrow \Delta \textit{ implies } \vdash_{\mathbf{S^M}} \Delta|a : A \Rightarrow \Delta.$$

Proof We prove the theorem by induction on the structure of formula A.

Assume that $\Delta \cup \{A\}$ is consistent.

Case $A = l$. Then, $\Delta \nvdash \neg l$, and by (M^+), $\Delta|a : l \Rightarrow \Delta, a : l$ is provable;

Case $A = A_1 \wedge A_2$. Then, $\Delta \cup \{a : A_1\}$ and $\Delta \cup \{a : A_1, a : A_2\}$ are consistent, and by induction assumption,

$$\vdash_{\mathbf{S^M}} \Delta|a : A_1 \Rightarrow \Delta, a : A_1$$
$$\vdash_{\mathbf{S^M}} \Delta, a : A_1|a : A_2 \Rightarrow \Delta, a : A_1, a : A_2$$

and by $(M^\wedge)$, $\Delta|a : A_1 \wedge A_2 \Rightarrow \Delta, a : A_1 \wedge A_2$ is provable;

Case $A = A_1 \vee A_2$. Then either $\Delta \cup \{a : A_1\}$ or $\Delta \cup \{a : A_2\}$ is consistent, and by induction assumption, either $\Delta|a : A_1 \Rightarrow \Delta, a : A_1$ or $\Delta|a : A_2 \Rightarrow \Delta, a : A_2$ are provable, and by $(M_1^\vee)$ or $(M_2^\vee)$, $\Delta|a : A_1 \vee A_2 \Rightarrow \Delta, a : A_1 \vee A_2$ is provable.

Case $A = \Box A_1$. Then, for any label c such that $R(a, c) \in \Delta$, $\Delta \cup \{c : A_1\}$ is consistent, and by induction assumption,

$$\vdash_{\mathbf{S^M}} \Delta|c : A_1 \Rightarrow \Delta, c : A_1,$$

and by $(M^\Box)$, $\vdash_{\mathbf{S^M}} \Delta|a : \Box A_1 \Rightarrow \Delta, a : \Box A_1$.

Case $A = \Diamond A_1$. Then, there is a label d such that $\Delta \cup \{d : A_1\}$ is consistent, and by induction assumption,

$$\vdash_{\mathbf{S^M}} \Delta|d : A_1 \Rightarrow \Delta, c : A_1,$$

and by $(M^\Diamond)$, $\vdash_{\mathbf{S^M}} \Delta|a : \Diamond A_1 \Rightarrow \Delta, a : \Diamond A_1$.

Similar for the case that $\Delta \cup \{A\}$ is inconsistent. □

Theorem 5.2.4 (Soundness theorem) *For any consistent formula set* Δ*, formula* A *and any label* a*, if* $\vdash_{\mathbf{S^M}} \Delta|a : A \Rightarrow \Delta, a : A$ *then* $\Delta \cup \{a : A\}$ *is consistent, i.e.,*

$$\vdash_{\mathbf{S^M}} \Delta|a : A \Rightarrow \Delta, a : A \textit{ implies } \models_{\mathbf{S^M}} \Delta|a : A \Rightarrow \Delta, a : A;$$

and if $\vdash_{\mathbf{S^M}} \Delta|a : A \Rightarrow \Delta$ *then* $\Delta \cup \{a : A\}$ *is inconsistent, i.e.,*

$$\vdash_{\mathbf{S^M}} \Delta|a : A \Rightarrow \Delta \textit{ implies } \models_{\mathbf{S^M}} \Delta|a : A \Rightarrow \Delta.$$

Proof Assume that $\Delta|A \Rightarrow \Delta$ is provable. We prove by induction on the structure of A, $\Delta \cup \{A\}$ is inconsistent.

Case $A = l$. By (M^-), $\Delta \vdash a : \neg l$, and hence, $\Delta \cup \{a : l\}$ is inconsistent;

Case $A = A_1 \wedge A_2$. By $(M_\wedge^1)$ and $(M_\wedge^2)$, either $\vdash_{\mathbf{S^M}} \Delta|a : A_1 \Rightarrow \Delta$ or $\vdash_{\mathbf{S^M}} \Delta a : A_1|a : A_2 \Rightarrow \Delta, a : A_1$, and by induction assumption, either $\Delta \cup \{a : A_1\}$ or $\Delta \cup \{a : A_1, a : A_2\}$ is inconsistent, and so is $\Delta \cup \{a : A_1 \wedge A_2\}$.

Case $A = A_1 \vee A_2$. By $(M_\vee)$,

$$\vdash_{\mathbf{S^M}} \Delta|a : A_1 \Rightarrow \Delta$$
$$\vdash_{\mathbf{S^M}} \Delta|a : A_2 \Rightarrow \Delta,$$

and by induction assumption, $\Delta \cup \{a : A_1\}$ and $\Delta \cup \{a : A_2\}$ are inconsistent, and so is $\Delta \cup \{a : A_1 \vee A_2\}$.

Case $A = \Box A_1$. By $(M_\Box)$, there is a label d such that $R(a, d) \in \Delta$ and $\vdash_{\mathbf{S^M}} \Delta|d : A_1 \Rightarrow \Delta$, and by induction assumption, $\Delta \cup \{d : A_1\}$ is inconsistent, and so is $\Delta \cup \{a : \Box A_1\}$.

Case $A = \Diamond A_1$. By $(M_\Diamond)$, there is a new label c such that $\vdash_{\mathbf{S^M}} \Delta|c : A_1 \Rightarrow \Delta$, and by induction assumption, $\Delta \cup \{c : A_1\}$ is inconsistent, and so is $\Delta \cup \{a : \Diamond A_1\}$.

Similar for case that $\Delta|A \Rightarrow \Delta, A$ is provable. □

R-calculus $\mathbf{S^M}$ for a theory

Let $\Gamma = (A_1, \ldots, A_n)$. Define

$$\Delta|\Gamma = (\cdots((\Delta|A_1)|A_2)\cdots)|A_n.$$

Definition 5.2.5 $\Delta|\Gamma \Rightarrow \Delta, \Theta$ is provable in $\mathbf{S^M}$, denoted by $\vdash_{\mathbf{S^M}} \Delta|\Gamma \Rightarrow \Delta, \Theta$, if there is a sequence $S_1, \ldots, S_m$ of statements such that

$$S_1 = \Delta|\Gamma \Rightarrow \Delta_1|\Gamma_1,$$
$$\ldots$$
$$S_m = \Delta_{m-1}|\Gamma_{m-1} \Rightarrow \Delta_m|\Gamma_m = \Delta, \Theta$$

and for each $i < m$, S_{i+1} is an axiom or is deduced from the previous statements by a deduction rule in $\mathbf{S^M}$.

Theorem 5.2.6 (Soundness Theorem) *For any consistent formula sets Θ, Δ and any finite consistent formula set Γ, if $\vdash_{\mathbf{S^M}} \Delta|\Gamma \Rightarrow \Delta, \Theta$ then Θ is a $\subseteq$-minimal change of Γ by Δ.*

□

Theorem 5.2.7 (Completeness Theorem) *For any consistent formula sets Θ, Δ and any finite consistent formula set Γ, if Θ is a $\subseteq$-minimal change of Γ by Δ then $\Delta|\Gamma \Rightarrow \Delta, \Theta$ is provable in $\mathbf{S^M}$.* □

Example 5.2.8 Let $\Delta = \{R(a, b), b : A_1, b : A_2\}$ and $a : A = a : \Box(\neg A_1 \wedge \neg A_2 \wedge A_3)$. We have the following deduction:

$$\Delta|b : \neg A_1 \wedge \neg A_2 \wedge A_3 \Rightarrow \Delta$$
$$\Delta|a : \Box(\neg A_1 \wedge \neg A_2 \wedge A_3) \Rightarrow \Delta.$$

Hence,

$$\vdash_{\mathbf{S}^{\mathrm{M}}} \Delta|a : \Box(\neg A_1 \wedge \neg A_2 \wedge A_3) \Rightarrow \Delta.$$

5.3 R-Calculus $\mathbf{T}^{\mathrm{M}}$ for $\preceq$-Minimal Change

Given any theories Δ and Γ, a theory Θ is a $\preceq$-minimal change of Γ by Δ, denoted by $\vdash_{\mathbf{T}^{\mathrm{M}}} \Delta|\Gamma \Rightarrow \Delta, \Theta$, if $\Theta \preceq \Gamma$; Θ is consistent with Δ, and for any theory Ξ with $\Theta \prec \Xi \preceq \Gamma$, Ξ is inconsistent with Δ.

R-calculus $\mathbf{T}^{\mathrm{M}}$ for $\preceq$-minimal change consists of the following axioms and deduction rules:

- **Axioms:**

$$(M^{\mathrm{con}})\ \frac{\Delta \nvdash a : \neg A}{\Delta|a : A, \Gamma \Rightarrow \Delta, a : A|\Gamma} \qquad (M^{\neg})\ \frac{\Delta \vdash a : \neg l}{\Delta|a : l, \Gamma \Rightarrow \Delta, \lambda|\Gamma}$$

- **Deduction rules:**

$$(M^{\wedge})\ \frac{\Delta|a : A_1, \Gamma \Rightarrow \Delta, a : C_1|\Gamma}{\Delta|a : A_1 \wedge A_2, \Gamma \Rightarrow \Delta, a : C_1|a : A_2, \Gamma}$$

$$(M_1^{\vee})\ \frac{\begin{array}{l}\Delta|a : A_1, \Gamma \Rightarrow \Delta, a : C_1|\Gamma \\ a : C_1 \neq \lambda\end{array}}{\Delta|a : A_1 \vee A_2 \Rightarrow \Delta, a : C_1 \vee A_2}$$

$$(M_2^{\vee})\ \frac{\begin{array}{l}\Delta|a : A_1, \Gamma \Rightarrow \Delta|\Gamma \\ \Delta|a : A_2, \Gamma \Rightarrow \Delta, a : C_2|\Gamma \\ a : C_2 \neq \lambda\end{array}}{\Delta|a : A_1 \vee A_2, \Gamma \Rightarrow \Delta, a : A_1 \vee C_2|\Gamma}$$

$$(M_3^{\vee})\ \frac{\begin{array}{l}\Delta|a : A_1, \Gamma \Rightarrow \Delta, \lambda|\Gamma \\ \Delta|a : A_2, \Gamma \Rightarrow \Delta, \lambda|\Gamma\end{array}}{\Delta|a : A_1 \vee A_2, \Gamma \Rightarrow \Delta, \lambda|\Gamma}$$

$$(M^{\Box})\ \frac{\begin{array}{l}\Delta \vdash R(a, d) \\ \Delta|d : A_1, \Gamma \Rightarrow \Delta|d : C_1, \Gamma\end{array}}{\Delta|a : \Box A_1, \Gamma \Rightarrow \Delta|a : \Box C_1, \Gamma}$$

$$(M^{\Diamond})\ \frac{\begin{array}{l}R(a, c) \in \Delta \\ \Delta|c : A_1, \Gamma \Rightarrow \Delta, c : C_1|\Gamma\end{array}}{\Delta|a : \Diamond A_1, \Gamma \Rightarrow \Delta, a : \Diamond C_1|\Gamma}$$

where if C is consistent then

$$\lambda \vee C \equiv C \vee \lambda \equiv C; \quad \lambda \wedge C \equiv C \wedge \lambda \equiv C; \quad \Delta, \lambda \equiv \Delta$$

and if C is inconsistent then

$$\lambda \vee C \equiv C \vee \lambda \equiv \lambda; \quad \lambda \wedge C \equiv C \wedge \lambda \equiv \lambda$$

and d is a label and c is a new label.

Definition 5.3.1 $\Delta|a : A \Rightarrow \Delta, a : C$ is provable in $\mathbf{T}^{\mathbf{M}}$, denoted by $\vdash_{\mathbf{T}^{\mathbf{M}}} \Delta|a : A \Rightarrow \Delta, a : C$, if there is a sequence $\theta_1, \ldots, \theta_m$ of statements such that

$$\begin{array}{l} \theta_1 = \Delta|a : A \Rightarrow \Delta|A_2(a), \\ \ldots \\ \theta_m = \Delta|A_m(a) \Rightarrow \Delta, a : C; \end{array}$$

and for each $j \leq m$, $\Delta|A_j(a) \Rightarrow \Delta|A_{j+1}(a)$ is an axiom or is deduced from the previous statements by a deduction rule in $\mathbf{T}^{\mathbf{M}}$, where $A_1 = A$, $A_{m+1} = C$.

Theorem 5.3.2 (Soundness theorem) *For any statement set Δ and statements $a : A, a : C$,*

$$\vdash_{\mathbf{T}^{\mathbf{M}}} \Delta|a : A \Rightarrow \Delta, a : C \text{ implies } \models_{\mathbf{T}^{\mathbf{M}}} \Delta|a : A \Rightarrow \Delta, a : C.$$

Proof We prove the theorem by induction on the structure of A. Assume that $\vdash_{\mathbf{T}^{\mathbf{M}}} \Delta|a : A \Rightarrow \Delta, a : C$.

Case $A = l$, where $l ::= p|\neg p$. Then, by $(M^{\mathbf{A}})$ and $(M_{\mathbf{A}})$,

$$a : C = \begin{cases} a : l & \text{if} A = l \text{ and } \Delta \nvdash \neg a : l \\ \neg a : l & \text{if } A = \neg l \text{ and } \Delta \nvdash a : l \\ \lambda & \text{otherwise}, \end{cases}$$

and it is clear that $a : C$ is a $\preceq^{\mathbf{M}}$-minimal change of $a : A$ by Δ.

Case $A = A_1 \wedge A_2$. Then, by $(M^{\wedge})$,

$$\begin{array}{l} \vdash_{\mathbf{T}^{\mathbf{M}}} \Delta|a : A_1 \Rightarrow \Delta, a : C_1, \\ \vdash_{\mathbf{T}^{\mathbf{M}}} \Delta, a : A_1|a : A_2 \Rightarrow \Delta, a : C_1, a : C_2, \end{array}$$

and by induction assumption, $a : C_1$ is a $\preceq^{\mathbf{M}}$-minimal change of $a : A_1$ by Δ, and $a : C_2$ is a $\preceq^{\mathbf{M}}$-minimal change of $a : A_2$ by $\Delta \cup \{a : A_1\}$. Therefore, $a : C_1 \wedge C_2$ is a $\preceq^{\mathbf{M}}$-minimal change of $a : A_1 \wedge A_2$ by Δ.

Case $A = A_1 \sqcup A_2$. Then, there are the following three cases:

(i) if $\vdash_{\mathbf{T}^{\mathbf{M}}} \Delta|a : A_1 \Rightarrow \Delta, a : C_1$ and $C_1 \neq \perp$ then by $(M_1^{\vee})$, $\vdash_{\mathbf{T}^{\mathbf{M}}} \Delta|a : A_1 \vee A_2 \Rightarrow \Delta, a : C_1 \vee A_2$, and by induction assumption, $a : C_1$ is a $\preceq^{\mathbf{M}}$-minimal change of $a : A_1$ by Δ, and $a : C_1 \vee A_2$ is a $\preceq^{\mathbf{M}}$-minimal change of $a : A_1 \vee A_2$ by Δ;

(ii) if

$$\begin{array}{l} \vdash_{\mathbf{T}^{\mathbf{M}}} \Delta|a : A_1 \Rightarrow \Delta, a : C_1, C_1 = \lambda \\ \vdash_{\mathbf{T}^{\mathbf{M}}} \Delta|a : A_2 \Rightarrow \Delta, a : C_2, C_2 \neq \lambda, \end{array}$$

then by $(M_2^{\vee})$, $\vdash_{\mathbf{T}^{\mathbf{M}}} \Delta|a : A_1 \vee A_2 \Rightarrow \Delta, a : A_1 \sqcup C_2$, and by induction assumption, $a : C_2$ is a $\preceq^{\mathbf{M}}$-minimal change of $a : A_2$ by Δ, and $a : A_1 \vee C_2$ is a $\preceq^{\mathbf{M}}$-minimal change of $a : A_1 \vee A_2$ by Δ.

(iii) if

$$\begin{array}{l}\vdash_{\mathbf{T^M}} \Delta|a : A_1 \Rightarrow \Delta, a : C_1,\\ \vdash_{\mathbf{T^M}} \Delta|a : A_2 \Rightarrow \Delta, a : C_2,\\ C_1 = C_2 = \lambda,\end{array}$$

then by $(M_3^\vee)$, $\vdash_{\mathbf{T^M}} \Delta|a : A_1 \vee A_2 \Rightarrow \Delta, \lambda$, and by induction assumption, λ is a $\preceq^{\mathbf{M}}$-minimal change of $a : A_1$ by Δ, and a $\preceq^{\mathbf{M}}$-minimal change of $a : A_2$ by Δ. Therefore, λ is a $\preceq^{\mathbf{M}}$-minimal change of $a : A_1 \vee A_2$ by Δ.

Case $A = a : \Box A_1$. Then, by $(M^\Box)$, there are labels $d_1, \ldots, d_n$ and formulas $A_2, \ldots, A_n, C$ such that $R(a, d_1), \ldots, R(a, d_n) \in \Delta$, and

$$\frac{\begin{array}{l}\Delta|d_1 : A_1 \Rightarrow \Delta|d_1 : A_2\\ \Delta|d_2 : A_2 \Rightarrow \Delta|d_2 : A_3\\ \ldots\\ \Delta|d_n : A_n \Rightarrow \Delta|d_n : C\end{array}}{\Delta|a : \Box A_1 \Rightarrow \Delta, a : \Box C.}$$

By induction assumption, for any $i \leq n+1$, $d_i : A_i$ is a $\preceq^\Delta$-minimal change of $d_i : A_{i-1}$, where $A_{n+1} = C$. Therefore, for any formula E with $a : \Box C \prec E \preceq a : \Box A_1$, there is a formula E' and $i \leq n-1$ such that $E = a : \Box E'$ and $d_{i+1} : A_i \prec d_{i+1} : E' \preceq d_{i+1} : A_{i-1}$. By induction assumption, Δ is inconsistent with $d_{i+1} : E'$ and $a : \Box E'$.

Case $A = a : \Diamond A_1$. Then, by $(M^\Diamond)$, there is a new label c and a formula C_1 such that $R(a, c) \in \Delta$, and

$$\frac{\Delta|c : A_1 \Rightarrow \Delta|c : C_1}{\Delta|a : \Diamond A_1 \Rightarrow \Delta, a : \Diamond C_1.}$$

By induction assumption, $c : C_1$ is a $\preceq^\Delta$-minimal change of $c : A_1$. Therefore, $a : \Diamond C_1$ is a $\preceq^\Delta$-minimal change of $a : \Diamond A_1$, because for any formula E with $a : \Diamond C_1 \prec E \preceq a : \Diamond A_1$, there is a formula E' such that $E = a : \Diamond E'$ and $c : C_1 \prec c : E' \preceq c : A_1$. By induction assumption, Δ is inconsistent with $c : E'$ and $a : \Box E'$. □

Theorem 5.3.3 (Completeness Theorem) *For any statement set Δ and statements $a : A, a : C$,*

$$\models_{\mathbf{T^M}} \Delta|a : A \Rightarrow \Delta, a : C \text{ implies } \vdash_{\mathbf{T^M}} \Delta|a : A \Rightarrow \Delta, a : C.$$

Proof We prove the theorem by induction on the structure of A. Assume that $\models_{\mathbf{T^M}} \Delta|a : A \Rightarrow \Delta, a : C$.

Case $A = l$, where $l ::= p|\neg p$. Then, by the definition of $\preceq^{\mathbf{M}}$-minimal change,

$$a : C = \begin{cases} a : l \text{ if } A = l \text{ and } \Delta \nvdash \neg a : l \\ \lambda \quad \text{otherwise,} \end{cases}$$

and by $(M^{\mathbf{A}})$ and $(M_{\mathbf{A}})$, we have

$$\vdash_{\mathbf{T^M}} \Delta|a : l \Rightarrow \Delta, a : C.$$

Case $A = A_1 \wedge A_2$. Then, there are formulas $a : C_1$ and $a : C_2$ such that $a : C_1$ is a $\preceq^{\mathbf{M}}$-minimal change of $a : A_1$ by Δ, and $a : C_2$ is a $\preceq^{\mathbf{M}}$-minimal change of $a : A_2$ by $\Delta \cup \{a : C_1\}$. By induction assumption, we have

$$\begin{aligned} &\vdash_{\mathbf{T^M}} \Delta | a : A_1 \Rightarrow \Delta, a : C_1, \\ &\vdash_{\mathbf{T^M}} \Delta, a : A_1 | a : A_2 \Rightarrow \Delta, a : A_1, a : A_2, \end{aligned}$$

and by $(M^{\wedge})$, we have

$$\vdash_{\mathbf{T^M}} \Delta | a : A_1 \wedge A_2 \Rightarrow \Delta, a : C_1 \wedge C_2.$$

Case $A = A_1 \vee A_2$. Then, either

(i) there is a formula $a : C_1 \neq \lambda$ such that $a : C_1$ is a $\preceq^{\mathbf{M}}$-minimal change of $a : A_1$ by Δ, and then $a : C_1 \vee A_2$ is a $\preceq^{\mathbf{M}}$-minimal change of $a : A_1 \vee A_2$ by Δ, or

(ii) there is a formula $a : C_2 \neq \lambda$ such that $a : C_2$ is a $\preceq^{\mathbf{M}}$-minimal change of $a : A_2$ by Δ, and then $a : A_1 \vee C_2$ is a $\preceq^{\mathbf{M}}$-minimal change of $a : A_1 \vee A_2$ by Δ, or

(iii) λ is a $\preceq^{\mathbf{M}}$-minimal change of $a : A_1 \vee A_2$ by Δ.

By induction assumption, we have one of the following

$$\begin{aligned} &\vdash_{\mathbf{T^M}} \Delta | a : A_1 \Rightarrow \Delta, a : C_1, C_1 \neq \lambda; \\ &\vdash_{\mathbf{T^M}} \Delta | a : A_2 \Rightarrow \Delta, a : C_2, C_1 = \lambda, C_2 \neq \lambda; \\ &\vdash_{\mathbf{T^M}} \Delta | a : A_1 \Rightarrow \Delta, \lambda; \ \vdash_{\mathbf{T^M}} \Delta | A_2 \Rightarrow \Delta, \lambda, \end{aligned}$$

and by one of $(M_1^{\vee})$, $(M_2^{\vee})$ and $(M_3^{\vee})$, we have

$$\vdash_{\mathbf{T^M}} \Delta | a : A_1 \vee A_2 \Rightarrow \Delta, a : C_1 \vee C_2.$$

Case $A = a : \Box A_1$. Then, there are formulas $C_1, \ldots, C_n, C$ and constants $d_1 = a, \ldots, d_n$ such that

(i) $R(a, d_1), \ldots, R(a, d_n) \in \Delta$;

(ii) $d_n : C$ is a $\preceq^{\mathbf{M}}$-minimal change of $d_n : C_n$ by Δ, and

(iii) for any $i \leq n$, $d_i : C_i$ is a $\preceq^{\mathbf{M}}$-minimal change of $d_i : C_{i-1}$.

By induction assumption, we have

$$\begin{aligned} &\vdash_{\mathbf{T^M}} \Delta | d_i : C_{i-1} \Rightarrow \Delta, d_i : C_i, \\ &\vdash_{\mathbf{T^M}} \Delta | a : \Box C_{i-1} \Rightarrow \Delta | a : \Box C_i, \end{aligned}$$

and by $(M^{\Box})$, we have

$$\vdash_{\mathbf{T^M}} \Delta | a : \Box A_1 \Rightarrow \Delta, a : \Box C.$$

Case $A = a : \Diamond A_1$. Then, there is a formula C_1 and a new label c such that

(i) $R(a, c) \in \Delta$; and

(ii) $c : C_1$ is a $\preceq^{\mathbf{M}}$-minimal change of $c : A_1$ by Δ.

By induction assumption, we have

$$\vdash_{\mathbf{T^M}} \Delta|c : A_1 \Rightarrow \Delta, c : C_1,$$

and by $(M^{\Diamond})$, we have

$$\vdash_{\mathbf{T^M}} \Delta|a : \Diamond A_1 \Rightarrow \Delta, a : \Diamond C.$$

□

Example 5.3.4 Let $\Delta = \{R(a, b_1), R(a, b_2), b_1 : A_1, b_2 : A_2\}$ and $A = \Box(\neg A_1 \wedge \neg A_2 \wedge A_3)$. We have the following deduction:

$$\begin{array}{r}\Delta|b_1 : (\neg A_1 \wedge \neg A_2 \wedge A_3) \Rightarrow \Delta, b_1 : (\neg A_2 \wedge A_3)\\ \Delta|b_2 : (\neg A_1 \wedge \neg A_2 \wedge A_3) \Rightarrow \Delta, b_2 : (\neg A_1 \wedge A_3)\\ \Delta|a : \Box(\neg A_1 \wedge \neg A_2 \wedge A_3) \Rightarrow \Delta, a : \Box.A_3.\end{array}$$

Because

$$a : \Box A_3 = a : \Box(\neg A_2 \wedge A_3) \curlywedge a : \Box(\neg A_1 \wedge A_3).$$

Hence,

$$\vdash_{\mathbf{T^M}} \Delta|a : \Box(\neg A_1 \wedge \neg A_2 \wedge A_3) \Rightarrow \Delta, a : \Box A_3.$$

Example 5.3.5 Let $\Delta = \{R(a, b_1), A_1(b_1), A_2(b_1)\}$ and $a : A = a : \Box(\neg A_1 \wedge (\neg A_2 \vee A_3))$. We have the following deduction:

$$\begin{aligned}\Delta|b_1 : \neg A_1 \wedge (\neg A_2 \vee A_3) &\Rightarrow \Delta|b_1 : (\neg A_2 \vee A_3)\\ &\Rightarrow \Delta, b_1 : A_3.\end{aligned}$$

Hence,

$$\vdash_{\mathbf{T^M}} \Delta|a : \Box(\neg A_1 \wedge (\neg A_2 \vee A_3)) \Rightarrow \Delta, a : \Box A_3,$$

and for any formula E with $a : \Box A_3 \prec a : E \preceq a : \Box(\neg A_1 \wedge (\neg A_2 \vee A_3))$, E has three cases:

$$\begin{array}{l}a : \Box(\neg A_2 \vee A_3),\\ a : \Box(\neg A_1 \wedge A_3),\\ a : \Box(\neg A_1 \wedge (\neg A_2 \vee A_3)).\end{array}$$

The last two are inconsistent with Δ, and for the first one, we have:

$$\begin{array}{l}\Delta, a : \Box A_3 \vdash a : \Box(\neg A_2 \vee A_3),\\ \Delta, a : \Box(\neg A_2 \vee A_3) \vdash a : \Box A_3.\end{array}$$

5.4 R-Modal Logic

Hoare logic [1, 5] is a logic for programming, where the statements are triples of form $\{A\}\alpha\{B\}$, where A, B are formulas in a first-order logical language and α is a program defined on the logical language. Soundness theorem and the relative completeness theorem are proved in [2].

Taking Δ as a program and Γ, Θ as formulas, $\Delta|\Gamma \Rightarrow \Delta, \Theta$ is represented as $\Gamma[\Delta]\Theta$, a statement in Hoare-typed logic, where $[\Delta]$ is a modality. Hence, we will give a Hoare-typed logic $\mathbf{H_R}$ for R-calculus of propositional logic such that $\mathbf{H_R}$ is sound and complete with respect to R-calculus $\mathbf{S}$. Precisely, for any formula set Δ and formula A, Δ is consistent with A if and only if $A[\Delta]A$ is provable in $\mathbf{H_R}$; and Δ is inconsistent with A if and only if $A[\Delta]\neg A$ is provable in $\mathbf{H_R}$.

A semantics will be given such that a model $\mathbf{M}$ is a pair (W, R), where W is the set of all the assignments (possible worlds), and for each formula Δ, $R_\Delta \subseteq W^2$ is the accessibility relation for modalities $[\Delta]$, such that for any possible worlds w and w', w' is accessible from w via R_Δ if and only if w' is nearest to w such that $w' \models \Delta$. Correspondingly, we define the validity of statements of forms

$$[A[\Delta]B], [A\langle\Delta\rangle B], \langle A[\Delta]B\rangle, \langle A\langle\Delta\rangle B\rangle,$$

where $\langle\Delta\rangle$ is the dual modality of $[\Delta]$, and prove soundness and completeness theorem for $\mathbf{H_R}$ in the following sense:

• **Soundness theorem**: If $\vdash A[\Delta]A$ then $\langle A\langle\Delta\rangle A\rangle$ is valid; and if $\vdash A[\Delta]\neg A$ then $[A[\Delta]\neg A]$ is valid, where the modality $\langle\cdot\rangle$ is lifted up to statement $A\langle\Delta\rangle B$ to form a new statement $\langle A\langle\Delta\rangle B\rangle$;

• **Completeness theorem**: If $\langle A\langle\Delta\rangle B\rangle$ is valid then $\vdash A[\Delta]A$; and if $[A[\Delta]\neg A]$ is valid then $\vdash A[\Delta]\neg A$, where the modality $[\cdot]$ is lifted up to statement $A[\Delta]B$ to form a new statement $[A[\Delta]B]$.

Belief revision is taken as a modal logic, dynamic-epistemic logic of information change, dynamic logic, doxastic logic, and dynamic doxastic logic. The difference between belief revision and dynamic logics lies in that belief revision is to change theories, and a dynamic logic is to change states, which indirectly changes theories. Usually, in Hoare logic, a program changing states is represented by a triple $\{A\}\alpha\{B\}$, where A is the precondition and B is the postcondition of program α. For the standard model of Peano arithmetic, we know that the behavior of α can be represented by a formula; and for the nonstandard model of Peano arithmetic, it cannot, which results in the difficulty of logically researching programs, and the relative completeness theorem [2] instead of completeness theorem of the Hoare logic.

R-modal logic given in this section has only one model $\mathbf{M} = (W, R)$, where W is the set of all the assignments, and R_Δ is uniquely determined by Δ. Hence, soundness theorem and completeness theorem are relative to this unique model.

5.4.1 A Logical Language of R-Modal Logic

Let the logical language contain the following symbols:

- propositional variables: $p_0, p_1, \ldots$;
- logical connectives: $\neg, \wedge, \vee$; and
- modal symbols: $[\cdot]$, $\langle \cdot \rangle$.

The formulas A are defined as follows:

$$A ::= p|\neg p|A_1 \wedge A_2|A_1 \vee A_2;$$

and the modalities are defined as

$$[\Delta]|\langle \Delta \rangle,$$

and the statements Φ are defined as

$$\Phi ::= A[\Delta]B|A\langle \Delta \rangle B|[A[\Delta]B]|[A\langle \Delta \rangle B]|\langle A[\Delta]B\rangle|\langle A\langle \Delta \rangle B\rangle,$$

where Δ is a set of formulas A.

A model $\mathbf{M} = (W, R)$ is a pair, where W is the set of all the assignments (possible worlds), and for each formula set Δ, $R_\Delta \subseteq W^2$ is an accessibility relation R_Δ for modalities $[\Delta]$ and $\langle \Delta \rangle$, where

$$R_\Delta = \{(w, w') : d(w, w') = \min\{d(w, w'') : w'' \models \Delta\}\},$$

where

$$d(w, w') = |\{p : w \models p\} - \{p' : w' \models p'\}| + |\{p' : w' \models p'\} - \{p : w \models p\}|.$$

Given a possible world $w \in W$, we say that A is satisfied at w, denoted by $w \models A$, if

$$\begin{cases} w(p) = 1 & \text{if } A = p \\ w(p) = 0 & \text{if } A = \neg p \\ w \models A_1 \& w \models A_2 & \text{if } A = A_1 \wedge A_2 \\ w \models A_1 \text{ or } w \models A_2 & \text{if} A = A_1 \vee A_2 \end{cases}$$

and a statement Φ is satisfied at w, denoted by $w \models \Phi$, if

$$\begin{cases} w \models A \Rightarrow \mathbf{A}w'((w, w') \in R_\Delta \Rightarrow w' \models B) \text{ if } \Phi = A[\Delta]B \\ w \models A \& \mathbf{E}w'((w, w') \in R_\Delta \& w' \models B) \quad \text{if} \Phi = A\langle \Delta \rangle B \end{cases}$$

Define

- $\mathbf{M} \models [A[\Delta]C]$ if for any w, $w \models A$ implies that for any w' with $(w, w') \in R_\Delta$, $w \models C$;
- $\mathbf{M} \models [A\langle \Delta \rangle C]$ if for any w, $w \models A$ implies that for some w' with $(w, w') \in R_\Delta$, $w \models C$;

- $\mathbf{M} \models \langle A[\Delta]C\rangle$ if there is w such that $w \models A$ and for any w' with $(w, w') \in R_\Delta$, $w \models C$;
- $\mathbf{M} \models \langle A\langle\Delta\rangle C\rangle$ if there is w such that $w \models A$ and for some w' with $(w, w') \in R_\Delta$, $w \models C$.

5.4.2 R-Modal Logic

A simple deduction system $\mathbf{H_R}$ consists of the following axioms:

$$\frac{\Delta \nvdash \neg A}{A[\Delta]A}, \qquad \frac{\Delta \vdash \neg A}{A[\Delta]\neg A},$$

where the axioms are decomposed into the following axioms and deduction rules.

$\mathbf{H_R}$ consists of the following axioms and deduction rules:

- **Axioms:**

$$(HR^{\mathbf{A}})\ \frac{\Delta \nvdash \neg l}{l[\Delta]l}\ (HR_{\mathbf{A}})\ \frac{\Delta \vdash \neg l}{l[\Delta]\neg l}$$

- **Deduction rules:**

$$(HR^{\wedge})\ \frac{A_1[\Delta]A_1 \quad A_2[\Delta, A_1]A_2}{A_1 \wedge A_2[\Delta]A_1 \wedge A_2} \qquad (HR^1_{\wedge})\ \frac{A_1[\Delta]\neg A_1}{A_1 \wedge A_2[\Delta]\neg(A_1 \wedge A_2)}$$

$$(HR^2_{\wedge})\ \frac{A_2[\Delta, A_1]\neg A_2}{A_1 \wedge A_2[\Delta]\neg(A_1 \wedge A_2)}$$

$$(HR^{\vee}_1)\ \frac{A_1[\Delta]A_1}{A_1 \vee A_2[\Delta]A_1 \vee A_2} \qquad (HR_{\vee})\ \frac{A_1[\Delta]\neg A_1 \quad A_2[\Delta]\neg A_2}{A_1 \vee A_2[\Delta]\neg(A_1 \vee A_2)}$$

$$(HR^{\vee}_2)\ \frac{A_2[\Delta]A_2}{A_1 \vee A_2[\Delta]A_1 \vee A_2}$$

where $(HR^{\mathbf{A}})$ and $(HR_{\mathbf{A}})$ are called the axioms and others are deduction rules.

Definition 5.4.1 A statement $A[\Delta]B$ is provable, denoted by $\vdash A[\Delta]B$, if there is a sequence $\{\Phi_1, \ldots, \Phi_n\}$ of statements such that for each $i \leq n$, either Φ_i is an axiom or is deduced from previous statements via the deduction rules.

Lemma 5.4.2 *If* $\vdash A[\Delta]\neg A$ *then* Δ, A *is inconsistent; and if* $\vdash A[\Delta]A$ *then* Δ, A *is consistent.*

Proof Assume that $\vdash A[\Delta]\neg A$. We prove that Δ, A is inconsistent by induction on the length of a proof of $\vdash A[\Delta]\neg A$.

If the last rule used is $(HR^{\neg})$ then $A = l$, $\Delta \vdash \neg l$ and Δ, l is inconsistent;

If the last rule used is $(HR^{\wedge}_1)$ then $\vdash A_1[\Delta]\neg A_1$. By the induction assumption, Δ, A_1 is inconsistent, and so is $\Delta, A_1 \wedge A_2$;

If the last rule used is $(HR^{\wedge}_2)$ then $\vdash A_2[\Delta, A_1]\neg A_2$. By the induction assumption, $\Delta \cup \{A_1\}, A_2$ is inconsistent, and so is $\Delta, A_1 \wedge A_2$;

If the last rule used is $(HR^{\vee})$ then $\vdash A_1[\Delta]\neg A_1$, and $\vdash A_2[\Delta]\neg A_2$. By the induction assumption, both Δ, A_1 and Δ, A_2 are inconsistent, and so is $\Delta, A_1 \vee A_2$.

Similar for case $\vdash A[\Delta]A$. □

Lemma 5.4.3 *If* Δ, A *is inconsistent then* $\mathbf{M} \models [A[\Delta]\neg A]$; *and if* Δ, A *is consistent then* $\mathbf{M} \models \langle A\langle\Delta\rangle A\rangle$.

Proof Assume that Δ, A is inconsistent. Then, for any $w \in W$, $w \models A$ iff $w \not\models \Delta$. Hence, for any $w \in W$, if $w \models A$ then for any possible world w' such that $(w, w') \in R_\Delta$, $w' \models \Delta$ and $w' \models \neg A$, that is, $w \models A[\Delta]\neg A$, i.e., $\mathbf{M} \models [A[\Delta]\neg A]$.

Assume that Δ, A is consistent. Then, there is a possible world $w \in W$ such that $w \models \Delta$ and $w \models A$, and hence, $(w, w) \in R_\Delta$. Therefore, $w \models A\langle\Delta\rangle A$, and $\mathbf{M} \models \langle A\langle\Delta\rangle A\rangle$. □

Theorem 5.4.4 *If* $\vdash A[\Delta]A$ *then* $\mathbf{M} \models \langle A\langle\Delta\rangle A\rangle$; *and if* $\vdash A[\Delta]\neg A$ *then* $\mathbf{M} \models [A[\Delta]A]$.

Proof By Theorem 5.3.2, if $\vdash A[\Delta]\neg A$ then Δ, A is inconsistent and by Theorem 5.3.3, $\mathbf{M} \models [A[\Delta]\neg A]$; and if $\vdash A[\Delta]A$ then Δ, A is consistent and by Theorem 5.3.3, $\mathbf{M} \models \langle A\langle\Delta\rangle A\rangle$. □

Lemma 5.4.5 *If* $\mathbf{M} \models \langle A\langle\Delta\rangle A\rangle$ *then* Δ, A *is consistent; and if* $\mathbf{M} \models [A[\Delta]\neg A]$ *then* Δ, A *is inconsistent.*

Proof Assume that $\mathbf{M} \models \langle A\langle\Delta\rangle A\rangle$. Then, by the definition, there is a possible world w such that $w \models A\langle\Delta\rangle A$, i.e., there is a possible world w' such that $w \models A$, $(w, w') \in R_\Delta$ and $w' \models A$. By the definition of R_Δ, $w' \models R_\Delta$, and Δ, A is consistent;

Assume that $\mathbf{M} \models [A[\Delta]\neg A]$. Then, by the definition, for any possible world w, if $w \models A$ then for any $w' \in W$ with $(w, w') \in R_\Delta$, $w' \models \neg A$, that is, for any possible world w, $(w, w) \notin R_\Delta$, i.e., Δ, A is inconsistent. □

Lemma 5.4.6 *If* Δ, A *is consistent then* $\vdash A[\Delta]A$; *and if* Δ, A *is inconsistent then* $\vdash A[\Delta]\neg A$.

Proof Assume that Δ, A is inconsistent. We prove by induction on the structure of A that $\vdash A[\Delta]\neg A$.

If $A = l$ then $\Delta \vdash \neg l$ and by $(HR^\neg)$, $\vdash l[\Delta]\neg l$;

If $A = A_1 \wedge A_2$ then either Δ, A_1 is inconsistent or $\Delta \cup \{A_1\}, A_2$ is inconsistent. If Δ, A_1 is inconsistent then by induction assumption, $\vdash A_1[\Delta]\neg A_1$ and by $(HR_1^\wedge)$, $\vdash A_1 \wedge A_2[\Delta]\neg(A_1 \wedge A_2)$; if $\Delta \cup \{A_1\}, A_2$ is inconsistent then by induction assumption, $\vdash A_2[\Delta, A_1]\neg A_2$ and by $(HR_1^\wedge)$, $\vdash A_1 \wedge A_2[\Delta]\neg(A_1 \wedge A_2)$.

Similar for case that Δ, A is consistent. □

Theorem 5.4.7 (Completeness theorem) *If* $\mathbf{M} \models \langle A\langle\Delta\rangle A\rangle$ *then* $\vdash A[\Delta]A$; *and if* $\mathbf{M} \models [A[\Delta]\neg A]$ *then* $\vdash A[\Delta]\neg A$.

Proof Assume that $\mathbf{M} \models \langle A\langle\Delta\rangle A\rangle$. Then, Δ, A is consistent, and by Theorem 5.3.3, $\vdash A[\Delta]A$;

Assume that $\mathbf{M} \models [A[\Delta]\neg A]$. Then, Δ, A is inconsistent, and by Theorem 5.3.3, $\vdash A[\Delta]\neg A$; □

Theorem 5.4.8 *For any formula set* Δ *and formula* A, *either* $\mathbf{M} \models \langle A\langle\Delta\rangle A\rangle$ *or* $\mathbf{M} \models [A[\Delta]\neg A]$.

Proof For any formula set Δ and formula A, Δ, A is either consistent or inconsistent, and by Theorem 5.3.3, either $\mathbf{M} \models \langle A\langle\Delta\rangle A\rangle$ or $\mathbf{M} \models [A[\Delta]\neg A]$. □

References

1. C.E. Alchourrón, P. Gärdenfors, D. Makinson, On the logic of theory change: partial meet contraction and revision functions. J. Symb. Logic **50**, 510–530 (1985)
2. K.R. Apt, Ten years of Hoares logic: a survey-part 1. ACM Trans. Program. Lang. Syst. **3**, 431–483 (1981)
3. M.J. Cresswell, Modal Logic, in Goble, Lou (eds.), *The Blackwell Guide to Philosophical Logic* (Basil Blackwell, Hoboken, 2001), pp. 136-58
4. R.W. Floyd, Assigning meanings to programs, in *Proceedings of the American Mathematical Society Symposia on Applied Mathematics*, vol. 19 (1967), pp. 19–31
5. C.A.R. Hoare, An axiomatic basis for computer programming. Commun. ACM **12**, 576–580 (1969)
6. G.E. Hughes, M.J. Cresswell, *A New Introduction to Modal Logic* (Routledge, Milton Park, 1996)

Chapter 6
R-Calculi for Logic Programming

This chapter will give R-calculi $\mathbf{S}^{\mathrm{LP}}$, $\mathbf{T}^{\mathrm{LP}}$ and $\mathbf{U}^{\mathrm{LP}}$ for logic programming [1, 3, 4, 6] such that they are sound and complete with respect to $\subseteq / \preceq / \vdash$-minimal changes, respectively [7]. Specially, for $\mathbf{U}^{\mathrm{LP}}$, we have the following conclusions:

- if $t|t' \Rightarrow t, t''$ is provable then t'' is a theory such that $t' \vdash t''$ and for any t_1 such that $t' \vdash t_1 \vdash t''$ and $t'' \nvdash t_1$, t_1 is inconsistent with t, that is, t'' is a $\vdash$-minimal change of t' by t; and
- $t|t' \Rightarrow t, t''$ is provable for any t'' such that t'' is a $\vdash$-minimal change of t' by t.

6.1 Logic Programming

Fix a first-order logical language. Literals, clauses and theories are defined as follows:

Definition 6.1.1 A *literal* l is an atom $p(t_1, ..., t_n)$ or the negation $\neg p(t_1, ..., t_n)$ of an atom. A *clause*

$$c = l_1 \vee \cdots \vee l_n$$

(denoted by $c = l_1; ...; l_n$) is a disjunction of literals, and a *theory*

$$t = c_1 \wedge \cdots \wedge c_m$$

is a conjunction of clauses. That is,

$$\begin{aligned} l &::= p(t_1, ..., t_n)|\neg p(t_1, ..., t_n) \\ c &::= l|l \vee c = l|l; c \\ t &::= c|c \wedge t = c|c, t \end{aligned}$$

where p is an n-ary predicate symbol.

W. Li and Y. Sui, *R-CALCULUS: A Logic of Belief Revision*, Perspectives in Formal Induction, Revision and Evolution,
https://doi.org/10.1007/978-981-16-2944-0_6

Let the base (language) L consist of finite sets of constant symbols, variable symbols, function symbols and predicate symbols. A term or formula is *ground* if there is no variable in it. A literal l is *positive* if l is an atom, otherwise, *negative*. A clause $\pi : l_1, ..., l_r \leftarrow l'_1, ..., l'_s$ is called

$$\begin{cases} a \quad rule \ if r \geq 1 ands \geq 1; \\ a \quad fact \ if r \geq 1 ands = 0; \\ a \quad goal \ if \ r = 0 \text{ and } s \geq 1, \end{cases}$$

where $l_1, ..., l_r$ is called the *rule head*, denoted by $head(\pi)$, and $l'_1, ..., l'_s$ is called the *rule body*, denoted by $body(\pi)$. Every l'_i is called a *subgoal*. We assume that the clauses are closed, i.e., every clause is a closed formula with universal quantification on variables in the clause. For example, $p(x) \leftarrow q(x)$ means that $\forall x[p(x) \leftarrow p(x)]$, where p, q are unary predicate symbols. A *logic program* is a set of clauses in some base L.

Definition 6.1.2 A clause $\pi : l_1, ..., l_r \leftarrow l'_1, ..., l'_s$ is called a *Horn clause* if $r \leq 1$ and every literal is positive. A logic program Π is *Horn* if every clause in Π is a Horn clause.

The *Herbrand universe* U_L of a logic program in L is the set of all ground terms in L. HB_L denotes the set of all the ground atoms in L, called the *Herbrand base*.

Definition 6.1.3 A *Herbrand interpretation* of a logic program Π is any subset I of HB_L. A Herbrand interpretation I of Π is a *Herbrand model of* Π if every clause is satisfied under I.

A Herbrand model I of a logic program Π is called *minimal* if there exists no subset $I' \subset I$ that is a model for Π; and called *least* if I is the unique minimal model of Π.

Theorem 6.1.4 *Let Π be a Horn program and let $HM(\Pi)$ denote the set of all Herbrand models of Π. Then the model intersection property holds, i.e., $\bigcap_{W \in HM(\Pi)} W$ is a Herbrand model of Π.*

□

6.1.1 Gentzen Deduction Systems

Let $t(x_1, ..., x_n)$ be a theory, where variables$x_1, ..., x_n$ occur in t.

A model$M$ is a subset of H. Theory t is satisfied in model M, denoted by $M \models t$, if

$$\begin{cases} l \in M & \text{if } \ t = l \\ M \models l \text{or} M \models c & \text{if } \ t = l; c \\ M \models c \& M \models t' & \text{if } \ t = c, t' \\ Aa_1, ..., a_n \in U(M \models t'(x_1/a_1, ..., x_n/a_n)) & \text{if } \ t = t'(x_1, ..., x_n); \end{cases}$$

A theory t is valid, denoted by $\models t$, if t is satisfied in any model.

A sequent is of form $t \Rightarrow t'$, where t, t' are theories. Sequent $t \Rightarrow t'$ is satisfied in models M, denoted by $M \models t \Rightarrow t'$, if $M \models t$ implies $M \models t'$. Sequent $t \Rightarrow t'$ is valid, denoted by $\models t \Rightarrow t'$, if $t \Rightarrow t'$ is satisfied in any model M.

Gentzen deduction system $\mathbf{G_3}$ for theories consists of the following axioms and deduction rules:

- **Axioms**:

$$\frac{t' \subseteq t}{t \Rightarrow t'},$$

where t, t' are sets of literals, i.e., $t = l_1, ..., l_n$.

- **Deduction rules**:

$$(;^L)\ \frac{t, l \Rightarrow t' \quad t, c \Rightarrow t'}{t, l; c \Rightarrow t'} \qquad (;^R_1)\ \frac{t \Rightarrow l, t'}{t \Rightarrow l; c, t'}$$

$$(;^R_2)\ \frac{t \Rightarrow c, t'}{t \Rightarrow l; c, t'}$$

$$(,^L_1)\ \frac{c \Rightarrow t'}{c, t \Rightarrow t'} \qquad (,^R)\ \frac{t \Rightarrow c \quad t \Rightarrow t'}{t \Rightarrow c, t'}$$

$$(,^L_2)\ \frac{t \Rightarrow t'}{c, t \Rightarrow t'}$$

$$(\forall^L)\ \frac{t(b) \Rightarrow t'}{t(x) \Rightarrow t'} \qquad (\forall^R)\ \frac{t \Rightarrow t'(a)}{t \Rightarrow t'(x)}$$

where b is a constant, and a is a new constant not occurring in t and t'.

Definition 6.1.5 A sequent $t \Rightarrow t'$ is provable in $\mathbf{G_3}$, denoted by $\vdash t \Rightarrow t'$, if there is a sequence $t_1 \Rightarrow t_1', ..., t_n \Rightarrow t_n'$ of sequents such that $t_n \Rightarrow t_n' = t \Rightarrow t'$, and for each $1 \leq i \leq n$, $t_i \Rightarrow t_i'$ is either an axiom or deduced from previous sequents by one deduction rule in $\mathbf{G_3}$.

Theorem 6.1.6 (Soundness theorem) *For any sequent $t \Rightarrow t'$, if $\vdash t \Rightarrow t'$ then $\models t \Rightarrow t'$.*

□

Theorem 6.1.7 (Completeness theorem) *For any sequent $t \Rightarrow t'$, if $\models t \Rightarrow t'$ then $\vdash t \Rightarrow t'$.*

□

6.1.2 Dual Gentzen Deduction System

Definition 6.1.8 A term $d = l_1 \wedge \cdots \wedge l_n$ (denoted by $c = l_1, ..., l_n$) is a disjunction of literals. A co-theory $s = d_1 \vee \cdots \vee c_m$ is a conjunction of terms. That is,

$$d ::= l|l \wedge d = l|l, c$$
$$s ::= d|d \vee s = d|d; s.$$

A co-theory s is satisfied in a model M, denoted by $M \models s$, if

$$\begin{cases} l \in M & \text{if } s = l \\ M \models l \& M \models d & \text{if } s = l, d \\ M \models d \text{ or } M \models s' & \text{if } s = d; s' \\ Aa_1, ..., a_n \in U(M \models s'(x_1/a_1, ..., x_n/a_n)) & \text{if } s = s'(x_1, ..., x_n). \end{cases}$$

A sequent is of form $s \Rightarrow s'$, where s, s' are co-theories, and is satisfied in a model M, denoted by $M \models s \Rightarrow s'$, if $M \models s$ implies $M \models s'$.

Gentzen deduction system $\mathbf{G}'_3$ for co-theories consists of the following axioms and deduction rules:

- **Axioms**:

$$\frac{s' \supseteq s}{s \Rightarrow s'},$$

where s, s' are sets of literals, i.e., $s = l_1; ...; l_n$.

- **Deduction rules:**

$$(;^L_1)\ \frac{s; l \Rightarrow s'}{s; l, d \Rightarrow s'} \qquad (;^R)\ \frac{s \Rightarrow l; s' \quad s \Rightarrow d; s'}{s \Rightarrow l, d; s'}$$
$$(;^L_2)\ \frac{s; d \Rightarrow s'}{s; l, d \Rightarrow s'}$$
$$(,^L)\ \frac{d \Rightarrow s' \quad s \Rightarrow s'}{d; s \Rightarrow s'} \qquad (,^R_1)\ \frac{s \Rightarrow d}{s \Rightarrow d; s'}$$
$$(,^R_2)\ \frac{s \Rightarrow s'}{s \Rightarrow d; s'}$$
$$(\forall^L)\ \frac{s(b) \Rightarrow s'}{s(x) \Rightarrow s'} \qquad (\forall^R)\ \frac{s \Rightarrow s'(a)}{s \Rightarrow s'(x)}$$

where b is a constant and a is a new constant not occurring in s and s'.

Theorem 6.1.9 (Soundness and completeness theorem) *For any sequent* $s \Rightarrow s'$, $\models s \Rightarrow s'$ *if and only if* $\vdash s \Rightarrow s'$.

□

6.1.3 Minimal Change

Definition 6.1.10 Given any consistent theories t' and t, a theory t'' is a $\subseteq$-minimal change of t' by t, denoted by $\models_{\mathrm{SLP}} t|t' \Rightarrow t, t''$, if (i) $t'' \subseteq t'$; (ii) t'' is consistent with t, and (iii) for any set t''' with $t'' \subset t''' \subseteq t'$, t''' is inconsistent with t.

Definition 6.1.11 A theory t'' is a $\preceq$-minimal change of t' by t, denoted by $\models_{\mathbf{T}^{\mathrm{LP}}} t|t' \Rightarrow t, t''$, if

(i) $t'' \preceq t'$, that is, for each clause $c \in t''$, there is a clause $c' \in t'$ such that $c \preceq c'$,
(ii) t'', t is consistent, and
(iii) for any theory t''' with $t'' \prec t''' \preceq t', t''', t$ is inconsistent.

Definition 6.1.12 Given three theories t', t and t'', t'' is a $\vdash$-minimal change of t' by t, denoted by $\models_{\mathbf{U}^{\mathrm{LP}}} t'|t \Rightarrow t, t''$, if

(i) $t' \vdash t''$,
(ii) t'', t is consistent, and
(iii) for any theory t''' with $t' \vdash t''' \vdash t''$ and $t'' \nvdash t''', t''', t$ is inconsistent.

6.2 R-Calculus $\mathbf{S}^{\mathrm{LP}}$ for $\subset$-Minimal Change

In this section, we will give an R-calculus for the logic programming, based on $\subseteq$-minimal change of Γ by Δ, and an Gentzen-typed deduction system $\mathbf{S}^{\mathrm{LP}}$ such that for any consistent theories t, t' and t'', $t|t' \Rightarrow t, t''$ is provable in $\mathbf{S}^{\mathrm{LP}}$ if and only if t'' is a $\subseteq$-minimal change of t' by t.

R-calculus $\mathbf{S}^{\mathrm{LP}}$ consists of the following axioms and deduction rules:

- **Axioms:**

$$(S^{\mathrm{A}})\ \frac{t \nvdash \neg l}{t|l \Rightarrow t, l} \quad (S_{\mathrm{A}})\ \frac{t \vdash \neg l}{t|l \Rightarrow t}$$

- **Deduction rules:**

$$(S^{,})\ \frac{\begin{array}{c} t|c_1 \Rightarrow t, c_1| \\ t, c_1|c_2 \Rightarrow t, c_1, c_2| \end{array}}{t|c_1, c_2 \Rightarrow t, c_1, c_2|} \quad (S^1_{,})\ \frac{t|c_1 \Rightarrow t|}{t|c_1, c_2 \Rightarrow t|}$$

$$(S^2_{,})\ \frac{t, c_1|c_2 \Rightarrow t, c_1}{t|c_1, c_2 \Rightarrow t}$$

$$(S^{;}_1)\ \frac{t|c_1 \Rightarrow t, c_1|}{t|c_1; c_2 \Rightarrow t, c_1; c_2} \quad (S_{;})\ \frac{t|c_1 \Rightarrow t \quad t|c_2 \Rightarrow t}{t|c_1; c_2 \Rightarrow t}$$

$$(S^{;}_2)\ \frac{t|c_2 \Rightarrow t, c_2|}{t|c_1; c_2 \Rightarrow t, c_1; c_2}$$

$$(S^{\forall})\ \frac{t|\theta t'(x) \Rightarrow t, \theta t'(x)|}{t|t'(x) \Rightarrow t|t'(x)} \quad (S_{\forall})\ \frac{t|\theta t'(x) \Rightarrow t}{t|t'(x) \Rightarrow t}$$

where the rules of the left-hand side are to put a formula of t' into t'', and the ones of the right-hand side are not to; and θ is a substitution of form $(x_1/b_1, ..., x_n/b_n)$, where $x_1, ..., x_n$ are variables and $b_1, ..., b_n$ are terms in first-order logic.

Definition 6.2.1 $t|t' \Rightarrow t, t''$ is provable in $\mathbf{S}^{\mathrm{LP}}$, denoted by $\vdash_{\mathbf{S}^{\mathrm{LP}}} t|t' \Rightarrow t, t''$, if there is a sequence $S_1, ..., S_m$ of statements such that

$$S_1 = t|t' \Rightarrow t|c_1',$$
$$\ldots$$
$$S_m = t|t_m \Rightarrow t, t'';$$

and for each $i < m$, S_{i+1} is an axiom or is deduced from the previous statements by a deduction rule in $\mathbf{S}^{\mathrm{LP}}$.

For example, the following are deductions in $\mathbf{S}^{\mathrm{LP}}$:

$$t(x, y)|\neg t(b, b') \Rightarrow t(x, y)$$
$$t(x, y)|\neg t(b, b') \vee \neg t(b, b') \Rightarrow t(x, y)$$
$$t(x, y)|\neg t(b, y) \vee \neg t(x, b') \Rightarrow t(x, y)$$

and

$$t(b, b')|\neg t(b, b') \Rightarrow t(b, b')t(b, b')|\neg t(x, y) \Rightarrow t(b, b')$$

where $\theta = (x/b, y/b')$.

Theorem 6.2.2 (Completeness theorem) *For any consistent theories t and t', if t, t' is consistent then $t|t' \Rightarrow t, t'$ is provable in $\mathbf{S}^{\mathrm{LP}}$, i.e.,*

$$\models_{\mathbf{S}^{\mathrm{LP}}} t|t' \Rightarrow t, t' \textit{ implies } \vdash_{\mathbf{S}^{\mathrm{LP}}} t|t' \Rightarrow t, t';$$

and if t, t' is inconsistent then $t|t' \Rightarrow t$ is provable in $\mathbf{S}^{\mathrm{LP}}$, i.e.,

$$\models_{\mathbf{S}^{\mathrm{LP}}} t|t' \Rightarrow t \textit{ implies } \vdash_{\mathbf{S}^{\mathrm{LP}}} t|t' \Rightarrow t.$$

Proof We prove the theorem by induction on the structure of formula t'.

Assume that t, t' is consistent.

If $t' = l$ then $t \nvdash \neg t'$, and by $(S^{\mathbf{A}})$, $t|t' \Rightarrow t, l$ is provable;

If $t' = c_1, c_2$ then t, c_1 and t, c_1, c_2 are consistent, and by induction assumption, $t|c_1 \Rightarrow t, c_1$ is provable and $t, c_1|c_2 \Rightarrow t, c_1, c_2$ is provable, and by $(S^{\wedge})$, $t|c_1, c_2 \Rightarrow t, c_1, c_2$ is provable;

If $t' = c_1; c_2$ then either t, c_1 or t, c_2 is consistent, and by induction assumption, either $t|c_1 \Rightarrow t, c_1$ or $t|c_2 \Rightarrow t, c_2$ are provable, and by $(S_1^{\vee})$ or $(S_2^{\vee})$, $t|c_1; c_2 \Rightarrow t, c_1; c_2$ is provable.

If $t' = t'(x)$ then for any substitution θ, $t, \theta t'(x)$ is consistent, and by induction assumption, $t|\theta t'(x) \Rightarrow t, \theta t'(x)$ is provable, and by $(S^{\forall})$, $t|t'(x) \Rightarrow t, t'(x)$ is provable.

Assume that t, t' is inconsistent.

If $t' = l$ then $t \vdash \neg t'$, and by $(S_{\mathbf{A}})$, $t|t' \Rightarrow t$ is provable;

If $t' = c_1, c_2$ then either t, c_1 or t, c_1, c_2 is inconsistent, and by induction assumption, either $t|c_1 \Rightarrow t$ is provable, or $t, c_1|c_2 \Rightarrow t, c_1|$ is provable, and by $(S_{\wedge}^1)$ and $(S_{\wedge}^2)$, $t|c_1, c_2 \Rightarrow t$ is provable;

If $t' = c_1; c_2$ then t, c_1 and t, c_2 are inconsistent, and by induction assumption, $t|c_1 \Rightarrow t$ and $t|c_2 \Rightarrow t$ are provable, and by $(S_{\vee})$, $t|c_1; c_2 \Rightarrow t$ is provable.

If $t' = t'(x)$ then for some substitution $\theta, t, \theta t'(x)$ is inconsistent, and by induction assumption, $t|\theta t'(x) \Rightarrow t$ is provable, and by $(S_\forall)$, $t|t'(x) \Rightarrow t$ is provable. □

Theorem 6.2.3 (Soundness theorem) *For any consistent theories t and t', if $t|t' \Rightarrow t, t'$ is provable in $\mathbf{S}^{\mathrm{LP}}$ then t, t' is consistent, i.e.,*

$$\vdash_{\mathbf{S}^{\mathrm{LP}}} t|t' \Rightarrow t, t' \textit{ implies } \models_{\mathbf{S}^{\mathrm{LP}}} t|t' \Rightarrow t, t';$$

and if $t|t' \Rightarrow t$ is provable in $\mathbf{S}^{\mathrm{LP}}$ then t, t' is inconsistent, i.e.,

$$\vdash_{\mathbf{S}^{\mathrm{LP}}} t|t' \Rightarrow t \textit{ implies } \models_{\mathbf{S}^{\mathrm{LP}}} t|t' \Rightarrow t.$$

Proof We prove the theorem by induction on the structure of theory t'.

Assume that $\vdash_{\mathbf{S}^{\mathrm{LP}}} t|t' \Rightarrow t, t'$.

If $t' = l$ then $t \nvdash \neg t'$, and t, t' is consistent;

If $t' = c_1, c_2$ then $\vdash_{\mathbf{S}^{\mathrm{LP}}} t|c_1 \Rightarrow t, c_1$ and $\vdash_{\mathbf{S}^{\mathrm{LP}}} t, c_1|c_2 \Rightarrow t, c_1, c_2$. By induction assumption, t, c_1 and t, c_1, c_2 are consistent, and so is t, c_1, c_2;

If $t' = c_1; c_2$ then either $\vdash_{\mathbf{S}^{\mathrm{LP}}} t|c_1 \Rightarrow t, c_1$ or $\vdash_{\mathbf{S}^{\mathrm{LP}}} t|c_2 \Rightarrow t, c_2$. By the induction assumption, either t, c_1 or t, c_2 is consistent, and so is $t, c_1; c_2$;

If $t' = t'(x)$ then for any substitution $\theta, \vdash_{\mathbf{S}^{\mathrm{LP}}} t|\theta t'(x) \Rightarrow t, \theta t'(x)$. By induction assumption, $t, \theta t'(x)$ is consistent, and so is $t, t'(x)$;

Assume that $\vdash_{\mathbf{S}^{\mathrm{LP}}} t|t' \Rightarrow t$.

If $t' = l$ then $t \vdash \neg t'$, and t, t' is inconsistent;

If $t' = c_1, c_2$ then either $\vdash_{\mathbf{S}^{\mathrm{LP}}} t|c_1 \Rightarrow t$ or $\vdash_{\mathbf{S}^{\mathrm{LP}}} t, c_1|c_2 \Rightarrow t, c_1$. By induction assumption, either t, c_1 or t, c_1, c_2 is inconsistent, and so is t, c_1, c_2;

If $t' = c_1; c_2$ then $\vdash_{\mathbf{S}^{\mathrm{LP}}} t|c_1 \Rightarrow t$ and $\vdash_{\mathbf{S}^{\mathrm{LP}}} t|c_2 \Rightarrow t$. By induction assumption, t, c_1 and t, c_2 are inconsistent, and so is $t, c_1; c_2$.

If $t' = t'(x)$ then for some substitution $\theta, \vdash_{\mathbf{S}^{\mathrm{LP}}} t|\theta t'(x) \Rightarrow t$. By induction assumption, $t, \theta t'(x)$ is inconsistent, and so is $t, t'(x)$; □

6.3 R-Calculus $\mathbf{T}^{\mathrm{LP}}$ for $\preceq$-Minimal Change

In this section, we will give an R-calculus $\mathbf{T}^{\mathrm{LP}}$ which is sound and complete with respect to $\preceq$-minimal changes of t' by t, that is, for any consistent theories t', t and $t'', t|t' \Rightarrow t, t''$ is provable in $\mathbf{T}^{\mathrm{LP}}$ if and only if t'' is a $\preceq$-minimal change of t' by t.

R-calculus $\mathbf{T}^{\mathrm{LP}}$ consists of the following axioms and deduction rules:

- **Axioms:**

$$(T^{\mathrm{A}}) \ \frac{t \nvdash \neg l}{t|l \Rightarrow t, l} \quad (T_{\mathrm{A}}) \ \frac{t \vdash \neg l}{t|l \Rightarrow t}$$

- **Deduction rules:**

$$(T^{,})\ \frac{t|c_1 \Rightarrow t, c_1}{t|c_1, c_2 \Rightarrow t, c_1|c_2} \qquad (T_1^{;})\ \frac{t|c_1 \Rightarrow t, c_1' \neq \lambda}{t|c_1; c_2 \Rightarrow t, c_1'; c_2}$$

$$(T_2^{;})\ \frac{\begin{array}{l} t|c_1 \Rightarrow t, \lambda \\ t|c_2 \Rightarrow t, c_2' \\ c_2' \neq \lambda \end{array}}{t|c_1; c_2 \Rightarrow t, c_1; c_2'}$$

$$(T_3^{;})\ \frac{t|c_1 \Rightarrow t, \lambda \ \ t|c_2 \Rightarrow t, \lambda}{t|c_1; c_2 \Rightarrow t, \lambda}$$

$$(T^{\forall})\ \frac{t|\theta t'(x) \Rightarrow t, \theta t''(x)}{t|t'(x) \Rightarrow t|t''(x)}$$

Definition 6.3.1 $t|t' \Rightarrow t, t''$ is provable in $\mathbf{T}^{\text{LP}}$, denoted by $\vdash_{\mathbf{T}^{\text{LP}}} t|t' \Rightarrow t, t''$, if there is a sequence $S_1, ..., S_m$ of statements such that

$$\begin{aligned} S_1 &= t|t' \Rightarrow t|t_1', \\ &\cdots \\ S_m &= t|t_m \Rightarrow t, t''; \end{aligned}$$

and for each $i < m$, S_{i+1} is an axiom or is deduced from the previous statements by a deduction rule in $\mathbf{T}^{\text{LP}}$.

Theorem 6.3.2 *For any consistent theories t and t', there is a theory t'' such that $t'' \preceq t'$ and $t|t' \Rightarrow t, t''$ is provable.*

Proof We prove the theorem by induction on the structure of theory t'.

Case $t' = l$. Then, if $t \vdash \neg l$ then let $t'' = \lambda$; if $t \nvdash \neg l$ then let $t'' = l$. Then, by $(T^{\mathbf{A}})$ and $(T_{\mathbf{A}})$, $t|t' \Rightarrow t, t''$ is provable.

Case $t' = c_1 \wedge c_2$. Then, by induction assumption, there are theories $c_1' \preceq c_1, c_2' \preceq c_2$ such that $t|c_1 \Rightarrow t, c_1'$ and $t, c_1|c_2 \Rightarrow t, c_1, c_2'$ are provable in $\mathbf{T}^{\text{LP}}$. Hence, so is $t|c_1, c_2 \Rightarrow t, c_1', c_2'$.

Case $c = c_1; c_2$. Then, by induction assumption, there are theories $c_1' \preceq c_1$ and $c_2' \preceq c_2$ such that $t|c_1 \Rightarrow t, c_1'$ and $t|c_2 \Rightarrow t, c_2'$ are provable. If $c_1' \neq \lambda$ then by $(T_1^{\vee})$, $\vdash_{\mathbf{T}^{\text{LP}}} t|c_1; c_2 \Rightarrow t, c_1'; c_2$; if $c_1' = \lambda$ and $c_2' \neq \lambda$ then by $(T_2^{\vee})$, $\vdash_{\mathbf{T}^{\text{LP}}} t|c_1; c_2 \Rightarrow c_1; c_2'$; if $c_1' = c_2' = \lambda$ then by $(T_3^{\vee})$, $\vdash_{\mathbf{T}^{\text{LP}}} t|c_1; c_2 \Rightarrow t$. Let

$$t'' = \begin{cases} c_1'; c_2 \text{ if } c_1' \neq \lambda \\ c_1; c_2' \text{ if } c_1' = \lambda \neq c_2' \\ \lambda \quad \text{ otherwise,} \end{cases}$$

and $\vdash_{\mathbf{T}^{\text{LP}}} t|c_1; c_2 \Rightarrow t, t''$.

Case $t' = t'(x)$. Assume that there are substitutions $\theta_1, ..., \theta_n$ and theories $t_1, ..., t_n$ such that $t_1 \preceq t_2 \preceq \cdots \preceq t_n$, and

$$\begin{array}{l} t|\theta_1 t'(x) \Rightarrow t, \theta_1 t_1(x), \\ t|\theta_2 t_1(x) \Rightarrow t, \theta_2 t_2(x), \\ \ldots \\ t|\theta_n t_{n-1}(x) \Rightarrow t, \theta_n t_n(x), \end{array}$$

are provable, and $t, t_n(x)$ is consistent, so that $t|t'(x) \Rightarrow t, t_n(x)$ is provable. Then, let $t'' = t_n(x)$. □

Lemma 6.3.3 *If t'' is a $\preceq$-minimal change of t' by t and t_2 is a $\preceq$-minimal change of t_1 by t, t'' then t_2 is a $\preceq$-minimal change of t', t_1 by t.*

Proof Assume that t'' is a $\preceq$-minimal change of t' by t and t_2 is a $\preceq$-minimal change of t_1 by t, t''. Then, by the definition, t_2 is a $\preceq$-minimal change of t', t_1 by t. □

Lemma 6.3.4 *If c'_1 is a $\preceq$-minimal change of c_1 by t and c'_2 is a $\preceq$-minimal change of c_2 by $t \cup \{c'_1\}$ then c'_1, c'_2 is a $\preceq$-minimal change of c_1, c_2 by t.*

Proof Assume that c'_1 is a $\preceq$-minimal changes of c_1 by t and c'_2 of c_2 by t, c'_1.

Then, $c'_1, c'_2 \preceq c_1, c_2$, and t, c'_1, c'_2 is consistent.

For any t_1 with $c'_1, c'_2 \prec t_1 \preceq c_1, c_2$, there are t_{11}, t_{12} such that $t_1 = t_{11}, t_{12}$, and

$$\begin{array}{l} c'_1 \preceq t_{11} \preceq c_1, \\ c'_2 \preceq t_{12} \preceq c_2. \end{array}$$

Then, if $c'_1 \prec t_{11}$ then t, t_{11} is inconsistent and so is t, t_{11}, t_{12}; if $c'_2 \prec t_{12}$ then t, t_{12} is inconsistent and so is t, t_{11}, t_{12}. □

Lemma 6.3.5 *If c'_1, c'_2 are $\preceq$-minimal changes of c_1 and c_2 by t, respectively, then c' is a $\preceq$-minimal change of $c_1; c_2$ by t, where*

$$c' = \begin{cases} c'_1; c_2 & \text{if } c'_1 \neq \lambda \\ c_1; c'_2 & \text{if } c'_1 = \lambda \text{ and } c'_2 \neq \lambda \\ \lambda & \text{if } c'_1 = c'_2 = \lambda \end{cases}$$

Proof It is clear that $c' \preceq c_1; c_2$, and t, c' is consistent.

For any t_1 with $c' \prec t_1 \preceq c_1; c_2$, there are t_{11}, t_{12} such that

$$\begin{array}{l} t_1 = t_{11}; t_{12}, \\ c'_1 \preceq t_{11} \preceq c_1, \\ c'_2 \preceq t_{12} \preceq c_2, \end{array}$$

and either $c'_1 \prec t_{11}$ or $t'_2 \prec t_{12}$, where c'_1 is either c_1 or c_1 and c'_2 is either c'_2 or c_2.

If $c'_1 \prec t_{11}$ then $c'_1 = c_1, c'_2 = c_2$, and by induction assumption, t, t_{11} is inconsistent and so is $t, t_{11}; c_2$; and if $c'_2 \prec t_{12}$ then $c'_2 = c_2, c'_1 = c_1$, and by induction assumption, t, t_{12} is inconsistent and so is $t, c_1; t_{12}$. □

Theorem 6.3.6 *For any formula theories t, t', t'', if $t|t' \Rightarrow t, t''$ is provable then t'' is a $\preceq$-minimal change of t' by t. That is,*

$$\vdash_{\mathbf{T}^{\mathrm{LP}}} t|t' \Rightarrow t, t'' \text{ implies } \models_{\mathbf{T}^{\mathrm{LP}}} t|t' \Rightarrow t, t''.$$

Proof We prove the theorem by induction on the structure of t'.

Case $t' = l$, a literal. Either $t|l \Rightarrow t, l$ or $t|l \Rightarrow t$ is provable. That is, either $t'' = l$ or $t'' = \lambda$, which is a $\preceq$-minimal change of t' by t.

Case $t' = c_1 \wedge c_2$. Then, if $t|c_1 \Rightarrow t, c_1'$ is provable then c_1' is a $\preceq$-minimal change of c_1 by t, and if $t, c_1'|c_2 \Rightarrow t, c_1', c_2'$ is provable then c_2' is a $\preceq$-minimal change of c_2 by t, c_1'. By Lemma 6.3.4, c_1', c_2' is a $\preceq$-minimal change of c_1, c_2 by t.

Case $t' = c_1; c_2$. Then t, c_1 and t, c_2 are inconsistent, and there are c_1', c_2' such that $t|c_1 \Rightarrow t, c_1'$ and $t|c_2 \Rightarrow t, c_2'$ are provable, and by induction assumption, c_1' and c_2' are $\preceq$-minimal changes of c_1 and c_2 by t, respectively. Then, $t|c_1; c_2 \Rightarrow t, c'$, and by Lemma 6.3.5, c' is a $\preceq$-minimal change of $c_1; c_2$ by t.

Case $t' = t'(x)$. Assume that there are substitutions $\theta_1, ..., \theta_n$ and theories $t_1, ..., t_n$ such that $t_1 \preceq t_2 \preceq \cdots \preceq t_n$, and

$$\begin{array}{l} t|\theta_1 t'(x) \Rightarrow t, \theta_1 t_1(x), \\ t|\theta_2 t_1(x) \Rightarrow t, \theta_2 t_2(x), \\ \ldots \\ t|\theta_n t_{n-1}(x) \Rightarrow t, \theta_n t_n(x), \end{array}$$

are provable, and $t, t_n(x)$ is consistent, so that $t|t'(x) \Rightarrow t, t_n(x)$ is provable. By induction assumption, for any $i \leq n$, $\theta_i t_i(x)$ is a $\preceq$-minimal change of $\theta_i t_{i-1}(x)$ by t, where $t_0 = t'$. Then, $t_n(x)$ is a $\preceq$-minimal change of $t'(x)$ by t. □

Theorem 6.3.7 *For any consistent theories t, t', t'', if t'' is a $\preceq$-minimal change of t' by t then $t|t' \Rightarrow t, t''$ is* $\mathbf{T}^{\mathrm{LP}}$*-provable. That is,*

$$\models_{\mathbf{T}^{\mathrm{LP}}} t|t' \Rightarrow, t, t'' \text{ implies } \vdash_{\mathbf{T}^{\mathrm{LP}}} t|t' \Rightarrow, t, t''.$$

Proof Let $t'' \preceq t'$ be a $\preceq$-minimal change of t' by t.

Case $t' = l$. Then $t'' = \lambda$ (if t, t' is inconsistent) or $t'' = t'$ (if t, t' is consistent), and $t, t' \Rightarrow t, t''$ is provable.

Case $t' = c_1, c_2$. Then there are c_1', c_2' such that $t'' = c_1', c_2'$, and c_1' and c_2' are $\preceq$-minimal changes of c_1 and c_2 by t and t, c_1', respectively. Hence, c_1', c_2' is a $\preceq$-minimal change of c_1, c_2 by t. By the induction assumption, $t|c_1 \Rightarrow t, c_1'$ and $t, c_1'|c_2 \Rightarrow t, c_1', c_2'$ are provable, and so is $t|c_1, c_2 \Rightarrow t, c_1', c_2'$.

Case $t' = c_1; c_2$. Then there are c_1' and c_2' such that $t'' = c'$, and c_1' and c_2' are $\preceq$-minimal changes of c_1 and c_2 by t, respectively, where

$$c' = \begin{cases} c_1'; c_2 \text{ if } & c_1' \neq \lambda \\ c_1; c_2' \text{ if } & c_1' = \lambda \text{ and } c_2' \neq \lambda \\ \lambda \quad\quad \text{ if } & c_1' = c_2' = \lambda \end{cases}$$

Then, c' is a $\preceq$-minimal change of $c_1; c_2$ by t. By the induction assumption, either (i) $t|c_1 \Rightarrow t, c_1'$ or (ii) $t|c_2 \Rightarrow t, c_2'$, or (iii) $t|c_1 \Rightarrow t$ and $t|c_2 \Rightarrow t$ are provable, and so is $t|c_1; c_2 \Rightarrow t, c'$, where if $c_1 \neq \lambda$ then $t|c_1; c_2 \Rightarrow t, c_1'; c_2$ is provable; if $c_1' = \lambda$ and $c_2' \neq \lambda$ then $t|c_1; c_2 \Rightarrow t, c_1; c_2'$ is provable; and if $c_1' = \lambda$ and $c_2' = \lambda$ then $t|c_1; c_2 \Rightarrow t$ is provable.

Case $t' = t'(x)$. Assume that there are substitutions $\theta_1, ..., \theta_n$ and theories $t_1, ..., t_n$ such that (i) $t_1 \preceq t_2 \preceq \cdots \preceq t_n$; (ii) for any $i \leq n$, $\theta_i t_i(x)$ is a $\preceq$-minimal change of $\theta_i t_{i-1}(x)$ by t, where $t_0 = t'$; and (iii) $t, t_n(x)$ is consistent. By induction assumption,

$$\begin{array}{l} t|\theta_1 t'(x) \Rightarrow t, \theta_1 t_1(x), \\ t|\theta_2 t_1(x) \Rightarrow t, \theta_2 t_2(x), \\ \cdots \\ t|\theta_n t_{n-1}(x) \Rightarrow t, \theta_n t_n(x), \end{array}$$

are provable. By $(T^{\forall})$, $t|t'(x) \Rightarrow t, t_n(x)$ is provable. □

References

1. C. Baral, M. Gelfond, Logic programming and knowledge representation. J. Logic Program. **19–20**, 73–148 (1994)
2. M. Brain, M. de Vos, Debugging logic programs under the answer set semantics, in *Answer Set Programming* (2005)
3. M. Dahr, *Deductive Databases: Theory and Applications* (International Thomson Computer Press, 1997)
4. M. Gelfond, V. Lifschitz, The stable model semantics for logic programming. ICLP/SLP **88**, 1070–1080 (1988)
5. R. Kowalski, The early years of logic programming. Commun. ACM **31**, 38–43 (1988)
6. J.W. Lloyd, *Foundations of Logic Programming*, 2nd edn. (Springer, Berlin, 1987)
7. W. Li, Y. Sui, An R-calculus for the propositional logic programming, in *Proceedings of International Conference on Computer Science and Information Technology* (2014), pp. 863–870
8. E. Shapiro, The family of concurrent logic programming languages. ACM Comput. Surv. **21**, 413–510 (1989)

Chapter 7
R-Calculi for First-Order Logic

Original R-calculus **R** [3] is a Gentzen-typed deduction system with cut-elimination rule. In this chapter we will give R-calculi $\mathbf{S}^{\mathrm{FOL}}$, $\mathbf{T}^{\mathrm{FOL}}$, $\mathbf{U}^{\mathrm{FOL}}$, which we call alternative R-calculi for first-order logic, where $\mathbf{S}^{\mathrm{FOL}}$, $\mathbf{T}^{\mathrm{FOL}}$, $\mathbf{U}^{\mathrm{FOL}}$ are sound and complete with respect to subset-minimal change, pseudo-subformula minimal change, deduction-based minimal change, respectively.

For first-order logic [1, 2, 4], because the undecidability of deduction relation we will give R-calculi without deciding whether $\Delta \nvdash \neg A$, which are sound and complete with respect to $\subseteq$ / $\preceq$ / $\vdash_{\preceq}$-minimal change, respectively [3].

In first-order logic, consistence has the following properties:

• $\Delta \cup \{\forall x A(x)\}$ is consistent iff for each constant c, $\Delta \cup \{A(c)\}$ is consistent; and, $\Delta \cup \{\forall x A(x)\}$ is inconsistent iff for some constant d, $\Delta \cup \{A(d)\}$ is inconsistent;

• $\Delta \cup \{\exists x A(x)\}$ is consistent iff for some constant d, $\Delta \cup \{A(d)\}$ is consistent; and, $\Delta \cup \{\exists x A(x)\}$ is inconsistent iff for each constant c, $\Delta \cup \{A(c)\}$ is inconsistent.

7.1 R-Calculus for ⊆-Minimal Change

For any theories Δ and Γ, a theory Θ is a $\subseteq$-minimal change of Γ by Δ, denoted by $\models_{\mathrm{S}}^{\mathrm{FOL}} \Delta|\Gamma \Rightarrow \Delta, \Theta$, if $\Theta \subseteq \Gamma$; Θ is consistent with Δ, and for any theory Ξ with $\Theta \subset \Xi \subseteq \Gamma$, Ξ is inconsistent with Δ.

7.1.1 R-Calculus $\mathbf{S}^{\mathrm{FOL}}$ for a Formula

R-calculus $\mathbf{S}^{\mathrm{FOL}}$ for a formula A consists of the following axioms and deduction rules:

W. Li and Y. Sui, *R-CALCULUS: A Logic of Belief Revision*, Perspectives in Formal Induction, Revision and Evolution,
https://doi.org/10.1007/978-981-16-2944-0_7

- **Axioms:**

$$(S^{\mathrm{con}})\ \frac{\Delta \nvdash \neg A}{\Delta|A \Rightarrow \Delta, A}\quad (S^{\neg})\ \frac{\Delta \vdash \neg l}{\Delta|l \Rightarrow \Delta}$$

- **Deduction rules:**

$$(S_1^{\wedge})\ \frac{\Delta|A_1 \Rightarrow \Delta}{\Delta|A_1 \wedge A_2 \Rightarrow \Delta}\qquad (S^{\vee})\ \frac{\Delta|A_1 \Rightarrow \Delta\ \ \Delta|A_2 \Rightarrow \Delta}{\Delta|A_1 \vee A_2 \Rightarrow \Delta}$$

$$(S_2^{\wedge})\ \frac{\Delta, A_1|A_2 \Rightarrow \Delta, A_1}{\Delta|A_1 \wedge A_2 \Rightarrow \Delta}$$

$$(S^{\forall})\ \frac{\Delta|A_1(t) \Rightarrow \Delta}{\Delta|\forall x A_1(x) \Rightarrow \Delta}\qquad (S^{\exists})\ \frac{\Delta|A_1(x) \Rightarrow \Delta}{\Delta|\exists x A_1(x) \Rightarrow \Delta}$$

For example, let $\Delta = \{swan(t) \wedge \neg white(t)\}$, and we have

$$\begin{aligned}
&\Delta|\neg swan(t) \Rightarrow swan(t) \wedge \neg white(t)\\
&\Delta|white(t) \Rightarrow swan(t) \wedge \neg white(t)\\
&\Delta|\neg swan(t) \vee white(t) \Rightarrow swan(t) \wedge \neg white(t)|\lambda\\
&\Delta|\forall x(swan(x) \to white(x)) \Rightarrow swan(t) \wedge \neg white(t)|\forall x(\lambda)\\
&\Delta|\forall x(swan(x) \to white(x)) \Rightarrow swan(t) \wedge \neg white(t)|\lambda.
\end{aligned}$$

Definition 7.1.1 $\Delta|A \Rightarrow \Delta, C$ is provable in $\mathbf{S}^{\mathrm{FOL}}$, denoted by $\vdash_{\mathbf{S}^{\mathrm{FOL}}} \Delta|A \Rightarrow \Delta, C$ if there is a sequence $S_1, ..., S_m$ of statements such that

$$\begin{aligned}
S_1 &= \Delta|A_1 \Rightarrow \Delta|A_1',\\
&\cdots\\
S_m &= \Delta|A_m \Rightarrow \Delta, A_m';\\
A_1 &= A,\\
A_m' &= C;
\end{aligned}$$

and for each $i < m$, S_{i+1} is an axiom or is deduced from the previous statements by a deduction rule in $\mathbf{S}^{\mathrm{FOL}}$.

Theorem 7.1.2 (Completeness theorem) *For any consistent formula set Δ and formula A, if $\Delta \cup \{A\}$ is consistent then $\Delta|A \Rightarrow \Delta, A$ is provable in $\mathbf{S}^{\mathrm{FOL}}$, i.e.,*

$$\models_{\mathbf{S}^{\mathrm{FOL}}} \Delta|A \Rightarrow \Delta, A \text{ implies } \vdash_{\mathbf{S}^{\mathrm{FOL}}} \Delta|A \Rightarrow \Delta, A;$$

and if $\Delta \cup \{A\}$ is inconsistent then $\Delta|A \Rightarrow \Delta$ is provable in $\mathbf{S}^{\mathrm{FOL}}$, i.e.,

$$\models_{\mathbf{S}^{\mathrm{FOL}}} \Delta|A \Rightarrow \Delta \text{ implies } \vdash_{\mathbf{S}^{\mathrm{FOL}}} \Delta|A \Rightarrow \Delta.$$

Proof Assume that $\Delta \cup \{A\}$ is consistent. Then, by (S^{con}), $\vdash_{\mathbf{S}^{\mathrm{FOL}}} \Delta|A \Rightarrow \Delta, A$;

Assume that $\Delta \cup \{A\}$ is inconsistent. We prove by induction on the structure of formula A that $\vdash_{\mathbf{S}^{\mathrm{FOL}}} \Delta | A \Rightarrow \Delta$.

Case $A = l$. Then, $\Delta \vdash \neg l$, and by $(S^{\neg})$, $\Delta | l \Rightarrow \Delta$ is provable;

Case $A = A_1 \wedge A_2$. Then, either $\Delta \cup \{A_1\}$ or $\Delta \cup \{A_1, A_2\}$ is inconsistent, and by induction assumption, either $\Delta | A_1 \Rightarrow \Delta$ is provable or $\Delta, A_1 | A_2 \Rightarrow \Delta, A_1$ is provable, and by $(S_1^{\wedge})$ or $(S_2^{\wedge})$, $\Delta | A_1 \wedge A_2 \Rightarrow \Delta$ is provable;

Case $A = A_1 \vee A_2$. Then $\Delta \cup \{A_1\}$ and $\Delta \cup \{A_2\}$ are inconsistent, and by induction assumption, $\Delta | A_1 \Rightarrow \Delta$ and $\Delta | A_2 \Rightarrow \Delta$ are provable. By $(F_{\vee})$, $\Delta | A_1 \vee A_2 \Rightarrow \Delta$ is provable.

Case $A = \forall x A_1(x)$. Then, there is a term t such that $\Delta \cup \{A_1(t)\}$ is inconsistent, and by induction assumption, $\Delta | A_1(t) \Rightarrow \Delta$ is provable. By $(S^{\forall})$, $\Delta | \forall x A_1(x) \Rightarrow \Delta$ is provable.

Case $A = \exists x A_1(x)$. Then, for any variable x not occurring in Δ, $\Delta \cup \{A_1(x)\}$ is inconsistent, and by induction assumption, $\Delta | A_1(x) \Rightarrow \Delta$ is provable. By $(S^{\exists})$, $\Delta | \exists x A_1(x) \Rightarrow \Delta$ is provable. □

Theorem 7.1.3 (Soundness theorem) *For any consistent formula set Δ and formula A, if $\Delta | A \Rightarrow \Delta, A$ is provable in $\mathbf{S}^{\mathrm{FOL}}$ then $\Delta \cup \{A\}$ is consistent; i.e.,*

$$\vdash_{\mathbf{S}^{\mathrm{FOL}}} \Delta | A \Rightarrow \Delta, A \text{ implies } \models_{\mathbf{S}^{\mathrm{FOL}}} \Delta | A \Rightarrow \Delta, A;$$

and if $\Delta | A \Rightarrow \Delta$ is provable in $\mathbf{S}^{\mathrm{FOL}}$ then $\Delta \cup \{A\}$ is inconsistent, i.e.,

$$\vdash_{\mathbf{S}^{\mathrm{FOL}}} \Delta | A \Rightarrow \Delta \text{ implies } \models_{\mathbf{S}^{\mathrm{FOL}}} \Delta | A \Rightarrow \Delta.$$

Proof If $\Delta | A \Rightarrow \Delta, A$ is provable then by (S^{con}), $\Delta \cup \{A\}$ is consistent.

Assume that $\Delta | A \Rightarrow \Delta$ is provable. We prove by induction on the structure of A that $\models_{\mathbf{S}^{\mathrm{FOL}}} \Delta | A \Rightarrow \Delta$.

Case $A = l$. Then, $\Delta \vdash \neg l$, and hence, $\Delta \cup \{l\}$ is inconsistent.

Case $A = A_1 \wedge A_2$. Then, either $\Delta | A_1 \Rightarrow \Delta$ or $\Delta, A_1 | A_2 \Rightarrow \Delta, A_1$ is provable, and by induction assumption, either $\Delta \cup \{A_1\}$ or $\Delta \cup \{A_1\} \cup \{A_2\}$ is inconsistent, and both imply that $\Delta \cup \{A_1 \wedge A_2\}$ is inconsistent.

Case $A = A_1 \vee A_2$. Then, $\Delta | A_1 \Rightarrow \Delta$ and $\Delta, A_2 \Rightarrow \Delta$ are provable, and by induction assumption, $\Delta \cup \{A_1\}$ and $\Delta \cup \{A_2\}$ are inconsistent, and so is $\Delta \cup \{A_1 \vee A_2\}$.

Case $A = \forall x A_1(x)$. Then, there is a term t such that $\Delta | A_1(t) \Rightarrow \Delta$ is provable, and by induction assumption, $\Delta \cup \{A_1(t)\}$ is inconsistent, and so is $\Delta \cup \{\forall x A_1(x)\}$.

Case $A = \exists x A_1(x)$. Then, for any variable x not occurring in Δ, $\Delta | A_1(x) \Rightarrow \Delta$ is provable, and by induction assumption, $\Delta \cup \{A_1(x)\}$ is inconsistent, and so is $\Delta \cup \{\exists x A_1(x)\}$. □

7.1.2 R-Calculus $\mathbf{S}^{\mathrm{FOL}}$ for a Theory

Let $\Gamma = (A_1, ..., A_n)$. Define

$$\Delta|\Gamma = (\cdots((\Delta|A_1)|A_2)\cdots)|A_n.$$

R-calculus $\mathbf{S}^{\mathrm{FOL}}$ for a theory Γ consists of the following axioms and deduction rules:

- **Axioms:**

$$(S^{\mathrm{con}})\ \frac{\Delta \nvdash \neg A}{\Delta|A, \Gamma \Rightarrow \Delta, A|\Gamma} \quad (S^{\neg})\ \frac{\Delta \vdash \neg l}{\Delta|l, \Gamma \Rightarrow \Delta|\Gamma}$$

- **Deduction rules:**

$$(S_1^{\wedge})\ \frac{\Delta|A_1, \Gamma \Rightarrow \Delta|\Gamma}{\Delta|A_1 \wedge A_2, \Gamma \Rightarrow \Delta|\Gamma} \quad (S^{\vee})\ \frac{\Delta|A_1, \Gamma \Rightarrow \Delta|\Gamma \quad \Delta|A_2, \Gamma \Rightarrow \Delta|\Gamma}{\Delta|A_1 \vee A_2, \Gamma \Rightarrow \Delta|\Gamma}$$

$$(S_2^{\wedge})\ \frac{\Delta, A_1|A_2, \Gamma \Rightarrow \Delta, A_1|\Gamma}{\Delta|A_1 \wedge A_2, \Gamma \Rightarrow \Delta|\Gamma}$$

$$(S^{\forall})\ \frac{\Delta|A_1(t), \Gamma \Rightarrow \Delta|\Gamma}{\Delta|\forall x A_1(x), \Gamma \Rightarrow \Delta|\Gamma} \quad (S^{\exists})\ \frac{\Delta|A_1(x), \Gamma \Rightarrow \Delta|\Gamma}{\Delta|\exists x A_1(x), \Gamma \Rightarrow \Delta|\Gamma}$$

Definition 7.1.4 $\Delta|\Gamma \Rightarrow \Delta, \Theta$ is provable in $\mathbf{S}^{\mathrm{FOL}}$, denoted by $\vdash_{\mathbf{S}^{\mathrm{FOL}}} \Delta|\Gamma \Rightarrow \Delta, \Theta$, if there is a sequence $S_1, ..., S_m$ of statements such that

$$\begin{aligned} S_1 &= \Delta|\Gamma \Rightarrow \Delta_1|\Gamma_1, \\ &\cdots \\ S_m &= \Delta_{m-1}|\Gamma_{m-1} \Rightarrow \Delta_m|\Gamma_m = \Delta, \Theta \end{aligned}$$

and for each $i < m$, S_{i+1} is an axiom or is deduced from the previous statements by a deduction rule in $\mathbf{S}^{\mathrm{FOL}}$.

Theorem 7.1.5 (Soundness theorem) *For any consistent formula sets Θ, Δ and any finite consistent formula set Γ, if $\Delta|\Gamma \Rightarrow \Delta, \Theta$ is provable in $\mathbf{S}^{\mathrm{FOL}}$ then Θ is a $\subseteq$-minimal change of Γ by Δ. That is,*

$$\vdash_{\mathbf{S}^{\mathrm{FOL}}} \Delta|\Gamma \Rightarrow \Delta, \Theta \textit{ implies } \models_{\mathbf{S}^{\mathrm{FOL}}} \Delta|\Gamma \Rightarrow \Delta, \Theta.$$

Proof We prove the theorem by induction on n.

Assume that $\Delta|\Gamma \Rightarrow \Delta, \Theta$ is provable in $\mathbf{S}^{\mathrm{FOL}}$.

Let $n = 1$. Then, either $C = A_1$ or $C = \lambda$. If $C = A_1$ then $\Delta \cup \{A_1\}$ is consistent, and C is a $\subseteq$-minimal change of A_1 by Δ; otherwise, $\Delta \cup \{A_1\}$ is inconsistent, and $C = \lambda$ is a $\subseteq$-minimal change of A_1 by Δ.

Assume that the theorem holds for n, that is, if $\Delta|\Gamma \Rightarrow \Delta, \Theta$ is provable in $\mathbf{S}^{\mathrm{FOL}}$ then Θ is a $\subseteq$-minimal change of Γ by Δ, where $\Gamma = (A_1, ..., A_n)$.

Let $\Gamma' = (\Gamma, A_{n+1}) = (A_1, ..., A_{n+1})$. Then, if $\Delta|\Gamma' \Rightarrow \Delta, \Theta'$ is provable then there is a Θ such that $\Delta|\Gamma \Rightarrow \Delta, \Theta$ and $\Delta, \Theta|A_{n+1} \Rightarrow \Delta, \Theta'$ are provable. By the case $n = 1$ and the induction assumption, Θ' is a $\subseteq$-minimal change of A_{n+1} by $\Delta \cup \Theta$, and Θ is a $\subseteq$-minimal change of Γ by Δ, therefore, Θ' is a $\subseteq$-minimal change of Γ' by Δ. □

Theorem 7.1.6 (Completeness theorem) *For any consistent formula sets Θ, Δ and any finite consistent formula set Γ, if Θ is a $\subseteq$-minimal change of Γ by Δ then $\Delta|\Gamma \Rightarrow \Delta, \Theta$ is provable in $\mathbf{S}^{\mathrm{FOL}}$. That is,*

$$\models_{\mathbf{S}^{\mathrm{FOL}}} \Delta|\Gamma \Rightarrow \Delta, \Theta \textit{ implies } \vdash_{\mathbf{S}^{\mathrm{FOL}}} \Delta|\Gamma \Rightarrow \Delta, \Theta.$$

Proof Assume that Θ is a $\subseteq$-minimal change of Γ by Δ. Then, there is an ordering $<$ of Γ such that $\Gamma = (A_1, A_2, ..., A_n)$, where $A_1 < A_2 < \cdots < A_n$, and Θ is a maximal subset of Γ such that $\Delta \cup \Theta$ is consistent.

We prove the theorem by induction on n.

Let $n = 1$. By the last theorem, if $\Theta = \{A_1\}$ then $\Delta|A_1 \Rightarrow \Delta, A_1$ is provable; and if $\Theta = \emptyset$ then $\Delta|A_1 \Rightarrow \Delta$ is provable.

Assume that the theorem holds for n, that is, if Θ is a $\subseteq$-minimal change of Γ by Δ then $\Delta|\Gamma \Rightarrow \Delta, \Theta$ is provable.

Let $\Gamma' = (\Gamma, A_{n+1}) = (A_1, ..., A_{n+1})$ and Θ' is a $\subseteq$-minimal change of Γ' by Δ. Then, Θ' is a $\subseteq$-minimal change of A_{n+1} by $\Delta \cup \Theta$, and $\Delta, \Theta|A_{n+1} \Rightarrow \Delta, \Theta'$ is provable. By the induction assumption, $\Delta|\Gamma \Rightarrow \Delta, \Theta$ is provable and $\Delta|\Gamma' \Rightarrow \Delta, \Theta|A_{n+1}$ is provable, and hence, $\Delta|\Gamma' \Rightarrow \Delta, \Theta'$ is provable. □

7.2 R-Calculus for ⪯-Minimal Change

In this section, we will give an R-calculus $\mathbf{T}^{\mathrm{FOL}}$ which is sound and complete with respect to $\preceq$-minimal changes of Γ by Δ, that is, for any consistent theories Γ, Δ and Θ, $\Delta|\Gamma \Rightarrow \Delta, \Theta$ is provable in $\mathbf{T}^{\mathrm{FOL}}$ if and only if Θ is a $\preceq$-minimal change of Γ by Δ.

7.2.1 R-Calculus $\mathbf{T}^{\mathrm{FOL}}$ *for a Formula*

R-calculus $\mathbf{T}^{\mathrm{FOL}}$ for a formula A consists of the following axioms and deduction rules:

- **Axioms:**

$$(T^{\mathrm{A}})\ \frac{\Delta \nvdash \neg A}{\Delta|A \Rightarrow \Delta, A} \quad (T_{\mathrm{A}})\ \frac{\Delta \vdash \neg l}{\Delta|l \Rightarrow \Delta}$$

- **Deduction rules:**

$$(T^{\wedge})\ \frac{\Delta|A_1 \Rightarrow \Delta, C_1}{\Delta|A_1 \wedge A_2 \Rightarrow \Delta, C_1|A_2} \qquad (T_1^{\vee})\ \frac{\Delta|A_1 \Rightarrow \Delta, C_1 \neq \lambda}{\Delta|A_1 \vee A_2 \Rightarrow \Delta, C_1 \vee A_2}$$

$$(T_2^{\vee})\ \frac{\begin{array}{l}\Delta|A_1 \Rightarrow \Delta \\ \Delta|A_2 \Rightarrow \Delta, C_2 \\ C_2 \neq \lambda\end{array}}{\Delta|A_1 \vee A_2 \Rightarrow \Delta, A_1 \vee C_2}$$

$$(T_3^{\vee})\ \frac{\Delta|A_1 \Rightarrow \Delta \quad \Delta|A_2 \Rightarrow \Delta}{\Delta|A_1 \vee A_2 \Rightarrow \Delta}$$

$$(T^{\forall})\ \frac{\Delta|A(t) \Rightarrow \Delta, C(t)}{\Delta|\forall x A(x) \Rightarrow \Delta|\forall x C(x)} \qquad (T^{\exists})\ \frac{\Delta|A(x) \Rightarrow \Delta, C(x)}{\Delta|\exists x A(x) \Rightarrow \Delta, \exists x C(x)}$$

Definition 7.2.1 $\Delta|A \Rightarrow \Delta, C$ is provable in $\mathbf{T}^{\text{FOL}}$, denoted by $\vdash_{\mathbf{T}}^{\text{FOL}} \Delta|A \Rightarrow \Delta, C$, if there is a sequence $\{S_1, ..., S_m\}$ of statements such that

$$\begin{aligned} S_1 &= \Delta|A \Rightarrow \Delta|A_1', \\ &\cdots \\ S_m &= \Delta|A_m \Rightarrow \Delta, C; \end{aligned}$$

and for each $i < m$, S_{i+1} is an axiom or is deduced from the previous statements by a deduction rule in $\mathbf{T}^{\text{FOL}}$.

Theorem 7.2.2 *For any consistent formula set Δ and formula A, there is a formula C such that $C \preceq A$ and $\Delta|A \Rightarrow \Delta, C$ is provable.*

Proof We prove the theorem by induction on the structure of formula A.

Case $A = l$. Then, if $\Delta \vdash \neg l$ then let $C = \lambda$; if $\Delta \nvdash \neg l$ then let $C = l$. Then, by $(T^{\mathbf{A}})$ and $(T_{\mathbf{A}})$, $\Delta|A \Rightarrow \Delta, C$ is provable.

Case $A = A_1 \wedge A_2$. Then, by induction assumption, there are formulas $C_1 \preceq A_1, C_2 \preceq A_2$ such that $\Delta|A_1 \Rightarrow \Delta, C_1$ and $\Delta, C_1|A_2 \Rightarrow \Delta, C_2$ are provable in $\mathbf{T}^{\text{FOL}}$. Hence, so is $\Delta|A_1 \wedge A_2 \Rightarrow \Delta, C_1 \wedge C_2$.

Case $A = A_1 \vee A_2$. Then, by induction assumption, there are formulas $C_1 \preceq A_1$ and $C_2 \preceq A_2$ such that $\Delta|A_1 \Rightarrow \Delta, C_1$ and $\Delta|A_2 \Rightarrow \Delta, C_2$ are provable. If $C_1 \neq \lambda$ then by $(T_1^{\vee})$, $\vdash_{\mathbf{T}}^{\text{FOL}} \Delta|A_1 \vee A_2 \Rightarrow \Delta, C_1 \vee A_2$; if $C_1 = \lambda$ and $C_2 \neq \lambda$ then by $(T_2^{\vee})$, $\vdash_{\mathbf{T}}^{\text{FOL}} \Delta|A_1 \vee A_2 \Rightarrow A_1 \vee C_2$; if $C_1 = C_2 = \lambda$ then by $(T_3^{\vee})$, $\vdash_{\mathbf{T}}^{\text{FOL}} \Delta|A_1 \vee A_2 \Rightarrow \Delta$. Let

$$C = \begin{cases} C_1 \vee A_2 \text{ if } C_1 \neq \lambda \\ A_1 \vee C_2 \text{ if } C_1 = \lambda \neq C_2 \\ \lambda \qquad\quad \text{otherwise,} \end{cases}$$

and $\vdash_{\mathbf{T}}^{\text{FOL}} \Delta|A_1 \vee A_2 \Rightarrow \Delta, C$.

Case $A = \forall x A_1(x)$. Then, by $(T^{\forall})$, there are constants $d_1, ..., d_n$ and formulas $A_2, ..., A_n$ such that

$$\frac{\begin{array}{l}\Delta|A_1(d_1) \Rightarrow \Delta|A_2(d_1)\\ \Delta|A_2(d_2) \Rightarrow \Delta|A_3(d_2)\\ \ldots\\ \Delta|A_n(d_n) \Rightarrow \Delta|C(d_n)\end{array}}{\Delta|\forall x A_1(x) \Rightarrow \Delta, \forall x C(x).}$$

By induction assumption, for any $i \le n+1$, $A_i(d_i)$ is a $\preceq^\Delta$-minimal change of $A_{i-1}(d_i)$, where $A_{n+1} = C$. Therefore, for any formula E with $\forall x C(x) \prec E \preceq \forall x A_1(x)$, there is a formula E' and $i \le n+1$ such that $E = \forall x E'(x)$ and $A_i(d_{i-1}) \prec E'(d_{i-1}) \preceq A_{i-1}(d_{i-1})$. By induction assumption, Δ is inconsistent with $E'(d_{i-1})$ and $\forall x E'(x)$.

Case $A = \exists x A_1(x)$. Then, by $(T^\exists)$, there is a constant c and a formula D_1 such that

$$\frac{\Delta|A_1(c) \Rightarrow \Delta|D_1(c)}{\Delta|\exists x A_1(x) \Rightarrow \Delta, \exists x D_1(x).}$$

By induction assumption, $D_1(c)$ is a $\preceq^\Delta$-minimal change of $A_1(c)$. Therefore, $\exists x D_1(x)$ is a $\preceq^\Delta$-minimal change of $\exists x A_1(x)$, because for any formula E with $\exists x D_1(x) \prec E \preceq \exists x A_1(x)$, there is a formula E' such that $E = \exists x E'(x)$ and $D_1(c) \prec E'(c) \preceq A_1(c)$. By induction assumption, Δ is inconsistent with $E'(c)$ and $\exists x E'(x)$. □

Lemma 7.2.3 *If there are constants $d_1, \ldots, d_n$ and formulas $A_2, \ldots, A_n, C_1$ such that*

$A_2(d_1)$ is a $\preceq$-minimal change of $A_1(d_1)$ by Δ
$A_3(d_2)$ is a $\preceq$-minimal change of $A_2(d_2)$ by Δ
...
$C_1(d_n)$ is a $\preceq$-minimal change of $A_n(d_n)$ by Δ

then $\forall x C_1(x)$ is a $\preceq$-minimal change of $\forall x A_1(x)$ by Δ.

Proof We prove the Lemma by induction on n. When $n = 1$ we assume that $C_1(d)$ is a $\preceq$-minimal changes of $A_1(d)$ by Δ.

Then, $\forall x C_1(x) \preceq \forall x A_1(x)$, and $\Delta \cup \{\forall x C_1(x)\}$ is consistent.

For any B with $\forall x C_1(x) \prec B \preceq \forall x A_1(x)$, there is a formula B_1 such that $B = \forall x B_1(x)$, and

$$C_1(d) \prec B_1(d) \preceq A_1(d).$$

Then, $\Delta \cup \{B_1(d)\}$ is inconsistent and so is $\Delta \cup \{\forall x B_1(x)\}$.

Assume that the Lemma holds for $n = k$. Let $n = k+1$. Then, $C_1(d_n)$ is a $\preceq$-minimal changes of $A_1(d_n)$ by Δ.

Then, $\forall x C_1(x) \preceq \forall x A_n(x)$, and $\Delta \cup \{\forall x C_1(x)\}$ is consistent.

For any B with $\forall x C_1(x) \prec B \preceq \forall x A_n(x)$, there is a formula B_1 such that $B = \forall x B_1(x)$, and

$$C_1(d_n) \prec B_1(d_n) \preceq A_n(d_n) \preceq A_i(d_n).$$

Then, $\Delta \cup \{B_1(d_n)\}$ is inconsistent and so is $\Delta \cup \{\forall x B_1(x)\}$. □

Lemma 7.2.4 *If there is a constant c and formula D_1 such that*

$$D_1(c) \text{ is a } \preceq\text{-minimal change of } A_1(c) \text{ by } \Delta$$

then $\exists x D_1(x)$ is a $\preceq$-minimal change of $\exists x A_1(x)$ by Δ.

Proof The proof is similar to that of the last lemma. □

Theorem 7.2.5 *For any formula set Δ and formulas A, B, if $\Delta|A \Rightarrow \Delta, B$ is provable then B is a $\preceq$-minimal change of A by Δ. That is,*

$$\vdash_{\mathbf{T}}^{\mathrm{FOL}} \Delta|A \Rightarrow \Delta, B \text{ implies } \models_{\mathbf{T}}^{\mathrm{FOL}} \Delta|A \Rightarrow \Delta, B.$$

Proof We prove the theorem by induction on the structure of A.

If $B = A$ then, A is consistent with Δ and B is a $\preceq$-minimal change of A by Δ.

Case $A = l$, a literal. Then, $\Delta|l \Rightarrow \Delta$ is provable. That is, $B = \lambda$ is a $\preceq$-minimal change of l by Δ.

Case $A = A_1 \wedge A_2$. Then, if $\Delta|A_1 \Rightarrow \Delta, C_1$ is provable then C_1 is a $\preceq$-minimal change of A_1 by Δ, and if $\Delta, C_1|A_2 \Rightarrow \Delta, C_1, C_2$ is provable then C_2 is a $\preceq$-minimal change of A_2 by $\Delta \cup \{C_1\}$. Then, $C_1 \wedge C_2$ is a $\preceq$-minimal change of $A_1 \wedge A_2$ by Δ.

Case $A = A_1 \vee A_2$. Then $\Delta \cup \{A_1\}$ and $\Delta \cup \{A_2\}$ are inconsistent, and there are C_1', C_2' such that $\Delta|A_1 \Rightarrow \Delta, C_1'$ and $\Delta|A_2 \Rightarrow \Delta, C_2'$ are provable, and by induction assumption, C_1' and C_2' are $\preceq$-minimal changes of A_1 and A_2 by Δ, respectively. Then, $\Delta|A_1 \vee A_2 \Rightarrow \Delta, C'$, and then, C' is a $\preceq$-minimal change of $A_1 \vee A_2$ by Δ.

Case $A = \forall x A_1(x)$. Then, there are formulas $A_2, ..., A_n$ and constants $d_1, ..., d_n$ such that

$$\begin{array}{l} \Delta|A_1(d_1) \Rightarrow \Delta, A_2(d_1) \\ \Delta|A_2(d_2) \Rightarrow \Delta, A_3(d_2) \\ \cdots \\ \Delta|A_n(d_n) \Rightarrow \Delta, C_1(d_n) \end{array}$$

are provable, and by induction assumption,

$$\begin{array}{l} A_2(d_1) \text{ is a } \preceq\text{-minimal change of } A_1(d_1) \text{ by } \Delta, \\ A_3(d_2) \text{ is a } \preceq\text{-minimal change of } A_2(d_2) \text{ by } \Delta, \\ \cdots \\ C_1(d_n) \text{ is a } \preceq\text{-minimal change of } A_n(d_n) \text{ by } \Delta, \end{array}$$

and by Lemma 7.2.3, $\forall x C_1(x)$ is a $\preceq$-minimal change of $\forall x A_1(x)$ by Δ.

Case $A = \exists x A_1(x)$. Similar to case $A = \forall x A_1(x)$. □

Theorem 7.2.6 *For any formula sets Δ and formulas A, B, if B is a $\preceq$-minimal change of A by Δ then there is a formula A' such that $A \simeq A'$ and $\Delta|A' \Rightarrow \Delta, B$ is* $\mathbf{T}^{\mathrm{FOL}}$*-provable. That is,*

$$\models^{\rm FOL}_{\bf T} \Delta|A \Rightarrow \Delta, B \textit{ implies } \vdash^{\rm FOL}_{\bf T} \Delta|A' \Rightarrow \Delta, B.$$

Proof Let $C \preceq A$ be a $\preceq$-minimal change of A by Δ.

Case $A = l$. Then $C = \lambda$ (if Δ, A is inconsistent) or $C = A$ (if Δ, A is consistent), and $\Delta, A \Rightarrow \Delta, C$ is provable.

Case $A = A_1 \wedge A_2$. Then there are C_1, C_2 such that $C = C_1 \wedge C_2$, and C_1 and C_2 are $\preceq$-minimal changes of A_1 and A_2 by Δ and Δ, C_1, respectively. Hence, $C_1 \wedge C_2$ is a $\preceq$-minimal change of $A_1 \wedge A_2$ by Δ. By the induction assumption, $\Delta|A_1 \Rightarrow \Delta, C_1$ and $\Delta, C_1|A_2 \Rightarrow \Delta, C_1, C_2$ are provable, and by $(T^\wedge)$, $\Delta|A_1 \wedge A_2 \Rightarrow \Delta, C_1 \wedge C_2$ is provable.

Case $A = A_1 \vee A_2$. Then there are C_1 and C_2 such that $C = C'$, and C_1 and C_2 are $\preceq$-minimal changes of A_1 and A_2 by Δ, respectively, where

$$C' = \begin{cases} C_1 \vee A_2 & \text{if } C_1 \neq \lambda \\ A_1 \vee C_2 & \text{if } C_1 = \lambda \text{ and } C_2 \neq \lambda \\ \lambda & \text{if } C_1 = C_2 = \lambda \end{cases}$$

Then, C' is a $\preceq$-minimal change of $A_1 \vee A_2$ by Δ. By the induction assumption, either (i) $\Delta|A_1 \Rightarrow \Delta, C_1$ or (ii) $\Delta|A_2 \Rightarrow \Delta, C_2$, or (iii) $\Delta|A_1 \Rightarrow \Delta$ and $\Delta|A_2 \Rightarrow \Delta$ are provable, and so is $\Delta|A_1 \vee A_2 \Rightarrow \Delta, C'$, where if $C_1 \neq \lambda$ then $\Delta|A_1 \vee A_2 \Rightarrow \Delta, C_1 \vee A_2$ is provable; if $C_1 = \lambda$ and $C_2 \neq \lambda$ then $\Delta|A_1 \vee A_2 \Rightarrow \Delta, A_1 \vee C_2$ is provable; and if $C_1 = \lambda$ and $C_2 = \lambda$ then $\Delta|A_1 \vee A_2 \Rightarrow \Delta$ is provable.

Case $A = \forall x A_1(x)$. Then there is a formula $C_1(t) \preceq A_1(t)$, and $C_1(t)$ is a $\preceq$-minimal changes of $A_1(t)$ by Δ. By the induction assumption, $\Delta|A_1(t) \Rightarrow \Delta, C_1(t)$ is provable, and by $(T^\forall)$, $\Delta|\forall x A_1(x) \Rightarrow \Delta, \forall x C_1(x)$ is provable.

Case $A = \exists x A_1(x)$. Then there is a formula $C_1(x) \preceq A_1(x)$, and $C_1(x)$ is a $\preceq$-minimal changes of $A_1(x)$ by Δ. By the induction assumption, $\Delta|A_1(x) \Rightarrow \Delta, C_1(x)$ is provable, and by $(T^\exists)$, $\Delta|\exists x A_1(x) \Rightarrow \Delta, \exists x C_1(x)$ is provable. □

7.2.2 *R-Calculus* $\mathbf{T}^{\rm FOL}$ *for a Theory*

Let $\Gamma = (A_1, ..., A_n)$. Define

$$\Delta|\Gamma = (\cdots((\Delta|A_1)|A_2)\cdots)|A_n.$$

R-calculus $\mathbf{T}^{\rm FOL}$ for a theory Γ consists of the following axioms and deduction rules:

- **Axioms:**

$$(T^{\rm A})\ \frac{\Delta \nvdash \neg A}{\Delta|A, \Gamma \Rightarrow \Delta, A|\Gamma} \quad (T_{\rm A})\ \frac{\Delta \vdash \neg l}{\Delta|l, \Gamma \Rightarrow \Delta|\Gamma}$$

• **Deduction rules:**

$$(T^\wedge)\ \frac{\Delta|A_1,\Gamma\Rightarrow\Delta,C_1|\Gamma}{\Delta|A_1\wedge A_2,\Gamma\Rightarrow\Delta,C_1|A_2,\Gamma}\qquad (T_1^\vee)\ \frac{\Delta|A_1,\Gamma\Rightarrow\Delta,C_1|\Gamma\ \ C_1\neq\lambda}{\Delta|A_1\vee A_2,\Gamma\Rightarrow\Delta,C_1\vee A_2|\Gamma}$$

$$(T_2^\vee)\ \frac{\begin{array}{l}\Delta|A_1,\Gamma\Rightarrow\Delta|\Gamma\\ \Delta|A_2,\Gamma\Rightarrow\Delta,C_2|\Gamma\\ C_2\neq\lambda\end{array}}{\Delta|A_1\vee A_2,\Gamma\Rightarrow\Delta,A_1\vee C_2|\Gamma}$$

$$(T_3^\vee)\ \frac{\Delta|A_1,\Gamma\Rightarrow\Delta|\Gamma\ \ \Delta|A_2,\Gamma\Rightarrow\Delta|\Gamma}{\Delta|A_1\vee A_2,\Gamma\Rightarrow\Delta|\Gamma}$$

$$(T^\forall)\ \frac{\Delta|A_1(t),\Gamma\Rightarrow\Delta,C_1(t)|\Gamma}{\Delta|\forall xA_1(x),\Gamma\Rightarrow\Delta|\forall xC_1(x),\Gamma}\qquad (T^\exists)\ \frac{\Delta|A_1(x),\Gamma\Rightarrow\Delta,C_1(x)|\Gamma}{\Delta|\exists xA_1(x),\Gamma\Rightarrow\Delta|\exists xC_1(x),\Gamma}$$

Theorem 7.2.7 (Soundness theorem) *For any formula sets* Θ, Δ *and any finite formula set* Γ, *if* $\Delta|\Gamma\Rightarrow\Delta,\Theta$ *is provable in* $\mathbf{T}^{\rm FOL}$ *then* Θ *is a* $\preceq$*-minimal change of* Γ *by* Δ. *That is,*

$$\vdash^{\rm FOL}_{\mathbf{T}}\Delta|\Gamma\Rightarrow\Delta,\Theta \textit{ implies } \models^{\rm FOL}_{\mathbf{T}}\Delta|\Gamma\Rightarrow\Delta,\Theta.$$

Theorem 7.2.8 (Completeness theorem) *For any formula sets* Θ, Δ *and* Γ, *if* Θ *is a* $\preceq$*-minimal change of* Γ *by* Δ *then* $\Delta|\Gamma\Rightarrow\Delta,\Theta$ *is provable in* $\mathbf{T}^{\rm FOL}$. *That is,*

$$\models^{\rm FOL}_{\mathbf{T}}\Delta|\Gamma\Rightarrow\Delta,\Theta \textit{ implies } \vdash^{\rm FOL}_{\mathbf{T}}\Delta|\Gamma\Rightarrow\Delta,\Theta.$$

□

References

1. J. Barwise, An introduction to first-order logic, in *Handbook of Mathematical Logic*. Studies in Logic and the Foundations of Mathematics, ed. by J. Barwise (North-Holland, Amsterdam, 1982)
2. W. Li,*Mathematical Logic, Foundations for Information Science*. Progress in Computer Science and Applied Logic, vol. 25 (Birkhäuser, 2010)
3. W. Li, R-calculus: an inference system for belief revision. The Computer J. **50**, 378–390 (2007)
4. G. Takeuti, Proof theory, in *Handbook of Mathematical Logic*. Studies in Logic and the Foundations of Mathematics. ed. by J. Barwise (Amsterdam, North-Holland, 1987)

Chapter 8
Nonmonotonicity of R-Calculus

Nonmonotonic logics [3, 4, 10] are different from monotonic logics in that the deduction is nonmonotonic. The traditional logics, such as propositional logic, first-order logic, modal logic, etc., are monotonic. The nontraditional logics, such as default logic, R-calculi, autoepistemic logic, circumscription, etc., are nonmonotonic.

The nonmonotonicity of a nonmonotonic logic follows from using a negation $\Delta \nvdash A$ of a monotonic deduction $\Delta \vdash A$ [3]. We found that $\Delta \nvdash A$ occurs in each nonmonotonic logic. For example, in default logic, a formula B is deducible (or in some extension of default theory $(\Delta, A \rightsquigarrow B)$) from a default theory $(\Delta, A \rightsquigarrow B)$ [1, 6–9, 11] if A is deducible in propositional logic from Δ and $\neg B$ is not, that is,

$$\Delta \vdash A \& \Delta \nvdash \neg B,$$

where $A \rightsquigarrow B$ is a normal default $\dfrac{A : B}{B}$.

It is obvious that the monotonicity of $\Delta \vdash A$ implies the nonmonotonicity of $\Delta \nvdash A$.

As a deduction relation, $\nvdash$ is contradictory to $\vdash$. Correspondingly, in Gentzen deduction systems, the validity of $\Gamma \Rightarrow \Delta$ is contradictory to the validity of $\Gamma \not\Rightarrow \Delta$ (denoted by $\Gamma \mapsto \Delta$), where $\Gamma \mapsto \Delta$ is valid if there is an assignment v such that v satisfies Γ and does not satisfy Δ, where v satisfies Γ if v satisfies each formula in Γ; and v satisfies Δ if v satisfies some formula in Δ.

Therefore, as a contradictory relation $\Gamma \mapsto \Delta$ of $\Gamma \Rightarrow \Delta$, there is a Gentzen deduction system $\mathbf{G}_2$ such that $\mathbf{G}_2$ is sound, complete and nonmonotonic, that is, for any sequent $\Gamma \mapsto \Delta$, if $\Gamma \mapsto \Delta$ is provable in $\mathbf{G}_2$ then $\Gamma \mapsto \Delta$ is valid; and conversely, if $\Gamma \mapsto \Delta$ is valid then $\Gamma \mapsto \Delta$ is provable in $\mathbf{G}_2$.

W. Li and Y. Sui, *R-CALCULUS: A Logic of Belief Revision*, Perspectives in Formal Induction, Revision and Evolution,
https://doi.org/10.1007/978-981-16-2944-0_8

8.1 Nonmonotonic Propositional Logic

This section gives a sound and complete Gentzen deduction system for nonmonotonic propositional logic.

8.1.1 Monotonic Gentzen Deduction System $\mathbf{G}_1'$

Gentzen deduction system $\mathbf{G}_1'$ consists of the following axioms and deductions:

- **Axioms**:

$$\frac{\mathrm{incon}(\Gamma) \text{ or } \mathrm{incon}(\Delta) \text{ or } \Gamma \cap \Delta \neq \emptyset}{\Gamma \Rightarrow \Delta},$$

where Γ, Δ are sets of literals, and $\mathrm{incon}(\Gamma)$ denotes that there is a literal l such that $l, \neg l \in \Gamma$.

- **Deduction rules**:

$$(\wedge_1^L)\ \frac{\Gamma, A_1 \Rightarrow \Delta}{\Gamma, A_1 \wedge A_2 \Rightarrow \Delta} \qquad (\wedge^R)\ \frac{\Gamma \Rightarrow B_1, \Delta \quad \Gamma \Rightarrow B_2, \Delta}{\Gamma \Rightarrow B_1 \wedge B_2, \Delta}$$

$$(\wedge_2^L)\ \frac{\Gamma, A_2 \Rightarrow \Delta}{\Gamma, A_1 \wedge A_2 \Rightarrow \Delta}$$

$$(\vee^L)\ \frac{\Gamma, A_1 \Rightarrow \Delta \quad \Gamma, A_2 \Rightarrow \Delta}{\Gamma, A_1 \vee A_2 \Rightarrow \Delta} \qquad (\vee_1^R)\ \frac{\Gamma \Rightarrow B_1, \Delta}{\Gamma \Rightarrow B_1 \vee B_2, \Delta}$$

$$(\vee_2^R)\ \frac{\Gamma \Rightarrow B_2, \Delta}{\Gamma \Rightarrow B_1 \vee B_2, \Delta}$$

$$(\neg\wedge^L)\ \frac{\Gamma, \neg A_1 \Rightarrow \Delta \quad \Gamma, \neg A_2 \Rightarrow \Delta}{\Gamma, \neg(A_1 \wedge A_2) \Rightarrow \Delta} \qquad (\neg\wedge_1^R)\ \frac{\Gamma \Rightarrow \neg B_1, \Delta}{\Gamma \Rightarrow \neg(B_1 \wedge B_2), \Delta}$$

$$(\neg\wedge_2^R)\ \frac{\Gamma \Rightarrow \neg B_2, \Delta}{\Gamma \Rightarrow \neg(B_1 \wedge B_2), \Delta}$$

$$(\neg\vee_1^L)\ \frac{\Gamma, \neg A_1 \Rightarrow \Delta}{\Gamma, \neg(A_1 \vee A_2) \Rightarrow \Delta} \qquad (\vee^R)\ \frac{\Gamma \Rightarrow \neg B_1, \Delta \quad \Gamma \Rightarrow \neg B_2, \Delta}{\Gamma \Rightarrow \neg(B_1 \vee B_2), \Delta}$$

$$(\neg\vee_2^L)\ \frac{\Gamma, \neg A_2 \Rightarrow \Delta}{\Gamma, \neg(A_1 \vee A_2) \Rightarrow \Delta}$$

Remark Condition

$$\mathrm{incon}(\Gamma) \text{ or } \mathrm{incon}(\Delta) \text{ or } \Gamma \cap \Delta \neq \emptyset$$

is equivalent to $\mathrm{incon}(\Gamma \cup \neg\Delta)$. □

Theorem 8.1.1 (Soundness and completeness theorem) *For any sequent* $\Gamma \Rightarrow \Delta$,

$$\vdash_{\mathbf{G}_1'} \Gamma \Rightarrow \Delta \textit{ iff } \models_{\mathbf{G}_1'} \Gamma \Rightarrow \Delta.$$

□

Proposition 8.1.2 *For any theories* $\Gamma, \Gamma', \Delta, \Delta'$ *and assignment* v,

$$\Gamma \subseteq \Gamma' \& v \models_{\mathbf{G}_1'} \Gamma \Rightarrow \Delta \textit{ imply } v \models_{\mathbf{G}_1'} \Gamma' \Rightarrow \Delta,$$
$$\Delta \subseteq \Delta' \& v \models_{\mathbf{G}_1'} \Gamma \Rightarrow \Delta \textit{ imply } v \models_{\mathbf{G}_1'} \Gamma \Rightarrow \Delta';$$

and

$$\Gamma \subseteq \Gamma' \& \vdash_{\mathbf{G}_1'} \Gamma \Rightarrow \Delta \textit{ imply } \vdash_{\mathbf{G}_1'} \Gamma' \Rightarrow \Delta,$$
$$\Delta \subseteq \Delta' \& \vdash_{\mathbf{G}_1'} \Gamma \Rightarrow \Delta \textit{ imply } \vdash_{\mathbf{G}_1'} \Gamma \Rightarrow \Delta'.$$

□

8.1.2 Nonmonotonic Gentzen Deduction System Logic $\mathbf{G}_2$

About the negation of $\Rightarrow$, we have the following deduction system $\mathbf{G}_2'$:

• **Axioms**: let Γ, Δ be sets of literals.

$$(\mathbf{A}^{\not\Rightarrow})\ \frac{\text{con}(\Gamma)\&\text{con}(\Delta)\&\Gamma \cap \Delta = \emptyset}{\Gamma \not\Rightarrow \Delta},$$

where con(Γ) means that theory Γ is consistent in propositional logic.

• **Deduction rules**:

$$(\not\Rightarrow\wedge^L)\ \frac{\Gamma, A_1 \not\Rightarrow \Delta \quad \Gamma, A_1 \not\Rightarrow \Delta}{\Gamma, A_1 \wedge A_2 \not\Rightarrow \Delta} \qquad (\not\Rightarrow\wedge_1^R)\ \frac{\Gamma \not\Rightarrow B_1, \Delta}{\Gamma \not\Rightarrow B_1 \wedge B_2, \Delta}$$
$$(\not\Rightarrow\wedge_2^R)\ \frac{\Gamma \not\Rightarrow B_2, \Delta}{\Gamma \not\Rightarrow B_1 \wedge B_2, \Delta}$$
$$(\not\Rightarrow\vee_1^L)\ \frac{\Gamma, A_1 \not\Rightarrow \Delta}{\Gamma, A_1 \vee A_2 \not\Rightarrow \Delta} \qquad (\not\Rightarrow\vee^R)\ \frac{\Gamma \not\Rightarrow B_1, \Delta \quad \Gamma \not\Rightarrow B_1, \Delta}{\Gamma \not\Rightarrow B_1 \vee B_2, \Delta}$$
$$(\not\Rightarrow\vee_2^L)\ \frac{\Gamma, A_2 \not\Rightarrow \Delta}{\Gamma, A_1 \vee A_2 \not\Rightarrow \Delta}$$
$$(\not\Rightarrow\neg\wedge_1^L)\ \frac{\Gamma, \neg A_1 \not\Rightarrow \Delta}{\Gamma, \neg(A_1 \wedge A_2) \not\Rightarrow \Delta} \qquad (\not\Rightarrow\neg\wedge^R)\ \frac{\Gamma \not\Rightarrow \neg B_1, \Delta \quad \Gamma \not\Rightarrow \neg B_2, \Delta}{\Gamma \not\Rightarrow B_1 \wedge B_2, \Delta}$$
$$(\not\Rightarrow\neg\wedge_2^L)\ \frac{\Gamma, \neg A_2 \not\Rightarrow \Delta}{\Gamma, \neg(A_1 \wedge A_2) \not\Rightarrow \Delta}$$
$$(\not\Rightarrow\neg\vee^L)\ \frac{\Gamma, \neg A_1 \not\Rightarrow \Delta \quad \Gamma, \neg A_2 \not\Rightarrow \Delta}{\Gamma, \neg(A_1 \vee A_2) \not\Rightarrow \Delta} \qquad (\not\Rightarrow\neg\vee_1^R)\ \frac{\Gamma \not\Rightarrow \neg B_1, \Delta}{\Gamma \not\Rightarrow \neg(B_1 \vee B_2), \Delta}$$
$$(\not\Rightarrow\neg\vee_2^R)\ \frac{\Gamma \not\Rightarrow \neg B_2, \Delta}{\Gamma \not\Rightarrow \neg(B_1 \vee B_2), \Delta}$$

Definition 8.1.3 A sequent $\Gamma \mapsto \Delta$ is valid, denoted by $\models_{\mathbf{G}_2} \Gamma \mapsto \Delta$, if there is an assignment v such that $v(\Gamma) = 1$ and $v(\Delta) = 0$, where $v(\Gamma) = 1$ if for each $A \in \Gamma$, $v(A) = 1$; and $v(\Delta) = 0$ if for each $B \in \Delta$, $v(B) = 0$.

A sequent $\Gamma \mapsto \Delta$ is not valid if $\Gamma \mapsto \Delta$ is unsatisfiable, i.e., there is no assignment v such that $v(\Gamma) = 1$ and $v(\Delta) = 0$.

Lemma 8.1.4 *Given two sets* Γ, Δ *of atoms or the negation of atoms,* $\models_{\mathbf{G}_2} \Gamma \mapsto \Delta$ *if and only if* Γ *and* Δ *are consistent, and* $\Gamma \cap \Delta = \emptyset$.

Proof Assume that $\mathrm{con}(\Gamma)$, $\mathrm{con}(\Delta)$ and $\Gamma \cap \Delta = \emptyset$. Then, there is an assignment v such that $v(\Gamma) = 1$ and $v(\Delta) = 0$. Define v such that for any propositional variable p,

$$v(p) = \begin{cases} 0 \text{ if } \neg p \in \Gamma \\ 1 \text{ if } p \in \Gamma \\ 0 \text{ if } p \in \Delta \\ 1 \text{ if } \neg p \in \Delta \end{cases}$$

Then, v is well-defined, $v(\Gamma) = 1$, and for any literal $l \in \Delta$, $v(l) = 0$.

Conversely, assume that $\models_{\mathbf{G}_2} \Gamma \mapsto \Delta$. Then, there is an assignment v such that $v(\Gamma) = 1$ and $v(\Delta) = 0$, which implies that Γ and Δ are consistent and $\Gamma \cap \Delta = \emptyset$. □

Intuitively,

$\models_{\mathbf{G}_2} \Gamma \mapsto \Delta$ iff there is an assignment v such that (i) that for each formula $A \in \Gamma$, $v(A) = 1$ and (2) that for each formula $B \in \Delta$, $v(B) = 0$; iff it is not true that for any assignment v, (1) that for each formula $A \in \Gamma$, $v(A) = 1$ implies (2) that for some formula $B \in \Delta$, $v(B) = 1$; iff it is not true that $\mathrm{incon}(\Gamma)$ or $\mathrm{incon}(\Delta)$ or $\Gamma \cap \Delta \neq \emptyset$, iff $\mathrm{con}(\Gamma)\&\mathrm{con}(\Delta)\&\Gamma \cap \Delta = \emptyset$.

Notice that $\mathrm{incon}(\Gamma)$ or $\mathrm{incon}(\Delta)$ or $\Gamma \cap \Delta \neq \emptyset$ is equivalent to $\mathrm{incon}(\Gamma \cup \neg\Delta)$, and $\mathrm{con}(\Gamma)\&\mathrm{con}(\Delta)\&\Gamma \cap \Delta = \emptyset$ is equivalent to $\mathrm{con}(\Gamma \cup \neg\Delta)$, where $\neg\Delta = \{\neg B : B \in \Delta\}$.

Gentzen deduction system $\mathbf{G}_2$ [3] consists of the following axioms and deductions:

- Axioms:

$$(_{\mapsto}\mathbf{A})\ \frac{\mathrm{con}(\Gamma)\&\mathrm{con}(\Gamma)\&\Gamma \cap \Delta = \emptyset}{\Gamma \mapsto \Delta},$$

where Δ, Γ are sets of literals.

- Deduction rules:

$$(_{\mapsto}\wedge^L)\ \frac{\Gamma, A_1 \mapsto \Delta\ \ \Gamma, A_2 \mapsto \Delta}{\Gamma, A_1 \wedge A_2 \mapsto \Delta} \qquad (_{\mapsto}\wedge_1^R)\ \frac{\Gamma \mapsto B_1, \Delta}{\Gamma \mapsto B_1 \wedge B_2, \Delta}$$

$$(_{\mapsto}\wedge_2^R)\ \frac{\Gamma \mapsto B_2, \Delta}{\Gamma \mapsto B_1 \wedge B_2, \Delta}$$

$$(_{\mapsto}\vee_1^L)\ \frac{\Gamma, A_1 \mapsto \Delta}{\Gamma, A_1 \vee A_2 \mapsto \Delta} \qquad (_{\mapsto}\vee^R)\ \frac{\Gamma \mapsto B_1, \Delta\ \ \Gamma \mapsto B_2, \Delta}{\Gamma \mapsto B_1 \vee B_2, \Delta}$$

$$(_{\mapsto}\vee_2^L)\ \frac{\Gamma, A_2 \mapsto \Delta}{\Gamma, A_1 \vee A_2 \mapsto \Delta}$$

$$(_{\mapsto}\neg\wedge_1^L)\ \frac{\Gamma, \neg A_1 \mapsto \Delta}{\Gamma, \neg(A_1 \wedge A_2) \mapsto \Delta} \qquad (_{\mapsto}\neg\wedge^R)\ \frac{\Gamma \mapsto \neg B_1, \Delta\ \ \Gamma \mapsto \neg B_2, \Delta}{\Gamma \mapsto \neg(B_1 \wedge B_2), \Delta}$$

$$(_{\mapsto}\neg\wedge_2^L)\ \frac{\Gamma, \neg A_2 \mapsto \Delta}{\Gamma, \neg(A_1 \wedge A_2) \mapsto \Delta}$$

$$(_{\mapsto}\neg\vee^L)\ \frac{\Gamma, \neg A_1 \mapsto \Delta\ \ \Gamma, \neg A_2 \mapsto \Delta}{\Gamma, \neg(A_1 \vee A_2) \mapsto \Delta} \qquad (_{\mapsto}\neg\vee_1^R)\ \frac{\Gamma \mapsto \neg B_1, \Delta}{\Gamma \mapsto \neg(B_1 \vee B_2), \Delta}$$

$$(_{\mapsto}\neg\vee_2^R)\ \frac{\Gamma \mapsto \neg B_2, \Delta}{\Gamma \mapsto \neg(B_1 \vee B_2), \Delta}$$

Definition 8.1.5 A sequent $\Gamma \mapsto \Delta$ is provable, denoted by $\vdash_{\mathbf{G}_2} \Gamma \mapsto \Delta$, if there is a sequence $\Gamma_1 \mapsto \Delta_1, ..., \Gamma_n \mapsto \Delta_n$ such that $\Gamma_n \mapsto \Delta_n = \Gamma \mapsto \Delta$, and for each $1 \leq i \leq n, \Gamma_i \mapsto \Delta_i$ is an axiom or is deduced from the previous sequents by one of the deduction rules in $\mathbf{G}_2$.

Theorem 8.1.6 (Soundness theorem) *For any sequent* $\Gamma \mapsto \Delta$,

$$\vdash_{\mathbf{G}_2} \Gamma \mapsto \Delta \textit{ implies } \models_{\mathbf{G}_2} \Gamma \mapsto \Delta.$$

Proof We prove that each axiom is valid and each deduction rule preserves the validity.

To verify the validity of the axiom, assume that $\text{con}(\Gamma \cup \neg\Delta)$. There is an assignment v such that $v(\Gamma) = 1$ and $v(\neg\Delta) = 1$, that is, for every $A \in \Gamma$, $v(A) = 1$; and for every $B \in \Delta$, $v(B) = 0$.

To verify that $(_{\mapsto}\wedge^L)$ preserves the validity, assume that there is an assignment v such that $v(\Gamma, A_1) = 1, v(\Gamma, A_2) = 1$ and $v(\Delta) = 0$. For this very assignment v, $v(\Gamma, A_1 \wedge A_2) = 1$ and $v(\Delta) = 0$.

To verify that $(_{\mapsto}\wedge_1^R)$ preserves the validity, assume that there is an assignment v such that $v(\Gamma) = 1$ and $v(\Delta, B_1) = 0$. For this very assignment v, $v(\Gamma) = 1$ and $v(\Delta, B_1 \wedge B_2) = 0$.

To verify that $(_{\mapsto}\vee_1^L)$ preserves the validity, assume that there is an assignment v such that $v(\Gamma, A_1) = 1$ and $v(\Delta) = 0$. For this very assignment v, $v(\Gamma, A_1 \vee A_2) = 1$ and $v(\Delta) = 0$.

To verify that $(_{\mapsto}\vee^R)$ preserves the validity, assume that there is an assignment v such that $v(\Gamma) = 1$ and $v(\Delta, B_1) = 0 = v(\Delta, B_2)$. For this very assignment v, $v(\Gamma) = 1$ and $v(\Delta, B_1 \vee B_2) = 0$.

Similar for other deduction rules. □

Theorem 8.1.7 (Completeness theorem) *For any sequent* $\Gamma \mapsto \Delta$,

$$\models_{\mathbf{G}_2} \Gamma \mapsto \Delta \textit{ implies } \vdash_{\mathbf{G}_2} \Gamma \mapsto \Delta.$$

Proof Given a sequent $\Gamma \mapsto \Delta$, we construct a tree T as follows:

- the root of T is $\Gamma \mapsto \Delta$;
- if a node $\Gamma' \mapsto \Delta'$ is such that Γ', Δ' are sets of literals then the node is a leaf; and
- if $\Gamma' \mapsto \Delta'$ is not a leaf of T then $\Gamma' \mapsto \Delta'$ has the direct child nodes

$$\begin{cases} \Gamma_1, A_1, A_2 \mapsto \Delta_1 & \text{if } \Gamma' \mapsto \Delta' = \Gamma_1, A_1 \wedge A_2 \mapsto \Delta_1 \\ \begin{cases} \Gamma_1 \mapsto B_1, \Delta_1 \\ \Gamma_1 \mapsto B_2, \Delta_1 \end{cases} & \text{if } \Gamma' \mapsto \Delta' = \Gamma_1 \mapsto B_1 \wedge B_2, \Delta_1 \\ \begin{cases} \Gamma_1, A_1 \mapsto \Delta_1 \\ \Gamma_1, A_2 \mapsto \Delta_1 \end{cases} & \text{if } \Gamma' \mapsto \Delta' = \Gamma_1, A_1 \vee A_2 \mapsto \Delta_1 \\ \Gamma_1 \mapsto B_1, B_2, \Delta_1 & \text{if } \Gamma' \mapsto \Delta' = \Gamma_1 \mapsto B_1 \vee B_2, \Delta_1, \end{cases}$$

and

$$\begin{cases} \begin{cases} \Gamma_1, \neg A_1 \mapsto \Delta_1 \\ \Gamma_1, \neg A_2 \mapsto \Delta_1 \end{cases} & \text{if } \Gamma' \mapsto \Delta' = \Gamma_1, \neg(A_1 \wedge A_2) \mapsto \Delta_1 \\ \Gamma_1 \mapsto \neg B_1, \neg B_2, \Delta_1 & \text{if } \Gamma' \mapsto \Delta' = \Gamma_1 \mapsto \neg(B_1 \wedge B_2), \Delta_1 \\ \Gamma_1, \neg A_1, \neg A_2 \mapsto \Delta_1 & \text{if } \Gamma' \mapsto \Delta' = \Gamma_1, \neg(A_1 \vee A_2) \mapsto \Delta_1 \\ \begin{cases} \Gamma_1 \mapsto B_1, \Delta_1 \\ \Gamma_1 \mapsto B_2, \Delta_1 \end{cases} & \text{if } \Gamma' \mapsto \Delta' = \Gamma_1 \mapsto \neg(B_1 \vee B_2), \Delta_1. \end{cases}$$

Lemma 8.1.8 *If there is a branch $\xi \subseteq T$ such that the leaf of ξ is an axiom in $\mathbf{G}_2$ then there is a proof of $\Gamma \mapsto \Delta$.*

Proof We prove the theorem by induction on the nodes of ξ.

Assume that the leaf of ξ is an axiom in $\mathbf{G}_2$. Then, let the leaf be $\Gamma' \mapsto \Delta'$, and $\text{con}(\Gamma' \cup \neg(\Delta'))$, that is, $\vdash_{\mathbf{G}_2} \Gamma' \mapsto \Delta'$.

We prove that for each node $\Gamma_1 \mapsto \Delta_1$ of $\xi, \vdash_{\mathbf{G}_2} \Gamma_1 \mapsto \Delta_1$. There are the following cases for $\Gamma_1 \mapsto \Delta_1$.

Case 1. $\Gamma_1 \mapsto \Delta_1 = \Gamma_2, A_1 \wedge A_2 \mapsto \Delta_2 \in \xi$. Then, $\Gamma_1 \mapsto \Delta_1$ has a direct child node $\Gamma_2, A_1, A_2 \mapsto \Delta_2 \in \xi$. By the assumption that $\vdash_{\mathbf{G}_2} \Gamma_2, A_1, A_2 \mapsto \Delta_2$ and $(_{\mapsto}\wedge^L)$ in $\mathbf{G}_2, \vdash_{\mathbf{G}_2} \Gamma_2, A_1 \wedge A_2 \mapsto \Delta_2$.

Case 2. $\Gamma_1 \mapsto \Delta_1 = \Gamma_2, A_1 \vee A_2 \mapsto \Delta_2 \in \xi$. Then, $\Gamma_1 \mapsto \Delta_1$ has two direct children nodes $\Gamma_2, A_1 \mapsto \Delta_2$ and $\Gamma_2, A_2 \mapsto \Delta_2$. There is an $i \in \{1, 2\}$ such that $\Gamma_2, A_i \mapsto \Delta_2 \in \xi$. By induction assumption, $\vdash_{\mathbf{G}_2} \Gamma_2, A_i \mapsto \Delta_2$, and by $(_{\mapsto}\vee_i^L)$ in $\mathbf{G}_2, \vdash_{\mathbf{G}_2} \Gamma_2, A_1 \vee A_2 \mapsto \Delta_2$.

Case 3. $\Gamma_1 \mapsto \Delta_1 = \Gamma_2 \mapsto B_1 \wedge B_2, \Delta_2 \in \xi$. Then, $\Gamma_1 \mapsto \Delta_1$ has two direct children nodes $\Gamma_2 \mapsto B_1, \Delta_2$ and $\Gamma_2 \mapsto B_2, \Delta_2$. There is an $i \in \{1, 2\}$ such that $\Gamma_2 \mapsto B_i, \Delta_2 \in \xi$. By induction assumption, $\vdash_{\mathbf{G}_2} \Gamma_2 \mapsto B_i, \Delta_2$, and by $(_{\mapsto}\wedge_i^R)$ in $\mathbf{G}_2, \vdash_{\mathbf{G}_2} \Gamma_2 \mapsto B_1 \wedge B_2, \Delta_2$.

Case 4. $\Gamma_1 \mapsto \Delta_1 = \Gamma_2 \mapsto B_1 \vee B_2, \Delta_2 \in \xi$. Then, $\Gamma_1 \mapsto \Delta_1$ has a direct child node $\Gamma_2 \mapsto B_1, B_2, \Delta_2 \in \xi$. By the assumption that $\vdash_{\mathbf{G}_2} \Gamma_2 \mapsto B_1, B_2, \Delta_2$ and $(_{\mapsto}\vee^R)$ in $\mathbf{G}_2, \vdash_{\mathbf{G}_2} \Gamma_2 \mapsto B_1 \vee B_2, \Delta_2$.

Similar for other cases. Hence, we have $\vdash_{\mathbf{G}_2} \Gamma \mapsto \Delta$. □

Lemma 8.1.9 *If each leaf $\Gamma' \mapsto \Delta'$ of T is not an axiom in $\mathbf{G}_2$ then T is a proof tree of $\Gamma \Rightarrow \Delta$ in $\mathbf{G}'_1$.*

Proof Assume that each leaf $\Gamma' \mapsto \Delta'$ of T is not an axiom in $\mathbf{G}_2$. Then, each leaf $\Gamma' \mapsto \Delta'$ of T is an axiom in $\mathbf{G}'_1$, because Γ' and Δ' are sets of literals, $\text{incon}(\Gamma' \cup \neg\Delta')$. Therefore, $\vdash_{\mathbf{G}'_1} \Gamma' \Rightarrow \Delta'$.

Given any node $\Gamma_1 \mapsto \Delta_1$, there are the following cases for $\Gamma_1 \mapsto \Delta_1$:

Case 5. $\Gamma_1 \mapsto \Delta_1 = \Gamma_2, A_1 \wedge A_2 \mapsto \Delta_2$. Then, $\Gamma_1 \mapsto \Delta_1$ has a direct child node $\Gamma_2, A_1, A_2 \mapsto \Delta_2$. By the assumption that $\vdash_{\mathbf{G}'_1} \Gamma_2, A_1, A_2 \Rightarrow \Delta_2$, and by $(\wedge^L)$ in $\mathbf{G}'_1, \vdash_{\mathbf{G}'_1} \Gamma_2, A_1 \wedge A_2 \Rightarrow \Delta_2$.

Case 6. $\Gamma_1 \mapsto \Delta_1 = \Gamma_2, A_1 \vee A_2 \mapsto \Delta_2$. Then, $\Gamma_1 \mapsto \Delta_1$ has two direct children nodes $\Gamma_2, A_1 \mapsto \Delta_2$ and $\Gamma_2, A_2 \mapsto \Delta_2$. By induction assumption, $\vdash_{\mathbf{G}'_1} \Gamma_2, A_1 \Rightarrow \Delta_2$, and $\vdash_{\mathbf{G}'_1} \Gamma_2, A_2 \Rightarrow \Delta_2$, and by $(\vee^L)$ in $\mathbf{G}'_1, \vdash_{\mathbf{G}'_1} \Gamma_2, A_1 \vee A_2 \Rightarrow \Delta_2$.

Case 7. $\Gamma_1 \mapsto \Delta_1 = \Gamma_2 \mapsto B_1 \wedge B_2, \Delta_2$. Then, $\Gamma_1 \mapsto \Delta_1$ has two direct children nodes $\Gamma_2 \mapsto B_1, \Delta_2$ and $\Gamma_2 \mapsto B_2, \Delta_2$. By induction assumption, $\vdash_{\mathbf{G}_1'} \Gamma_2 \Rightarrow B_1, \Delta_2$, and $\vdash_{\mathbf{G}_1'} \Gamma_2 \Rightarrow B_2, \Delta_2$, and by $(\wedge^R)$ in $\mathbf{G}_1'$, $\vdash_{\mathbf{G}_1'} \Gamma_2 \Rightarrow B_1 \wedge B_2, \Delta_2$.

Case 8. $\Gamma_1 \mapsto \Delta_1 = \Gamma_2 \mapsto B_1 \vee B_2, \Delta_2$. Then, $\Gamma_1 \mapsto \Delta_1$ has a direct child node $\Gamma_2 \mapsto B_1, B_2, \Delta_2$. By the assumption that $\vdash_{\mathbf{G}_1'} \Gamma_2 \Rightarrow B_1, B_2, \Delta_2$, and by $(\vee^R)$ in $\mathbf{G}_1'$, $\vdash_{\mathbf{G}_1'} \Gamma_2 \Rightarrow B_1 \vee B_2, \Delta_2$.

Similar for other cases.

□

8.1.3 Nonmonotonicity of $\mathbf{G}_2$

Definition 8.1.10 A deduction system **X** is nonmonotonic in Γ if for any formula sets Γ, Γ' and Δ,

$$\vdash_{\mathbf{X}} \Gamma \mapsto \Delta \& \Gamma' \supseteq \Gamma \text{ may not imply } \vdash_{\mathbf{X}} \Gamma' \mapsto \Delta.$$

X is nonmonotonic in Δ if for any formula sets Γ, Δ and Δ',

$$\vdash_{\mathbf{X}} \Gamma \mapsto \Delta \& \Delta' \supseteq \Delta \text{ may not imply } \vdash_{\mathbf{X}} \Gamma \mapsto \Delta'.$$

Theorem 8.1.11 (Nonmonotonicity theorem) $\mathbf{G}_2$ *is nonmonotonic in both* Γ *and* Δ, *that is, for any formula sets* Γ, Γ', Δ *and* Δ',

$$\begin{array}{l}\Gamma \subseteq \Gamma' \& \vdash_{\mathbf{G}_2} \Gamma \mapsto \Delta \textit{ may not imply } \vdash_{\mathbf{G}_2} \Gamma' \mapsto \Delta;\\ \Delta \subseteq \Delta' \& \vdash_{\mathbf{G}_2} \Gamma \mapsto \Delta \textit{ may not imply } \vdash_{\mathbf{G}_2} \Gamma \mapsto \Delta'.\end{array}$$

Proof We prove that the axiom is nonmonotonic and each deduction rule preserves the monotonicity.

Assume that $\mathrm{con}(\Gamma \cup \neg\Delta)$. There is a superset $\Gamma' \supseteq \Gamma$ such that $\Gamma' \cup \neg\Delta$ is inconsistent; and there is a superset $\Delta' \supseteq \Delta$ such that $\Gamma \cup \neg\Delta'$ is inconsistent. Hence, $\mathbf{G}_2$ is nonmonotonic in both Γ and Δ.

To show that $(\wedge^L)$ preserves the monotonicity of Γ, assume that $\Gamma, A_1, A_2 \mapsto \Delta$ is monotonic with respect to Γ. By $(_{\mapsto}\wedge^L)$, from $\Gamma, A_1, A_2 \mapsto \Delta$ we infer $\Gamma, A_1 \wedge A_2 \mapsto \Delta$. Then, for any $\Gamma' \supseteq \Gamma, \Gamma', A_1, A_2 \mapsto \Delta$; and by $(_{\mapsto}\wedge^L)$, from $\Gamma', A_1, A_2 \mapsto \Delta$, infer $\Gamma', A_1 \wedge A_2 \mapsto \Delta$. Hence, $\Gamma, A_1 \wedge A_2 \mapsto \Delta$ implies $\Gamma', A_1 \wedge A_2 \mapsto \Delta$, that is, $\Gamma, A_1 \wedge A_2 \mapsto \Delta$ is monotonic with respect to Γ.

To show that $(\wedge^L)$ preserves the nonmonotonicity of Γ, assume that $\Gamma, A_1, A_2 \mapsto \Delta$ is nonmonotonic with respect to Γ. By $(_{\mapsto}\wedge^L)$, from $\Gamma, A_1, A_2 \mapsto \Delta$ we infer $\Gamma, A_1 \wedge A_2 \mapsto \Delta$. Then, for some $\Gamma' \supseteq \Gamma, \Gamma, A_1, A_2 \mapsto \Delta$ does not imply $\Gamma', A_1, A_2 \not\mapsto \Delta$; and by $(_{\mapsto}\wedge^L)$, $\Gamma, A_1 \wedge A_2 \mapsto \Delta$ does not imply $\Gamma', A_1 \wedge A_2 \mapsto \Delta$, that is, $\Gamma, A_1 \wedge A_2 \mapsto \Delta$ is nonmonotonic with respect to Γ.

To show that $(\wedge^L)$ preserves the monotonicity of Δ, assume that $\Gamma, A_1, A_2 \mapsto \Delta$ is monotonic with respect to Δ. By $(_{\mapsto}\wedge^L)$, from $\Gamma, A_1, A_2 \mapsto \Delta$ we infer

$\Gamma, A_1 \wedge A_2 \mapsto \Delta$. Then, for any $\Delta' \supseteq \Delta, \Gamma, A_1, A_2 \mapsto \Delta'$; and by $(_\mapsto\wedge^L)$, from $\Gamma, A_1, A_2 \mapsto \Delta'$, infer $\Gamma, A_1 \wedge A_2 \mapsto \Delta'$. Hence, $\Gamma, A_1 \wedge A_2 \mapsto \Delta$ implies $\Gamma, A_1 \wedge A_2 \mapsto \Delta'$, that is, $\Gamma, A_1 \wedge A_2 \mapsto \Delta$ is monotonic with respect to Δ.

To show that $(\wedge^L)$ preserves the nonmonotonicity of Δ, assume that $\Gamma, A_1, A_2 \mapsto \Delta$ is nonmonotonic with respect to Δ'. By $(_\mapsto\wedge^L)$, from $\Gamma, A_1, A_2 \mapsto \Delta$ we infer $\Gamma, A_1 \wedge A_2 \mapsto \Delta$. Then, for some $\Delta' \supseteq \Delta, \Gamma, A_1, A_2 \mapsto \Delta$ does not imply $\Gamma, A_1, A_2 \not\mapsto \Delta'$; and by $(\wedge^L)$, $\Gamma, A_1 \wedge A_2 \mapsto \Delta$ does not imply $\Gamma, A_1 \wedge A_2 \mapsto \Delta'$, that is, $\Gamma, A_1 \wedge A_2 \mapsto \Delta$ is nonmonotonic with respect to Δ.

Similar for other cases. □

By soundness and completeness theorems, we have that for any formula sets Γ, Γ', Δ and Δ',

$$\begin{array}{l}\Gamma \subseteq \Gamma' \& \models_{\mathbf{G}_2} \Gamma \mapsto \Delta \text{ may not imply } \models_{\mathbf{G}_2} \Gamma' \mapsto \Delta, \\ \Delta \subseteq \Delta' \& \models_{\mathbf{G}_2} \Gamma \mapsto \Delta \text{ may not imply } \models_{\mathbf{G}_2} \Gamma \mapsto \Delta'.\end{array}$$

The comparison of deduction systems $\mathbf{G}_1'$ and $\mathbf{G}_2$:

	monotonic $\mathbf{G}_1'$	nonmonotonic $\mathbf{G}_2$
(A)	$\dfrac{\text{incon}(\Gamma) \vee \text{incon}(\Delta) \vee \Gamma \cap \Delta \neq \emptyset}{\Gamma \Rightarrow \Delta}$	$\dfrac{\text{con}(\Gamma) \wedge \text{con}(\Delta) \wedge \Gamma \cap \Delta = \emptyset}{\Gamma \mapsto \Delta}$

(notice here we abuse $\vee, \wedge$ for *or*, &) and

	monotonic	nonmonotonic
$(\wedge^L)$	$\dfrac{\Gamma, A_1 \Rightarrow \Delta}{\Gamma, A_1 \wedge A_2 \Rightarrow \Delta}$ $\dfrac{\Gamma, A_2 \Rightarrow \Delta}{\Gamma, A_1 \wedge A_2 \Rightarrow \Delta}$	$\dfrac{\Gamma, A_1, A_2 \mapsto \Delta}{\Gamma, A_1 \wedge A_2 \mapsto \Delta}$
$(\wedge^R)$	$\dfrac{\Gamma \Rightarrow B_1, \Delta \quad \Gamma \Rightarrow B_2, \Delta}{\Gamma \Rightarrow B_1 \wedge B_2, \Delta}$	$\dfrac{\Gamma \mapsto B_1, \Delta}{\Gamma \mapsto B_1 \wedge B_2, \Delta}$ $\dfrac{\Gamma \mapsto B_2, \Delta}{\Gamma \mapsto B_1 \wedge B_2, \Delta}$
$(\vee^L)$	$\dfrac{\Gamma, A_1 \Rightarrow \Delta \quad \Gamma, A_2 \Rightarrow \Delta}{\Gamma, A_1 \vee A_2 \Rightarrow \Delta}$	$\dfrac{\Gamma, A_1 \mapsto \Delta}{\Gamma, A_1 \vee A_2 \mapsto \Delta}$ $\dfrac{\Gamma, A_2 \mapsto \Delta}{\Gamma, A_1 \vee A_2 \mapsto \Delta}$
$(\vee^R)$	$\dfrac{\Gamma \Rightarrow B_1, \Delta}{\Gamma \Rightarrow B_1 \vee B_2, \Delta}$ $\dfrac{\Gamma \Rightarrow B_2, \Delta}{\Gamma \Rightarrow B_1 \vee B_2, \Delta}$	$\dfrac{\Gamma \mapsto B_1, B_2, \Delta}{\Gamma \mapsto B_1 \vee B_2, \Delta}$

In monotonic propositional logic,

$$\frac{\Gamma, A_1, A_2 \Rightarrow \Delta}{\Gamma, A_1 \wedge A_2 \Rightarrow \Delta}$$

is equivalent to both

$$\frac{\Gamma, A_1 \Rightarrow \Delta}{\Gamma, A_1 \wedge A_2 \Rightarrow \Delta} \& \frac{\Gamma, A_1 \Rightarrow \Delta}{\Gamma, A_1 \wedge A_2 \Rightarrow \Delta};$$

and in nonmonotonic propositional logic,

$$\frac{\Gamma, A_1, A_2 \mapsto \Delta}{\Gamma, A_1 \wedge A_2 \mapsto \Delta}$$

is equivalent to

$$\frac{\Gamma, A_1 \mapsto \Delta \ \Gamma, A_2 \mapsto \Delta}{\Gamma, A_1 \wedge A_2 \mapsto \Delta},$$

not equivalent to

$$\frac{\Gamma, A_1 \mapsto \Delta}{\Gamma, A_1 \wedge A_2 \mapsto \Delta} \& \frac{\Gamma, A_2 \mapsto \Delta}{\Gamma, A_1 \wedge A_2 \mapsto \Delta}.$$

8.2 Involvement of $\Gamma \nvdash A$ in a Nonmonotonic Logic

R-calculus is nonmonotonic, because

• For any consistent theories Δ, Δ' and Γ, if $\Delta|\Gamma \Rightarrow \Delta, \Theta$ is provable in R-calculus and $\Delta' \supset \Delta$, then $\Delta'|\Gamma \Rightarrow \Delta', \Theta$ may not be provable in R-calculus;

• For any consistent theories Δ, Γ and Γ', if $\Delta|\Gamma \Rightarrow \Delta, \Theta$ is provable in R-calculus and $\Gamma' \supset \Gamma$, then $\Delta|\Gamma' \Rightarrow \Delta, \Theta$ may not be provable in R-calculus;

• Moreover, if $\Delta'|\Gamma \Rightarrow \Delta, \Theta'$ and $\Delta|\Gamma' \Rightarrow \Delta, \Theta'$ are provable in R-calculus, then Θ' may not be comparable with Θ, with respect to relation $\subseteq$.

In R-calculus, there is a decision whether $\Delta \nvdash \neg A$, and if $\Delta \nvdash \neg A$ then $\vdash_{\mathbf{R}} \Delta|A \Rightarrow \Delta, A$; otherwise, $\vdash_{\mathbf{R}} \Delta|A \Rightarrow \Delta$. $\Delta \nvdash \neg A$ makes R-calculus nonmonotonic. Given a theory Δ and a formula A, assume that $\Delta \nvdash \neg A$. Then, $\Delta|A \Rightarrow \Delta, A$, and set $\Theta = \{\Delta, A\}$. Let $\Delta' = \Delta \cup \{\neg A\}$. Then, $\Delta' \vdash \neg A$, and $\Delta'|A \Rightarrow \Delta'$, and set $\Theta' = \Delta'$. Hence, even though $\Delta \subset \Delta'$, $\Theta \nsubseteq \Theta'$.

Each nonmonotonic logic involves $\Delta \nvdash A$.

• In default logic [1, 9, 11], to compute an extension E of a default theory (Δ, D), for a normal default $A \rightsquigarrow B \in D$, if $E \vdash A$ and $E \nvdash \neg B$ then $E \vdash B$;

• In circumscription [4, 10], there is the second order formula $\neg\exists P[A(P) \wedge P < p]$ which states that it is impossible to find a predicate P satisfying certain properties;

• In autoepistemic logic [4, 10], an autoepistemic theory E is stable if (i) $\mathrm{Cn}(E) \subseteq E$; (ii) positive introspection: $A \in E \Rightarrow \mathbf{K}A \in E$; and (iii) negative introspection: $A \notin E \Rightarrow \neg\mathbf{K}A \in E$. In another words, $E \nvdash A$ implies $E \vdash \neg\mathbf{K}A$.

• In logic programming with negation as failure [4], *not* l is satisfied in an answer set S if $\neg l \notin S$, i.e., $S \nvdash \neg l$.

8.2.1 Default Logic

A *default* δ is an expression of the form

$$\frac{A : B_1, ..., B_n}{C},$$

where $A, B_1, ..., B_n, C$ are all first order formulas and $n \geq 1$. Here A is called the *prerequisite*, $B_1, ..., B_n$ the *justifications*, and C the *consequent* of δ. δ is *normal* if it has the form $A : B/B$ (denoted by $A \rightsquigarrow B$). Intuitively, a default can be interpreted as follows: if A is known, and if it is consistent to assume $B_1, ..., B_n$, then conclude C.

A *default theory* is a pair (Δ, D), where Δ is a set of closed formulas and D is a set of defaults. (Δ, D) is a *normal default theory* iff every default of D is normal.

Given a default theory (Δ, D), an extension E of (Δ, D) can be derived by applying as many defaults consistently as possible. Define

$$\begin{aligned} &E_0 = \Delta, \\ &E_{i+1} = \mathrm{Th}(E_i) \cup \{B : A \rightsquigarrow B \in D, A \in E_i, \neg B \notin E_i\}; \\ &E = \bigcup_{i \in \omega} E_i. \end{aligned}$$

Then, E is an extension of (Δ, D).

A normal default theory has at least one extension. If E and E' are two distinct extensions of a normal default theory, then E and E' are incompatible.

Proposition 8.2.1 *Let E be an extension of a default theory (Δ, D). Then, E is $\subseteq$-maximal, that is, there is no consistent superset $E' \supseteq E$ such that each formula B is produced by formulas in Δ and defaults in D.*

8.2.2 Circumscription

Circumscription was introduced by McCarthy [5].

Let p, q be two predicate symbols of the same arity n. Define

$$\begin{aligned} &p = q \text{ denotes } \forall x_1, ..., x_n(p(x_1, ..., x_n) \equiv q(x_1, ..., x_n)) \\ &p \leq q \text{ denotes } \forall x_1, ..., x_n(p(x_1, ..., x_n) \to q(x_1, ..., x_n)) \\ &p < q \text{ denotes } p \leq q \wedge \neg(p = q). \end{aligned}$$

In semantics, $p = q$ is true iff p, q have the same extent; $p \leq q$ is true iff the extent of p is a subset of the extent of q; and $p < q$ is true iff the extent of p is a proper subset of the extent of q.

Let $A(p)$ be a sentence containing a predicate symbol p. Let P be a predicate variable of the same arity as p. The circumscription of p in $A()$, denoted by $circ[A(p); p]$, is the following second-order sentence:

$$circ[A(p); p] = A(p) \wedge \neg\exists P[A(P) \wedge P < p].$$

Intuitively, the second order formula $\neg\exists P[A(P) \wedge P < p]$ says: it is not possible to find a predicate P such that

- P satisfies what is said in $A(p)$ about p, and
- the extent of P is a proper subset of the extent of p.

In other words: the extent of p is minimal satisfying condition $A(p)$.

For example,

$A(P)$	$circ[A(P); P]$
$P(a)$	$\forall x(P(x) \equiv x = a)$
$P(a) \wedge P(b)$	$\forall x(P(x) \equiv x = a \vee x = b)$
$P(a) \vee P(b)$	$\forall x(P(x) \equiv x = a) \vee \forall x(P(x) \equiv x = b)$
$\neg P(a)$	$\forall x \neg P(x)$
$\forall x(Q(x) \to P(x))$	$\forall x(Q(x) \equiv P(x))$

Definition 8.2.2 Let M_1 and M_2 be structures, and p a predicate symbol. M_1 is at least as p-preferred as M_2, denoted $M_1 \leq^p M_2$, whenever the following conditions hold:

(i) $|M_1| = |M_2|$,

(ii) $M_1[\![q]\!] = M_2[\![q]\!]$ for every constant $q \neq p$, and

(iii) $M_1[\![p]\!] \subseteq M_2[\![p]\!]$.

Proposition 8.2.3 *M is a model of* $circ[A; P; Z]$ *if and only if M is* $\leq^{P;Z}$*-minimal among the models of* A.

8.2.3 Autoepistemic Logic

Mooly [5] introduced the autoepistemic logic. Autoepistemic logic is a modal logic with modality **K**, where modal logic with **K** is monotonic and the autoepistemic logic with stable theories is not.

An autoepistemic theory E is stable if

(i) $\mathrm{Cn}(E) \subseteq E$;

(ii) positive introspection: $A \in E \Rightarrow \mathbf{K}A \in E$; and

(iii) negative introspection: $A \notin E \Rightarrow \neg\mathbf{K}A \in E$.

Hence, assume that $E' \supseteq E$. If $A \in E' - E$ then $\neg\mathbf{K}A \in E$ and $\mathbf{K}A \in E'$, i.e., $\neg\mathbf{K}A \notin E'$.

Given a consistent theory E, the stability of E implies that

$$A \in E \Leftrightarrow \mathbf{K}A \in E,$$
$$A \notin E \Leftrightarrow \neg\mathbf{K}A \in E.$$

Given a theory D, a theory E is a stable (autoepistemic) expansion of D if for each $A \in E$,

$$D \cup \{\mathbf{K}A : A \in E\} \cup \{\neg\mathbf{K}A : A \notin E\} \vdash A.$$

Hence, E is a stable expansion of D if it is a fixed point of

$$E = \mathrm{Cn}(D \cup \{\mathbf{K}A : A \in E\} \cup \{\neg\mathbf{K}A : A \notin E\}).$$

The correspondence between autoepistemic logic and default logic:

$$\frac{A : B}{C}$$

corresponds to

$$\mathbf{K}A \wedge \neg\mathbf{K}\neg B \to C.$$

Therefore, for

$$\begin{cases} \mathbf{K}p \wedge \neg\mathbf{K}\neg q \to q \\ \mathbf{K}p \wedge \neg\mathbf{K}\neg q \to \mathbf{K}q \end{cases}$$

we have

$$\mathbf{K}A \wedge \neg\mathbf{K}\neg B \to B$$

corresponds to

$$\frac{A : B}{B};$$

and

$$\mathbf{K}A \wedge \mathbf{K}\neg B_1 \wedge \neg\mathbf{K}\neg(B_1 \wedge B_2) \to B_2$$

corresponds to

$$\frac{A, \neg B_1 : B_1 \wedge B_2}{B_2}.$$

8.2.4 Logic Programming with Negation as Failure

Let Π be a logic program, a set of clauses.

The answer set of Π is the smallest sets S of literals satisfying the following conditions:

- either $l_1 \in S$ or $l_2 \in S$ implies $l_1 \vee l_2 \in S$; and
- $l_1 \in S$ and $l_2 \in S$ imply $l_1 \wedge l_2 \in S$.

If there is a literal l such that $l \in S$ and $\neg l \in S$ then S is contradictory, and $S =$ the set of all the formulas composed of the literals and propositional connectives. The reasoning in S is classical.

If a clause is Horn then $l_1 \in S, l_2 \in S, ..., l_n \in S$ imply that $l \in S$, denoted by

$$\frac{l_1 \in S, l_2 \in S, ..., l_n \in S}{l \in S}.$$

We design a language to describe the membership of S. Let $\mathcal{B}$ be the set of all the ground atoms. A string P of symbols is a proposition if either $P \in \mathcal{B}$, or there is a $Q \in \mathcal{B}$ such that $P = \neg Q$, or there are two $Q, R \in \mathcal{B}$ such that $P = Q \vee R$ or $Q \wedge R$.

Assume that $l_1, ..., l_n, l$ are literals.

The classical negation

If $l \leftarrow \neg l_1, l_2, ..., l_n$, then let $\Gamma = \{l_2 \in S, ..., l_n \in S\}$, and if Γ and $\neg l_1 \in S$ then $l \in S$, denoted by $\frac{\Gamma, \neg l_1 \in S}{l \in S}$.

If $\neg l \leftarrow l_1, l_2, ..., l_n$, then let $\Gamma = \{l_1 \in S, l_2 \in S, ..., l_n \in S\}$, and if Γ then $\neg l \in S$, denoted by $\frac{\Gamma}{\neg l \in S}$.

The negation as failure

If $l \leftarrow not\ l_1, l_2, ..., l_n$ then let $\Gamma = \{l_2 \in S, ..., l_n \in S\}$, and if Γ and $l_1 \notin S$ then $l \in S$, denoted by $\frac{\Gamma, l_1 \notin S}{l \in S}$.

If $not\ l \leftarrow l_1, l_2, ..., l_n$, then let $\Gamma = l_1 \in S, l_2 \in S, ..., l_n \in S$, and if Γ then $l \notin S$, denoted by $\frac{\Gamma}{l \notin S}$, which is not equivalent to

$$\frac{\Gamma}{\neg l \in S}.$$

Because the former asserts that $l \notin S$, and it may be that $\neg l \in S$ or there is no information about $\neg l \in S$; and the latter asserts that $\neg l \in S$.

The correspondence between autoepistemic logic and logic programming with negation as failure:

$$l \leftarrow l_1, ..., l_n, not\ l'_1, ..., not\ l'_m$$

corresponds to

$$l_1 \wedge \cdots \wedge l_n \wedge \neg \mathbf{K} l'_1 \wedge \cdots \wedge \neg \mathbf{K} l'_m \rightarrow l.$$

8.3 Correspondence Between R-Calculus and Default Logic

A normal default $A \rightsquigarrow B$ is called simple if $A = \top$. A normal default theory is called simple if each default in the default theory is simple and normal.

There are transformations between those known nonmonotonic logics [5]. This section will give two transformations from the simple normal default logic into R-calculus and vice versa. Precisely,

• given a simple normal default theory (Δ, D) and an extension E of (Δ, D), there is a transformation σ_E such that $\sigma_E(\Delta, D) = \Delta|\Gamma'$ and $\Delta|\Gamma' \Rightarrow E$ is provable in R-calculus;

• there is a transformation τ such that given an R-configuration $\Delta|\Gamma$, $\tau(\Delta|\Gamma) = (\Delta, D)$ is a simple normal default theory, and for any theory Θ, if $\Delta|\Gamma \Rightarrow \Delta, \Theta$ is provable in R-calculus then $\Theta \cup \Delta$ is an extension of (Δ, D).

Hence, we have the following table:

	first-order logic	R-calculus	default logic
semi-decidable	$\Gamma \vdash A$	$\mathrm{incon}(\Delta \cup \{A\})$	$\Delta \vdash A \& \Delta \vdash \neg B$
undecidable	$\Gamma \nvdash A$	$\mathrm{con}(\Delta \cup \{A\})$	$\Delta \vdash A \& \Delta \nvdash \neg B$

8.3.1 *Transformation from R-Calculus to Default Logic*

There is a mapping τ from R-calculus to the simple normal default logic such that for any configuration $\Delta|\Gamma$, $\tau(\Delta|\Gamma) = (\Delta, D)$ is a default theory; and for any $\subseteq$-minimal change Θ of Γ by Δ, $\tau(\Theta \cup \Delta)$ is an extension of (Δ, D), where $D = \{\lambda \rightsquigarrow A : A \in \Gamma\}$.

Hence, we have the following commutative diagram:

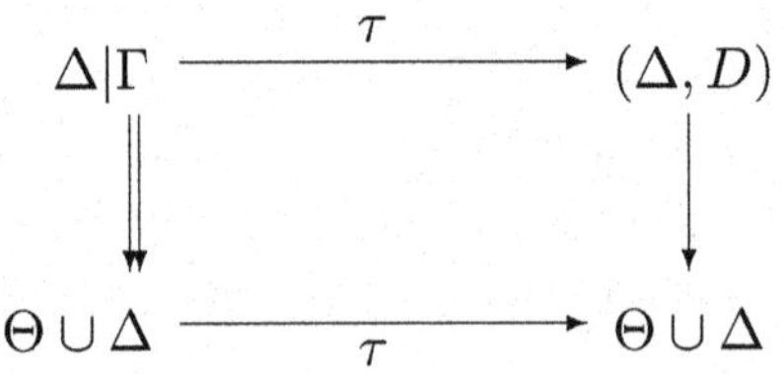

R-calculus $\mathbf{S}^{\mathrm{FOL}}$ is transformed into the following deduction system $\mathbf{D}$ for the simple normal default logic:

$$(D^{\mathrm{con}})\ \frac{\Delta \nvdash \neg A}{(\Delta, \top \rightsquigarrow A, D) \Rightarrow (\Delta \cup \{A\}, D)}$$

$$(D^{\mathbf{A}})\ \frac{\Delta \vdash \neg l}{(\Delta, \top \rightsquigarrow l, D) \Rightarrow (\Delta, D)}$$

$$(D^{\wedge})\ \frac{\begin{array}{c}(\Delta, \top \rightsquigarrow A_1, D) \Rightarrow (\Delta, D)\\ (\Delta, \top \rightsquigarrow A_2, D) \Rightarrow (\Delta, D)\end{array}}{(\Delta, \top \rightsquigarrow A_1 \wedge A_2, D) \Rightarrow (\Delta, D)}$$

$$(D_1^{\vee})\ \frac{(\Delta, \top \rightsquigarrow A_1, D) \Rightarrow (\Delta, D)}{(\Delta, \top \rightsquigarrow A_1 \vee A_2, D) \Rightarrow (\Delta, D)}$$

$$(D_2^{\vee})\ \frac{\begin{array}{c}(\Delta, \top \rightsquigarrow A_1, D)\Gamma \Rightarrow (\Delta \cup \{A_1\}, D)\\ (\Delta \cup \{\neg A_1\}, \top \rightsquigarrow A_2, D)\Gamma \Rightarrow (\Delta \cup \{\neg A_1\}, D)\end{array}}{(\Delta, \top \rightsquigarrow A_1 \vee A_2, D) \Rightarrow (\Delta, D)}$$

$$(D^{\forall})\ \frac{(\Delta, \top \rightsquigarrow A_1(t), D) \Rightarrow (\Delta, D)}{(\Delta, \top \rightsquigarrow \forall x A_1(x), D) \Rightarrow (\Delta, D)}$$

$$(D^{\exists})\ \frac{(\Delta, \top \rightsquigarrow A_1(x), D) \Rightarrow (\Delta, D)}{(\Delta, \top \rightsquigarrow \exists x A_1(x), D) \Rightarrow (\Delta, D)}$$

Definition 8.3.1 $(\Delta, D) \Rightarrow E$ is provable in **D**, denoted by $\vdash_{\mathbf{D}} (\Delta, D) \Rightarrow E$, if there is a sequence $\{(\Delta_i, D_i) : i \in \omega\}$ such that $(\Delta_1, D_1) = (\Delta, D)$, $E = \lim_{i\to\infty} \Delta_i$, and for each $i \in \omega$, (Δ_i, D_i) is an axiom or is deduced from the previous statements by the deduction rules in **D**.

Theorem 8.3.2 *Given a default theory* $(\Delta, \top \rightsquigarrow A)$, *there is a formula* C *such that* $\vdash_{\mathbf{D}} (\Delta, \top \rightsquigarrow A) \Rightarrow \Delta \cup \{C\}$.

Proof Define

$$C = \begin{cases} A \text{ if } \Delta \nvdash \neg A \\ \lambda \text{ otherwise.} \end{cases}$$

Then, we prove by induction on structure of A that $\vdash_{\mathbf{D}} (\Delta, \top \rightsquigarrow A) \Rightarrow \Delta \cup \{C\}$.

If $\Delta \nvdash \neg A$ then by (D^{con}), $\vdash_{\mathbf{D}} (\Delta, \top \rightsquigarrow A) \Rightarrow \Delta \cup \{A\}$.

Assume that $\Delta \vdash \neg A$.

Case $A = l$. $\Delta \vdash \neg l$, and by $(D^{\mathbf{A}})$, $\vdash_{\mathbf{D}} (\Delta, \top \rightsquigarrow l) \Rightarrow \Delta$.

Case $A = A_1 \wedge A_2$. Then, $\Delta \vdash \neg A_1$ and $\Delta, A_1 \vdash \neg A_2$, and by induction assumption,

$$\vdash_{\mathbf{D}} (\Delta, \top \rightsquigarrow A_1) \Rightarrow \Delta$$

or

$$\vdash_{\mathbf{D}} (\Delta, \top \rightsquigarrow A_2) \Rightarrow \Delta,$$

and by $(D^{\wedge})$,

$$\vdash_{\mathbf{D}} (\Delta, \top \rightsquigarrow A_1 \wedge A_2) \Rightarrow \Delta.$$

Case $A = A_1 \vee A_2$. Then, either $\Delta \vdash \neg A_1$ or $\Delta, \neg A_1 \vdash \neg A_2$, and by induction assumption, either

$$\vdash_{\mathbf{D}} (\Delta, \top \rightsquigarrow A_1) \Rightarrow \Delta$$

or

$$\vdash_{\mathbf{D}} (\Delta \cup \{\neg A_1\}, \top \rightsquigarrow A_2) \Rightarrow \Delta \cup \{\neg A_1\},$$

and by $(D_1^{\vee})$ or $(D_2^{\vee})$, $\vdash_{\mathbf{D}} (\Delta, \top \rightsquigarrow A_1 \vee A_2) \Rightarrow \Delta$.

Case $A = \forall x A_1(x)$. Then, there is a term t such that $\Delta \vdash \neg A_1(t)$, and by induction assumption,

$$\vdash_{\mathbf{D}} (\Delta, \top \rightsquigarrow A_1(t)) \Rightarrow \Delta,$$

and by $(D^{\forall})$, $\vdash_{\mathbf{D}} (\Delta, \top \rightsquigarrow \forall x A_1(x)) \Rightarrow \Delta$.

Case $A = \exists x A_1(x)$. Then, there is a variable x not occurring in Δ such that $\Delta \vdash \neg A_1(x)$, and by induction assumption,

$$\vdash_{\mathbf{D}} (\Delta, \top \rightsquigarrow A_1(x)) \Rightarrow \Delta,$$

and by $(D^{\exists})$, $\vdash_{\mathbf{D}} (\Delta, \top \rightsquigarrow \exists x A_1(x)) \Rightarrow \Delta$. □

Therefore, we have the following

Theorem 8.3.3 $\vdash_{\mathbf{D}} (\Delta, \top \rightsquigarrow A) \Rightarrow \Delta \cup \{A\}$ *if and only if* $\Delta \nvdash \neg A$; *and* $\vdash_{\mathbf{D}} (\Delta, \top \rightsquigarrow A) \Rightarrow \Delta$ *if and only if* $\Delta \vdash \neg A$. □

Proposition 8.3.4 $\vdash_{\mathbf{D}} (\Delta, \top \rightsquigarrow A) \Rightarrow \Delta \cup \{C\}$ *iff* $\Delta \cup \{C\}$ *is an extension of* $(\Delta, \top \rightsquigarrow A)$. □

Inductively we have the following

Theorem 8.3.5 (Soundness theorem) *Given a statement* $(\Delta, D) \Rightarrow E$, *if* $(\Delta, D) \Rightarrow E$ *is provable in* **D** *then* E *is an extension of* (Δ, D). □

Theorem 8.3.6 (Completeness theorem) *Given a statement* $(\Delta, D) \Rightarrow E$, *if* E *is an extension of* (Δ, D) *then there is an ordering* $<$ *of* D *such that* $(\Delta, D^{<}) \Rightarrow E$ *is provable in* **D**. □

Theorem 8.3.7 *For any theory* Θ, $\Delta|\Gamma \Rightarrow \Theta$ *is provable in* $\mathbf{S}^{\mathrm{FOL}}$ *if and only if* $(\Delta, D) \Rightarrow \Theta$ *is provable in* **D**.

Proof Assume that $\Gamma = \{A_0, A_1, ...\}$. By the deduction rules of $\mathbf{S}^{\mathrm{FOL}}$, if $\Delta_i|A_i, A_{i+1}, ... \Rightarrow \Delta_{i+1}|A_{i+1}, ...$, where

$$\Delta_{i+1} = \begin{cases} \Delta_i & \text{if } \Delta_i \vdash \neg A_i \\ \Delta_i \cup \{A_i\} & \text{otherwise} \end{cases}$$

then $E_{i+1} = \Delta_{i+1}$, that is, $A_i \in \Theta$ if and only if A_i is in the extension E of (Δ, D). □

Therefore, given a maximal consistent theory Θ of Γ by Δ, if $\Gamma|\Delta \Rightarrow \Theta$ is provable in $\mathbf{S}^{\mathrm{FOL}}$ then Θ is an extension of (Δ, D).

8.3.2 Transformation from Default Logic to R-Calculus

Given a simple normal default theory (Δ, D) and a consistent theory G, assume that there is a well-founded ordering $\prec$ on D and $D^{\prec, G} = \{A \rightsquigarrow B \in D : G \vdash A\}$. Define

$$\begin{aligned} E_0 &= \Delta, \\ E_{i+1} &= E_i \cup \{B_j\}, \\ E &= \bigcup_{i \in \omega} E_i, \end{aligned}$$

where j is the least such that $E_i \nvdash \neg B_j$ and $E_i \nvdash B_j$. Then, E is an extension of (Δ, D), and denoted by $f_1(\Delta, D^{\prec}, G)$.

There is a mapping σ from default logic into R-calculus such that for any default theory (Δ, D) and an extension E of (Δ, D), $\sigma_E(\Delta, D)$ is a configuration $\Delta|^E D'$; and $\sigma(E)$ is an irreducible configuration and $\Delta|D' \Rightarrow \sigma(E)$ is provable in $\mathbf{S}^{\mathrm{FOL}}$, so that we have the following commutative diagram:

$$\begin{array}{ccc} (\Delta, D) & \xrightarrow{\quad \sigma_E \quad} & \Delta|_E D' \\ \downarrow & & \Downarrow \\ E & \xrightarrow[\quad \sigma_E \quad]{} & E \end{array}$$

Given a default theory (Δ, D), define

$$\begin{aligned} &\sigma_E(\Delta, D) = \Delta|_E D', \\ &D' = \{B : A \rightsquigarrow B \in D, E \vdash A\}, \\ &\sigma_E(E) = E. \end{aligned}$$

Theorem 8.3.8 *For any theory Θ, $\Delta|D' \Rightarrow \Theta$ is provable in $\mathbf{S}^{\mathrm{FOL}}$ if and only if $(\Delta, D) \Rightarrow \Theta$ is provable in* **D**.

Proof By the definition of E, $E = \{B : A \rightsquigarrow B \in D, E \vdash A\}$.

Assume that $D' = \{B_0, B_1, ...\}$. Then, for each $i \in \omega$,

$$\Delta_i | B_i, B_{i+1}, ... \Rightarrow \Delta_{i+1} | B_{i+1}, ...,$$

where $\Delta_{i+1} = \begin{cases} \Delta_i & \text{if } \Delta_i \vdash \neg B_i \\ \Delta_i \cup \{B_i\} & \text{otherwise} \end{cases}$

Therefore, $B_i \in \Delta_{i+1}$ if and only if $B_i \in E$. Hence, $\Delta|D' \Rightarrow E$ is provable in R-calculus $\mathbf{S}^{\mathrm{FOL}}$.

By soundness and completeness theorem of R-calculus $\mathbf{S}^{\mathrm{FOL}}$, E is a $\subseteq$-minimal change of D' by Δ. □

Therefore, given an extension E of (Δ, D), there is an ordering $\prec$ such that $\Delta|\Gamma \Rightarrow E$ is provable in $\mathbf{S}^{\mathrm{FOL}}$.

Theorem 8.3.9 *For any default theory* (Δ, D),

$$\tau \circ \sigma_E(\Delta, D) = (\Delta, D)^E;$$

and for any R-configuration $\Delta|\Gamma$,

$$\sigma_\Theta \circ \tau(\Delta|\Gamma) = \Delta|\Gamma.$$

Proof By the definition of σ_E, given an extension E of (Δ, D), $\sigma_E(\Delta, D) = \Delta|\Gamma$, where $\Gamma = \{B : A \rightsquigarrow B \in D, E \vdash A\}$; and by the definition of τ, $\tau(\Delta|\Gamma) = (\Delta, D')$, where $D' = \{\gamma \rightsquigarrow B : B \in \Theta\}$, and $\Theta = D$. By notation, $(\Delta, D') = (\Delta, D)^E$.

Conversely, given an R-configuration $\Delta|\Gamma$, $\tau(\Delta|\Gamma) = (\Delta, D)$, where $D = \{\lambda \rightsquigarrow A : A \in \Gamma\}$; and given an extension E of (Δ, D), by the definition of σ_E, $\sigma_E(\Delta, D) = \Delta|\Gamma$. □

References

1. G. Antoniou, A tutorial on default logics. ACM Comput. Surv. **31**, 337–359 (1999)
2. H. Arló-Costa, Epistemic conditionals, snakes and stars, in *Conditionals: from Philosophy to Computer Science, Studies in Logic and Computation*, vol. 5, ed. by G. Crocco, L. Fariás del Cerro, A. Herzig (Oxford University Press, Oxford, 1995)
3. C. Cao, Y. Sui, Y. Wang, The nonmonotonic propositional logics. Artif. Intell. Res. **5**, 111–120 (2016)
4. K. Clark, Negation as failure, in *Readings in Nonmonotonic Reasoning* (Morgan Kaufmann Publishers, San Francisco, 1987), pp. 311–325
5. M.L. Ginsberg (ed.), *Readings in Nonmonotonic Reasoning* (Morgan Kaufmann, San Francisco, 1987)
6. J.P. Delgrande, T. Schaub, W.K. Jackson, Alternative approaches to default logic. Artif. Intell. **70**, 167–237 (1994)
7. M. Denecker, V.W. Marek, M. Truszczynski, Uniform semantic treatment of default and autoepistemic logics. Artif. Intell. **143**, 79–122 (2003)
8. W. Lukaszewicz, Considerations on default logic: an alternative approach. Comput. Intell. **4**, 1–16 (1988)
9. M. Kaminski, J. Rubin-Mosin, Default theories over monadic languages. Theoret. Comput. Sci. **364**, 241–253 (2006)
10. W. Marek, M. Truszczynski, *Nonmonotonic Logics: Context-Dependent Reasoning* (Springer, Berlin, 1993)
11. R. Reiter, A logic for default reasoning. Artif. Intell. **13**, 81–132 (1980)

Chapter 9
Approximate R-Calculus

In R-calculus for first-order logic, there is a rule

$$\frac{\Delta \nvdash \neg A}{\Delta | A \Rightarrow \Delta, A},$$

which cannot be reduced to the atomic form:

$$\frac{\Delta \nvdash \neg l}{\Delta | l \Rightarrow \Delta, l},$$

because $\Delta \nvdash \neg l$ is undecidable and $\Delta \vdash \neg l$ is semi-decidable.

A set A is decidable [9, 10] if there is an algorithm to decide whether a given x is in A, such that

(i) if $x \in A$ then the algorithm gives yes, and
(ii) if $x \notin A$ then the algorithm gives no.

A set A is semi-decidable if there is an algorithm to decide whether a given x is in A, such that

(i) if $x \in A$ then the algorithm gives yes, and
(ii) if $x \notin A$ then the algorithm may not terminate.

Therefore, a set A is decidable if and only if A and its complement $\bar{A}$ are semi-decidable.

For semi-decidable sets A, there is a limit lemma in recursion theory [9, 10] which says that for any semi-decidable set A, there is a sequence $\{A_s : s \in \omega\}$ of finite (decidable) sets A_s such that $A = \lim_{s \to \infty} A_s$, where $x \in \lim_{s \to \infty} A_s$ if there is a stage s_x such that for any $s \geq s_x$, $x \in A_s$.

W. Li and Y. Sui, *R-CALCULUS: A Logic of Belief Revision*, Perspectives in Formal Induction, Revision and Evolution,
https://doi.org/10.1007/978-981-16-2944-0_9

We use the decidable approximate deduction $\Delta \nvdash_s \neg A$ in R-calculus with the cost of finitely many injuries (mistakes) to have a computable R-calculus $\mathbf{F}^{\mathrm{app}}$ for first-order logic.

9.1 Finite Injury Priority Method

Finite injury priority method was firstly given by Friedberg [1] and Muchnik [7], who solved the Post problem independently. To construct a recursively enumerable set, the conditions the recursively enumerable set should satisfy are represented by an infinite set of requirements which are decomposed into the positive ones (putting elements in the set) and the negative ones (restraining elements from entering the set). The requirements are ordered by a priority ranking, so that the satisfaction of a requirement may injure the satisfaction of requirements with lower priority and cannot injure the satisfaction of requirements with higher priority [9, 10].

9.1.1 Post's Problem

Let $\{e\}$ be the computable function computed by the e-th Turing machine under a coding. If $\{e\}$ with input x halts then we say $\{e\}(x)$ converges, denoted by $\{e\}(x) \downarrow$; otherwise, disconverges, denoted by $\{e\}(x) \uparrow$.

Let $\{e\}_s(x)$ be the approximate computation of $\{e\}(x)$ at stage s.

Let $\{e\}^A$ be the function computed by the e-th Turing machine with oracle A under a coding, and $\{e\}_s^{A_s}(x)$ its approximation at stage s. Let $u(A, e, x, s)$ be the use function of $\{e\}^A(x)$, the maximal number of A used in computing $\{e\}_s^{A_s}(x)$.

The recursive sets are decidable sets. The recursively enumerable sets are the domain of recursive functions (equivalently, the ranges of the total recursive functions).

Notation: $W_e = \mathrm{dom}(\{e\})$, the e-th recursively enumerable set.

The halting problem $K_0 = \{e : \{e\}(e) \downarrow\}$ is undecidable. Let

$$K = \{(e, x) : \{e\}(x) \downarrow\}.$$

Then, K is undecidable.

Let $A, B \subseteq \omega$ be sets of natural numbers. A is recursive in (Turing reducible to) B, denoted by $A \leq_{\mathrm{T}} B$, if $A = \{e\}^B$ for some e.

The Turing degree of a set A is

$$\mathbf{a} = \deg(A) = \{B : B \leq_{\mathrm{T}} A \& A \leq_{\mathrm{T}} B\}.$$

Here, the least Turing degree is that of the recursive (decidable) sets, denoted by $\mathbf{0}$, and the Turing degree of the halting problem is denoted by $\mathbf{0}'$. That is,

$$\begin{aligned}\mathbf{0} &= \deg(\emptyset)\\ &= \{B : B \leq_T \emptyset \& \emptyset \leq_T B\}\\ &= \{B : B \text{ is recursive}\};\\ \mathbf{0}' &= \deg(K)\\ &= \{B : B \leq_T K \& K \leq_T B\}\end{aligned}$$

Proposition 9.1.1 $\mathbf{0} <_T \mathbf{0}'$ □

Post's problem I. Whether there exists any degree **a** between **0** and **0**′, i.e.,

$$\mathbf{0} <_T \mathbf{a} <_T \mathbf{0}'.$$

□

9.1.2 Construction with Oracle

Theorem 9.1.2 (Kleene-Post, [10], p 93) *There exist degrees* **a**, **b** $\leq \mathbf{0}'$ *such that* **a** *is incomparable with* **b**.

Proof We construct sets $A, B \leq_T \emptyset'$ in stages by a finite extension $\emptyset'$-oracle construction so that $\chi_A = \bigcup_s f_a$ and $\chi_B = \bigcup_s g_s$, where f_s and g_s are finite strings in $2^{<\omega}$ of length $\geq s$ viewed as initial segments of χ_A and χ_B. Hence, χ_A is the characteristic function of A, i.e., for any x,

$$\chi_A(x) = \begin{cases} 1 \text{ if } x \in A \\ 0 \text{ if } x \notin A. \end{cases}$$

Since the construction of f_s and g_s at stage s is recursive in $\emptyset'$, $\{f_s\}_{s\in\omega}$ and $\{g_s\}_{s\in\omega}$ are $\emptyset'$-recursive sequences, and so $A, B \leq_T \emptyset'$.

It suffices to meet for each e the following requirements:

$$\begin{aligned} R_e &: A \neq \{e\}^B \\ S_e &: B \neq \{e\}^A \end{aligned}$$

to ensure that $A \nleq_T B$ and $B \nleq_T A$, hence $\mathbf{a} = \deg(A)$ and $\mathbf{b} = \deg(B)$ are incomparable.

The construction

Stage $s = 0$. Define $f_0 = g_0 = \emptyset$.

Stage $s + 1 = 2e + 1$. Given $f_s, g_s \in 2^{<\omega}$ of length $\geq s$. Let $n = lh(f_s) = \mu x(x \notin dom(f_s))$.

Using a $\emptyset'$-oracle we test whether

$$\exists t \exists \sigma (\sigma \supset g_s \& \{e\}_t^{\sigma}(n) \downarrow).$$

Case 1: affirmative. Enumerate the recursive set $\{(\sigma, t) : \{e\}_t^{\sigma}(n) \downarrow\}$, choose the least (σ', t'), and define

$$g_{s+1} = \sigma',$$
$$f_{s+1}(n) = 1 - \{e\}_{t'}^{\sigma'}(n).$$

Case 2: negative. Define

$$f_{s+1}(lh(f_s)) = 0,$$
$$g_{s+1}(lh(g_s)) = 0.$$

For any e, $f(n) \neq \{e\}^g(n)$.

Stage $s + 1 = 2e + 2$. Interchange f and g.

□

9.1.3 Finite Injury Priority Method

Because both **0** and **0**$'$ are recursively enumerable, the recursively enumerable version of Post's problem is

Post's problem II. Whether there exists a recursively enumerable degree **a** between **0** and **0**$'$, i.e.,

$$\mathbf{0} <_{\mathrm{T}} \mathbf{a} <_{\mathrm{T}} \mathbf{0}'.$$

□

Definition 9.1.3 A set A is simple if A is infinite, recursively enumerable and $\bar{A}$ contains no infinite recursively enumerable set.

Theorem 9.1.4 (Friedberg-Muchnik, [10], p 111) *There is a simple set A which is low ($A' \equiv_{\mathrm{T}} \emptyset'$). Hence, $\mathbf{0} <_{\mathrm{T}} \deg(A) <_{\mathrm{T}} \mathbf{0}'$.*

Proof It suffices to construct a coinfinite recursively enumerable set A to meet for all e the requirements:

$$P_e : W_e \text{ infinite} \Rightarrow W_e \cap A \neq \emptyset;$$
$$N_e : \exists^{\infty} s(\{e\}_s^{A_s}(e) \downarrow) \Rightarrow \{e\}^A(e) \downarrow.$$

Let A_s consist of the elements enumerated in A by the end of stage s, and $A = \bigcup_s A_s$.

The priority ranking:

$$N_0, P_0, N_1, P_1, ...$$

The requirements $\{N_e\}_{e \in \omega}$ guarantee $A' \leq_{\mathrm{T}} \emptyset'$. Define the recursive function g by

$$g(e, s) = \begin{cases} 1 \text{ if } \{e\}_s^{A_s}(e) \downarrow \\ 0 \text{ otherwise.} \end{cases}$$

If requirement N_e is satisfied for all e then $\hat{g}(e) = \lim_s g(e, s)$ exists for all e, and $\hat{g} \leq_T \emptyset'$. $\hat{g} = \chi_{A'}$, and $A' \leq_T \emptyset'$.

The restraint function is defined by

$$r(e, s) = u(A_s; e, e, s).$$

To meet N_e we attempt to restrain with priority N_e any elements $x \leq r(e, s)$ from entering A_{s+1}.

Construction of A.

Stage $s = 0$. Let $A_0 = \emptyset$.

Stage $s + 1$. Given A_s we have $r(e, s)$ for all e. Choose the least $i \leq s$ such that

(1) $W_{i,s} \cap A_s = \emptyset$;
(2) $\exists x(x \in W_{i,s} \& x > 2i \& \forall e \leq i(r(e, s) < x))$.

If i exists, choose the least x satisfying (2). Enumerate x in A_{s+1}, and say that requirement P_i receives attention. Hence, $W_{i,s} \cap A_{s+1} \neq \emptyset$, so P_i is satisfied, (1) fails for stages $> s + 1$, and P_i never again receives attention. If i does not exist, do nothing, so $A_{s+1} = A_s$.

Let $A = \bigcup_s A_s$.

The ends the construction.

We say that x injures N_e at stage $s + 1$ if $x \in A_{s+1} - A_s$ and $x \leq r(e, s)$. Define the injury set for N_e

$$\begin{aligned} I_e &= \{x : \exists s(x \in A_{s+1} - A_s \& x \leq r(e, s))\} \\ &= \{x : x \text{ injures } N_e \text{ at some stage } s + 1\}. \end{aligned}$$

Lemma 9.1.5 *For any e, I_e is finite.*

Proof Each positive requirement P_i contributes at most one element to A by (1). By (2), N_e can be injured by P_i only if $i < e$. Hence, $|I_e| \leq e$. □

Lemma 9.1.6 *For every e, requirement N_e is met and $r(e) = \lim_s r(e, s)$ exists.*

Proof Fix e. By lemma 9.1.5, choose stage s_e such that N_e is not injured at any stage $s > s_e$. However, if $\{e\}_s^{A_s}(e)$ converges for $s > s_e$ then by induction on $t \geq s$, $r(e, t) = r(e, s)$ and $\{e\}_t^{A_t}(e) = \{e\}_s^{A_s}(e)$ for all $t \geq s$, so $A_s \upharpoonright r(e, s) = A \upharpoonright r(e, s)$, and hence $\{e\}^A(e)$ is defined. □

Lemma 9.1.7 *For every i, requirement P_i is met.*

Proof Fix i such that W_i is infinite. By lemma 9.1.6, choose s such that

$$\forall t \geq s \forall e \leq i(r(e, t) = r(e)).$$

Choose $s' \geq s$ such that no P_j, $j < i$, receives attention after stage s'.
Choose $t > s'$ such that

$$\exists x(x \in W_{i,t} \& x > 2i \& \forall e \leq i(r(e) < x)).$$

Now either $W_{i,t} \cap A_t \neq \emptyset$ or else P_i receives attention at stage $t+1$. In either case $W_{i,t} \cap A_{t+1} \neq \emptyset$, so P_i is met by the end of stage $t+1$.

$\bar{A}$ is infinite by (2), hence A is simple and low. □

9.2 Approximate Deduction

By the limit lemma, a semi-decidable set A is the limit of a sequence $\{A_0, A_1, ...\}$ of decidable (finite) sets. The deduction $\vdash$ in first-order logic is semidecidable. Therefore, we have the following table:

	first-order logic	R-calculus
semi-decidable	$\Gamma \vdash A$	$\Delta\|A \Rightarrow \Delta$
semi-undecidable	$\Gamma \nvdash A$	$\Delta\|A \Rightarrow \Delta, A$

where the semi-decidable $\Gamma \vdash A$ can be approximated by $\Gamma \vdash_s A$; and $\Delta|A \Rightarrow \Delta$ by $\Delta|A \Rightarrow_s \Delta$, i.e., we have the following table:

	first-order logic	R-calculus
monotonic in s	$\Delta \vdash_s A$	$\Delta\|A \Rightarrow_s \Delta$
nonmonotonic in s	$\Delta \nvdash_s A$	$\Delta\|A \Rightarrow_s \Delta, A$

where

- $\Delta \vdash_s A$ is monotonic in s, i.e., for any stage s, $\Delta \vdash_s A$ implies that for any $t \geq s$, $\Delta \vdash_t A$; and
- $\Delta \nvdash_s A$ is nonmonotonic in s, i.e., for any stage s, $\Delta \nvdash_s A$ does not imply that for any $t \geq s$, $\Delta \nvdash_t A$;

and correspondingly,

- $\Delta|A \Rightarrow_s \Delta$ is monotonic in s, i.e., for any stage s, $\Delta|A \Rightarrow_s \Delta$ implies that for any $t \geq s$, $\Delta|A \Rightarrow_t \Delta$; and
- $\Delta|A \Rightarrow_s \Delta, A$ is nonmonotonic in s, i.e., for any stage s, $\Delta|A \Rightarrow_s \Delta, A$ does not imply that for any $t \geq s$, $\Delta|A \Rightarrow_t \Delta, A$.

The rules of R-calculus $\mathbf{S}^{\mathrm{FOL}}$ (denoted by $\mathbf{F}$ in this chapter) can be reduced to the following rules:

$$(F^{\mathrm{incon}})\ \frac{\Delta \vdash \neg A}{\Delta|A \Rightarrow \Delta}\quad (F^{\mathrm{con}})\ \frac{\Delta \nvdash \neg A}{\Delta|A \Rightarrow \Delta, A}.$$

Because $\Delta \vdash \neg A$ is semi-decidable, we cannot recursively decide whether $\Delta \nvdash \neg A$. By the limit lemma in recursion theory, we can recursively decide the approximate deduction $\Delta \nvdash_s \neg A$ to approximate deciding $\Delta \nvdash \neg A$.

In this way, for the approximate deduction, the rules of R-calculus are represented by

$$(F_s^{\text{incon}})\ \frac{\Delta \vdash_s \neg A}{\Delta | A \Rightarrow_s \Delta}\quad (F_s^{\text{con}})\ \frac{\Delta \nvdash_s \neg A}{\Delta | A \Rightarrow_s \Delta, A}.$$

9.2.1 Approximate Deduction System for First-Order Logic

Gentzen deduction system $\mathbf{G}^{\text{app}}$ [3] for approximate reasoning consists of the following axioms and deduction rules:

- **Axiom**:

$$(\mathbf{A})\ \Gamma, A \Rightarrow_0 A, \Delta$$

- **Deduction rules**:

$$(\neg^L)\ \frac{\Gamma \Rightarrow_s A, \Delta}{\Gamma, \neg A \Rightarrow_{s+1} \Delta}\qquad (\neg^R)\ \frac{\Gamma, A \Rightarrow_s \Delta}{\Gamma \Rightarrow_{s+1} \neg A, \Delta}$$

$$(\wedge_1^L)\ \frac{\Gamma, A \Rightarrow_s \Delta}{\Gamma, A \wedge B \Rightarrow_{s+1} \Delta}\qquad (\wedge^R)\ \frac{\Gamma \Rightarrow_s A, \Delta\ \ \Gamma \Rightarrow_s B, \Delta}{\Gamma \Rightarrow_{s+1} A \wedge B, \Delta}$$

$$(\wedge_2^L)\ \frac{\Gamma, B \Rightarrow_s \Delta}{\Gamma, A \wedge B \Rightarrow_{s+1} \Delta}$$

$$(\vee^L)\ \frac{\Gamma, A \Rightarrow_s \Delta\ \ \Gamma, B \Rightarrow_s \Delta}{\Gamma, A \vee B \Rightarrow_{s+1} \Delta}\qquad (\vee_1^R)\ \frac{\Gamma \Rightarrow_s A, \Delta}{\Gamma \Rightarrow_{s+1} A \vee B, \Delta}$$

$$(\vee_1^R)\ \frac{\Gamma \Rightarrow_s B, \Delta}{\Gamma \Rightarrow_{s+1} A \vee B, \Delta}$$

$$(\forall^L)\ \frac{\Gamma, A(t) \Rightarrow_s \Delta}{\Gamma, \forall x A(x) \Rightarrow_{s+1} \Delta}\qquad (\forall^R)\ \frac{\Gamma \Rightarrow_s A(x), \Delta}{\Gamma \Rightarrow_{s+1} \forall x A(x), \Delta}$$

$$(\exists^L)\ \frac{\Gamma, A(x) \Rightarrow_s \Delta}{\Gamma, \exists x A(x) \Rightarrow_{s+1} \Delta}\qquad (\exists^R)\ \frac{\Gamma \Rightarrow_s A(t), \Delta}{\Gamma \Rightarrow_{s+1} \exists x A(x), \Delta}$$

where in quantifier rules, t is an arbitrary term, and x does not occur in the lower sequent.

Definition 9.2.1 A sequent $\Gamma \Rightarrow \Delta$ is s-deducible, denoted by $\vdash_s \Gamma \Rightarrow \Delta$, if there is a sequence $\Gamma \Rightarrow_0 \Delta, ..., \Gamma_{s+1} \Rightarrow_{s+1} \Delta_{s+1}$ of sequents which is a proof in $\mathbf{G}^{\text{app}}$ and $\Gamma_{s+1} = \Gamma, \Delta_{s+1} = \Delta$.

Proposition 9.2.2 *(i) For any sequent* $\Gamma \Rightarrow \Delta$, *if* $\vdash_s \Gamma \Rightarrow \Delta$ *then* $\vdash \Gamma \Rightarrow \Delta$.

(ii) For any sequent $\Gamma \Rightarrow \Delta$, *if* $\vdash \Gamma \Rightarrow \Delta$ *then there is an* $s \in \omega$ *such that* $\vdash_s \Gamma \Rightarrow \Delta$. □

9.3 R-Calculus $\mathbf{F}^{\mathrm{app}}$ and Finite Injury Priority Method

We will give a recursive R-calculus $\mathbf{F}^{\mathrm{app}}$ to approximate $\mathbf{S}^{\mathrm{FOL}}$ with finitely many injuries.

9.3.1 Construction with Oracle

Fix two consistent theories Δ, Γ. We will construct a theory Θ , satisfying the following conditions:

(i) $\Theta \subseteq \Gamma$;

(ii) Δ is a $\subseteq$-minimal change of Γ by Δ.

Let $\Gamma = \{A_0, A_1, ...\}$ and for each i, $\Gamma_i = \{A_0, A_1, ..., A_i\}$. We will construct a sequence $\{\Theta_0, \Theta_1, ...\}$ of theories such that $\Theta_0 = \Delta$; for each i, $\Theta_{i+1} \subseteq \Gamma_i$ is a $\subseteq$-minimal change of Γ_i by Δ, and $\Theta = \bigcup_i \Theta_i$ satisfies conditions (i) and (ii).

The construction is in stages with oracle whether $\Theta_i \cup \{A_i\}$ is consistent.

It suffices to meet for each e the following requirements:

$$P_e : \mathrm{con}(\Theta_e \cup \{A_e\}) \Rightarrow A_e \in \Theta_{e+1};$$
$$N_e : \mathrm{incon}(\Theta_e \cup \{A_e\}) \Rightarrow A_e \notin \Theta_{e+1}.$$

If $\{\Theta_i : i \in \omega\}$ is a sequence satisfying all the requirements then $\Theta = \bigcup_i \Theta_i$ satisfies (i) and (ii).

The construction:

`Stage s = 0. Define` $\Theta_0 = \emptyset$, `and for each i` ≥ 0, $\Theta_i = \emptyset$.

`Stage s + 1. If` $\Delta \cup \Theta_s \cup \{A_s\}$ `is consistent then let` $\Theta_{s+1} = \Theta_s \cup \{A_s\}$; `and if` $\Delta \cup \Theta_s \cup \{A_s\}$ `is inconsistent then let` $\Theta_{s+1} = \Theta_s$.

`This ends the construction.`

Lemma 9.3.1 *For each e, Θ_{e+1} is consistent with Δ.*

Proof By induction on e. Assume that Θ_e is consistent with Δ. Then,

$$\Theta_{e+1} = \begin{cases} \Theta_e \cup \{A_e\} \text{ if } \Delta \cup \Theta_e \cup \{A_e\} \text{ is consistent} \\ \Theta_e \quad\quad\quad \text{otherwise} \end{cases}$$

is consistent with Δ. □

Lemma 9.3.2 *For each e, $\Theta_{e+1} \subseteq \Gamma_e$.*

Proof By the construction. □

Lemma 9.3.3 *For each e, Θ_{e+1} is a $\subseteq$-minimal change of Γ_e by Δ.*

Proof By the construction,

$$\Theta_{e+1} = \begin{cases} \Theta_e \cup \{A_e\} \text{ if } \Delta \cup \Theta_e \cup \{A_e\} \text{ is consistent} \\ \Theta_e \qquad\qquad \text{otherwise.} \end{cases}$$

By induction on e, if Θ_e is a $\subseteq$-minimal change of Γ_{e-1} by Δ then Θ_{e+1} is a $\subseteq$-minimal change of Γ_e by Δ. □

By soundness and completeness theorem of R-calculus, we have that Θ is a $\subseteq$-minimal change of Γ by Δ, and hence, $\Delta|\Gamma \Rightarrow \Theta$ is provable in R-calculus **F**.

The procedure corresponding to the construction:

proxcheme R-calculus (Δ, Γ) — $\%\Delta$: a reversing theory
$\%\Gamma = \{A_0, A_1, ...\}$: theory to be revised

input: Δ, Γ
local variables: X_i, Y_i — $\%X_i$: set of formulas which may be in Θ,
$\% \ Y_i$: set of formulas which may not be in Θ

output: $\Theta = \lim_{i\to\infty} X_i$

$X_0 = \Delta, Y_0 = \emptyset$. — % initialization
For each i, do
begin
  If $X_i \vdash \neg A_i$ then set
    $X_{i+1} = X_i$
    $Y_{i+1} = Y_i \cup \{A_i\}$
  If $X_i \nvdash \neg A_i$ then set
    $X_{i+1} = X_i \cup \{A_i\}$
    $Y_{i+1} = Y_i$
end

Let $\Theta = \lim_{i\to\infty} X_i$. Then, $\Delta|\Gamma \Rightarrow \Theta$ is provable in R-calculus **F**.

9.3.2 Approximate Deduction System $\mathbf{F}^{\text{app}}$

Let $\Gamma_{i,s}$ be the formulas in $\Gamma_i = \{A_0, ..., A_{i-1}\}$ which are not deleted at the end of stage s and for each $j < i$, $A_j \in \Gamma_{i,s}$ iff $\Delta, \Gamma_{i,s} \nvdash \neg A_j$, and

$$\Gamma'_{i+2,s+1} = \{A_{i+1}, A_{i+2}, ...\}.$$

At stage $s + 1$, we consider A_i.

Gentzen deduction system $\mathbf{F}^{\text{app}}$ for approximate R-calculus consists of the following axioms and deduction rules:

- **Axioms:**

$$(F^{\mathrm{con,app}})\ \frac{\Delta,\Gamma_{i,s}\nvdash\neg A_i}{\Delta,\Gamma_{i,s}|A_i,\Gamma_{i+2,s}\Rightarrow_s\Delta,\Gamma_{i,s+1}|\Gamma_{i+2,s+1}};$$

$$(F^{\neg,\mathrm{app}})\ \frac{\Delta,\Gamma_{i,s}\vdash\neg p}{\Delta,\Gamma_{i,s}|p,\Gamma_{i+2,s}\Rightarrow\Delta,\Gamma_{i,s+1}|\Gamma_{i+2,s+1}};$$

- **Deduction rules:**

$$(F_1^{\wedge,\mathrm{app}})\ \frac{\Delta,\Gamma_{i,s}|A_{i1},\Gamma_{i+1,s}\Rightarrow\Delta,\Gamma_{i,s}|\Gamma_{i+2,s}}{\Delta,\Gamma_{i,s}|A_{i1}\wedge A_{i2},\Gamma_{i+2,s}\Rightarrow\Delta,\Gamma_{i,s+1}|\Gamma_{i+2,s+1}};$$

$$(F_2^{\wedge,\mathrm{app}})\ \frac{\begin{array}{l}\Delta,\Gamma_{i,s}|A_{i1},\Gamma_{i+2,s}\Rightarrow\Delta,\Gamma_{i,s},A_{i1}|\Gamma_{i+2,s}\\ \Delta,\Gamma_{i,s},A_{i1}|A_{i2},\Gamma_{i+2,s}\Rightarrow\Delta,\Gamma_{i,s},A_{i1}|\Gamma_{i+2,s}\end{array}}{\Delta,\Gamma_{i,s}|A_{i1}\wedge A_{i2},\Gamma_{i+2,s}\Rightarrow\Delta,\Gamma_{i,s+1}|\Gamma_{i+2,s}};$$

$$(F^{\vee,\mathrm{app}})\ \frac{\begin{array}{l}\Delta,\Gamma_{i,s}|A_{i1},\Gamma_{i+2,s}\Rightarrow\Delta,\Gamma_{i,s}|\Gamma_{i+2,s}\\ \Delta,\Gamma_{i,s}|A_{i2},\Gamma_{i+2,s}\Rightarrow\Delta,\Gamma_{i,s}|\Gamma_{i+2,s}\end{array}}{\Delta,\Gamma_{i,s}|A_{i1}\vee A_{i2},\Gamma_{i+2,s}\Rightarrow\Delta,\Gamma_{i,s+1}|\Gamma_{i+2,s+1}};$$

$$(F_1^{\forall,\mathrm{app}})\ \frac{\Delta,\Gamma_{i,s}|A_i(t),\Gamma_{i+1,s}\Rightarrow\Delta,\Gamma_{i,s}|\Gamma_{i+2,s}}{\Delta,\Gamma_{i,s}|\forall xA_i(x),\Gamma_{i+2,s}\Rightarrow\Delta,\Gamma_{i,s+1}|\Gamma_{i+2,s+1}};$$

$$(F_1^{\exists,\mathrm{app}})\ \frac{\Delta,\Gamma_{i,s}|A_i(x),\Gamma_{i+1,s}\Rightarrow\Delta,\Gamma_{i,s}|\Gamma_{i+2,s}}{\Delta,\Gamma_{i,s}|\exists xA_i(x),\Gamma_{i+2,s}\Rightarrow\Delta,\Gamma_{i,s+1}|\Gamma'_{i+2,s+1}},$$

where

$$\Gamma_{i,s+1}=\begin{cases}\Gamma_{i,s}\cup\{A_i\} & \text{if } \Delta,\Gamma_{i,s}\nvdash_s\neg A_i\\ \Gamma_{i,s} & \text{otherwise;}\end{cases}$$

and

$$\Gamma_{i+2,s+1}=\Gamma_{i+2,s}.$$

By soundness and completeness theorem [5] of R-calculus **F**, for any $i\in\omega$, there is a stage s_i such that

$$\Gamma_{i,s_i}=\lim_{s\to\infty}\Gamma_{i,s},$$

and $\Delta|\Gamma\Rightarrow\Sigma$ is provable in $\mathbf{F}^{\mathrm{app}}$, where

$$\Sigma=\Delta\cup\bigcup_i\Gamma_{i,s_i}.$$

Definition 9.3.4 $\Delta|\Gamma\Rightarrow\Delta,\Theta$ is provable in $\mathbf{F}^{\mathrm{app}}$, denoted by $\vdash_{\mathbf{F}^{\mathrm{app}}}\Delta|\Gamma\Rightarrow\Delta,\Theta$, if there is a sequence $\Delta_1|\Gamma_1,\ldots,\Delta_n|\Gamma_n,\ldots$ such that $\Delta_1|\Gamma_1=\Delta|\Gamma$, $\Theta=\lim_{n\to\infty}\Delta_n-\Delta$, and for each $j<n$, $\Delta_j|\Gamma_j$ is an axiom or is deduced from the previous statements by the deduction rules in $\mathbf{F}^{\mathrm{app}}$.

Therefore, we have

Theorem 9.3.5 *Given two theories* Γ *and* Δ, $\vdash_{\mathbf{F}^{\mathrm{app}}}\Delta|\Gamma\Rightarrow\Sigma$; *and conversely, if* $\vdash_{\mathbf{F}^{\mathrm{app}}}\Delta|\Gamma\Rightarrow\Delta,\Theta$ *then* $\Delta\cup\Theta=\Sigma$. □

9.3.3 Recursive Construction

By using the approximative deduction

$$(F_s^{\text{incon}})\ \frac{\Delta \vdash_s \neg A}{\Delta|A \Rightarrow_s \Delta}\ (F_s^{\text{con}})\ \frac{\Delta \nvdash_s \neg A}{\Delta|A \Rightarrow_s \Delta, A},$$

there are the following two cases:

- $\Delta \nvdash_s \neg A$, and for any $t \geq s$, $\Delta \nvdash_t \neg A$. Then for any $t \geq s$, $\Delta|A \Rightarrow_t \Delta, A$, and

$$\Delta|A \Rightarrow \Delta, A;$$

- $\Delta \nvdash_s \neg A$, and for some $t \geq s$, $\Delta \vdash_t \neg A$. Then for this t, $\Delta|A \Rightarrow_t \Delta$, and

$$\Delta|A \Rightarrow \Delta.$$

For the first case, there is no problem; and for the second case, canceling A at t may result in that some B canceled before t is consistent with Δ, which makes B put back in Γ. So, we call that the cancelation of A at t injures the one of B's before t. We will give a recursive construction such that for each B in Γ, its cancelation is injured only finitely often, and after finitely many stages, either B is canceled eventually, or B never be canceled.

Let

$$\begin{array}{l}[A] : A \rightsquigarrow \Delta \\ \langle A \rangle : A \not\rightsquigarrow \Delta\end{array}$$

represent that A is enumerated in Δ and extracted out of Δ, respectively. Assume that at stage s, we have

$$\begin{array}{l}\text{if } \Delta \nvdash_s A_1 \text{ then } \Delta|A_1 \Rightarrow_s \Delta|[A_1], \\ \text{if } \Delta \vdash_s \neg A_2 \text{ then } \Delta|A_2 \Rightarrow_s \Delta|\langle A_2 \rangle,\end{array}$$

and at some stage $t > s$, we have

$$\begin{array}{l}\text{if } \Delta \nvdash_s A_1 \text{ and } \Delta, A_1 \vdash_s \neg A_2 \text{ then } \Delta|A_1 \Rightarrow_s \Delta|[A_1], \text{ and} \\ \qquad\qquad\qquad\qquad\qquad\qquad \Delta|[A_1], A_2 \Rightarrow_s \Delta|[A_1], \langle A_2 \rangle \\ \text{if } \Delta \vdash_t \neg A_1 \text{ and } \Delta \nvdash_t \neg A_2 \text{ then} \Delta|[A_1], \langle A_2 \rangle \Rightarrow_t \Delta|\langle A_1 \rangle, [A_2],\end{array}$$

where $\langle A_2 \rangle \rightsquigarrow [A_2]$ at stage t only if there is a formula $A_1 \prec A_2$ such that $[A_1] \rightsquigarrow \langle A_1 \rangle$ at stage t.

In this case, $[A_2] \rightsquigarrow \langle A_2 \rangle$ at stage s is injured by $[A_1] \rightsquigarrow \langle A_1 \rangle$ at stage t.

The following gives the recursive construction for R-calculus $\mathbf{F}$.

Let $\Gamma = \{A_0, A_1, ...\}$ and for each i, $\Gamma_i = \{A_0, A_1, ..., A_i\}$. We will construct in stages a sequence $\{\Theta_0, \Theta_1, ...\}$ of theories such that

(i) $\Theta_0 = \emptyset$;
(ii) for each i, $\Theta_{i+1} \subseteq \Gamma_i$ is a $\subseteq$-minimal change of Γ_i by Δ, and
(iii) $\Theta = \bigcup_i \Theta_i$ satisfies the following conditions:

(a) $\Theta \subseteq \Gamma$;
(b) Θ is a $\subseteq$-minimal change of Γ by Δ.

The construction is in stages with approximate deduction $\Delta, \Theta_{i,s} \vdash_s \neg A_i$, where $\Gamma \vdash_s A$ if there is a sequence $\{\theta_0, ..., \theta_n\}$ of formulas with $n \leq s$ which is a proof of $\Gamma \vdash A$.

It suffices to meet for each e the following requirements:

$$P_e : \text{con}(\Delta \cup \Theta_e \cup \{A_e\}) \Rightarrow A_e \in \Theta_{e+1};$$
$$N_e : \text{incon}(\Delta \cup \Theta_e \cup \{A_e\}) \Rightarrow A_e \notin \Theta_{e+1}.$$

The priority ranking of requirements are

$$P_0, N_0, P_1, N_1, ..., P_e, N_e, ...$$

If $\{\Theta_i : i \in \omega\}$ is a sequence satisfying all the requirements then $\Theta = \bigcup_i \Theta_i$ satisfies (a) and (b).

A requirement N_e *requires attention* at stage $s+1$ if $\Delta, \Theta_{e,s} \vdash_{s+1} \neg A_e$ and $A_e \in \Theta_{e+1,s}$.

A requirement N_e *is satisfied* at stage $s+1$ if $\Delta, \Theta_{e,s} \vdash_{s+1} \neg A_e$ and $A_e \notin \Theta_{e+1,s}$.

The construction:

Stage $s = 0$. Define $\Theta_{0,0} = \emptyset$, and for each $i \geq 0$, $\Theta_{i,0} = \Gamma_i$.

Stage $s+1$. Find the least e such that N_e requires attention. Define

$$\Theta_{e+1,s+1} = \Theta_{e+1,s} - \{A_e\},$$

and for each $e' \geq e+1$,

$$\Theta_{e',s+1} = \Theta_{e',0},$$

and we say that N_e receives attention.

Define for each e,

$$\Theta_e = \lim_{s\to\infty} \Theta_{e,s}.$$

This ends the construction.

We say that N_e is injured at stage $s+1$ if there is $i < e$ such that $A_i \in \Theta_{i+1,s} - \Theta_{i+1,s+1}$.

Define the injury set for N_e as follows:

$$I_e = \{i : \exists s(A_i \in \Theta_{i+1,s} - \Theta_{i+1,s+1})\};$$
$$I_{e,s} = \{i : \exists s' \leq s(A_i \in \Theta_{i+1,s'} - \Theta_{i+1,s'+1})\}.$$

Lemma 9.3.6 *For each e, I_e is finite.*

Proof By induction on e. Assume that $I_{e'}(e' < e)$ is finite. Then, there is a stage s_e such that for any $s \geq s_e$, $N_{e'}$ never requires attention, and $I_e = I_{e,s_e}$ is finite. □

Lemma 9.3.7 *For every e, $\Theta_e = \lim_{s\to\infty} \Theta_{e,s+1}$ exists, and requirement N_e is met.*

Proof Fix e. By Lemma 9.3.6, there is a stage s_e such that N_e is not injured at any stage $s > s_e$.

If N_e never requires attention then $\Theta_e = \Theta_{e,s_e}$;

If N_e requires attention at some stage $s > s_e$ then, N_e is satisfied at s and never requires attention. Hence, $\Theta_e = \Theta_{e,s}$. □

Lemma 9.3.8 *For every e, requirement P_e is met.*

Proof If N_e is satisfied at stage s then so is P_e. □

By soundness and completeness theorem of R-calculus **F**, we have that Θ is a $\subseteq$-minimal change of Γ by Δ, and hence, $\Delta|\Gamma \Rightarrow \Delta, \Theta$ is provable in R-calculus **F**.

Example 9.3.9 Assume that

$\Delta\|$	A_0	A_1	A_2	$\cdots A_e \cdots$
	$\Theta_{0,s}$	$\Theta_{1,s}$	$\Theta_{2,s}$	$\cdots \Theta_{e,s} \cdots$
$s=0$	$\Delta \cup \{A_0\}$			
$s=1$	$\Theta_{0,0} \nvdash_1 \neg A_0$			
	$\Delta \cup \{A_0\}$	$\Delta \cup \{A_0, A_1\}$		
$s=2$	$\Theta_{0,1} \nvdash_2 \neg A_0$	$\Theta_{1,1} \nvdash_2 \neg A_1$		
	$\Delta \cup \{A_0\}$	$\Delta \cup \{A_0, A_1\}$	$\Delta \cup \{A_0, A_1, A_2\}$	
$s=3$	$\Theta_{0,2} \nvdash_3 \neg A_0$	$\Theta_{1,2} \nvdash_3 \neg A_1$	$\Theta_{2,2} \vdash_3 \neg A_2$	
	$\Delta \cup \{A_0\}$	$\Delta \cup \{A_0, A_1\}$	$\Delta \cup \{A_0, A_1\}$	
$s=4$	$\Theta_{0,3} \nvdash_4 \neg A_0$	$\Theta_{1,3} \vdash_3 \neg A_1$		
	$\Delta \cup \{A_0\}$	$\Delta \cup \{A_0\}$	$\Delta \cup \{A_0, A_2\}$	
$s=5$	$\Theta_{0,4} \nvdash_5 \neg A_0$		$\Theta_{2,4} \vdash_5 \neg A_2$	
	$\Delta \cup \{A_0\}$	$\Delta \cup \{A_0\}$	$\Delta \cup \{A_0\}$	
$s=6$	$\Theta_{0,5} \vdash_6 \neg A_0$			
	Δ	$\Delta \cup \{A_1\}$	$\Delta \cup \{A_1, A_2\}$	
$s=7$		$\Theta_{1,6} \nvdash_7 \neg A_1$	$\Theta_{2,6} \vdash_7 \neg A_2$	
	Δ	$\Delta \cup \{A_1\}$	$\Delta \cup \{A_1\}$	

and hence, $\Delta|\Gamma \Rightarrow \Delta \cup \{A_1\}$.

Let

$$\Delta = \{p_1, p_2, p_3, p_4\},$$
$$\Gamma = \{A_1, A_2, A_3\},$$

where

$$A_1 = p_1 \to (p_2 \to (p_3 \to \neg p_4)),$$
$$A_2 = p_1 \to (p_2 \to \neg p_3),$$
$$A_3 = (p_1 \to (p_2 \to \neg p_3)) \to \neg p_1\}.$$

Then,

Δ	A_0	A_1	A_2
	$\Theta_{0,s}$	$\Theta_{1,s}$	$\Theta_{2,s}$
$s=0$	$\Delta \cup \{A_0\}$	$\Delta \cup \{A_0, A_1\}$	$\Delta \cup \{A_0, A_1, A_2\}$
$s=1$	$\Theta_{0,0} \nvdash_1 \neg A_0$	$\Theta_{1,0} \nvdash_1 \neg A_1$	$\Theta_{0,1} \vdash_1 \neg A_2$
	$\Delta \cup \{A_0\}$	$\Delta \cup \{A_0, A_1\}$	$\Delta \cup \{A_0, A_1\}$
$s=2$	$\Theta_{0,1} \nvdash_2 \neg A_0$	$\Theta_{1,1} \vdash_2 \neg A_1$	
	$\Delta \cup \{A_0\}$	$\Delta \cup \{A_0\}$	$\Delta \cup \{A_0, A_2\}$
$s=3$	$\Theta_{0,2} \vdash_3 \neg A_0$		
	Δ	$\Delta \cup \{A_1\}$	$\Delta \cup \{A_1, A_2\}$
$s=4$		$\Theta_{1,3} \vdash_4 \neg A_1$	
	Δ	Δ	$\Delta \cup \{A_2\}$

and hence, $\Delta|\Gamma \Rightarrow \Delta \cup \{A_2\}$.

The algorithm corresponding to the construction:

algorithm $\mathbf{R}^{\text{rec}}$-calculus (Δ, Γ) %Δ : a reversing theory
%$\Gamma = \{A_0, A_1, ...\}$: theory to be revised
input: Δ, Γ
local variables: X_s, Y_s %X_s : set of elements which may be in Θ,
% Y_s : set of elements which may not be in Θ
output: $\Theta = \lim_{s\to\infty} X_s$

$X_0 = \Delta, Y_0 = \emptyset.$ % initialization
For each s, do
begin
 If there is no i satisfying
  $\Delta, X_s \upharpoonright i \vdash_{s+1} \neg A_i \& A_i \in X_s$
 do
  $X_{s+1} = X_s \cup \{A_{s+1}\}$
  $Y_{s+1} = Y_s;$
 Otherwise, let i be the least satisfying
 $\Delta, X_s \upharpoonright i \vdash_{s+1} \neg A_i \& A_i \in X_s$
 do
  $X_{s+1} = X_s \upharpoonright i \cup \{A_{i+1}, ..., A_{s+1}\}$ %$X_s \upharpoonright i = \{A_{i'} \in X_s : i' < i\}$
  $Y_{s+1} = Y_s \upharpoonright i \cup \{A_i\}$ %$Y_s \upharpoonright i = \{A_{i'} \in Y_s : i' < i\}$.
end

Let $\Theta = \lim_{s\to\infty} X_s$. Then, $\Delta|\Gamma \Rightarrow \Theta$ is provable in R-calculus.

9.3.4 *Approximate R-Calculus* $\mathbf{F}^{\text{rec}}$

Let $\Gamma_{i,s}$ be the formulas in $\Gamma_i = \{A_0, ..., A_{i-1}\}$ which are not deleted at the end of stage s and for each $j < i$, $A_j \in \Gamma_{i,s}$ iff $\Delta, \Gamma_{i,s} \nvdash \neg A_j$.

The approximate R-calculus $\mathbf{F}^{\text{rec}}$ consists of the following two sets of rules: one set is not to eliminate formulas in Γ, and another is to eliminate.

- **Axiom:**

$$(F^{\mathrm{con,rec}})\ \frac{\Delta, \Gamma_{i,s} \nvdash_s \neg p}{\Delta, \Gamma_{i,s}|p, \Gamma_{i+2,s} \Rightarrow_i \Delta, \Gamma_{i,s}, p|\Gamma_{i+2,s+1}};$$

- **Deduction rules:**

$$(F_{\wedge,\mathrm{rec}})\ \frac{\begin{array}{l}\Delta, \Gamma_{i,s}|A_{i1}, \Gamma_{i+2,s} \Rightarrow \Delta, \Gamma_{i,s}, A_{i1}|\Gamma_{i+2,s}\\ \Delta, \Gamma_{i,s}|A_{i2}, \Gamma_{i+2,s} \Rightarrow \Delta, \Gamma_{i,s}, A_{i2}|\Gamma_{i+2,s}\end{array}}{\Delta, \Gamma_{i,s}|A_{i1} \wedge A_{i2}, \Gamma_{i+2,s} \Rightarrow \Delta, \Gamma_{i+1,s+1}|\Gamma_{i+2,s+1}},$$

$$(F^1_{\vee,\mathrm{rec}})\ \frac{\Delta, \Gamma_{i,s}|A_{i1}, \Gamma_{i+2,s} \Rightarrow \Delta, \Gamma_{i,s}, A_{i1}|\Gamma_{i+2,s}}{\Delta, \Gamma_{i,s}|A_{i1} \vee A_{i2}, \Gamma_{i+2,s} \Rightarrow \Delta, \Gamma_{i+1,s+1}|\Gamma_{i+2,s+1}};$$

$$(F^2_{\vee,\mathrm{rec}})\ \frac{\Delta, \Gamma_{i,s}|A_{i2}, \Gamma_{i+2,s} \Rightarrow \Delta, \Gamma_{i,s}, A_{i2}|\Gamma_{i+2,s}}{\Delta, \Gamma_{i,s}|A_{i1} \vee A_{i2}, \Gamma_{i+2,s} \Rightarrow \Delta, \Gamma_{i+1,s+1}|\Gamma_{i+2,s+1}};$$

$$(F_{\forall,\mathrm{rec}})\ \frac{\Delta, \Gamma_{i,s}|A_i(t), \Gamma_{i+2,s} \Rightarrow \Delta, \Gamma_{i,s}, A_i(t)|\Gamma_{i+2,s}}{\Delta, \Gamma_{i,s}|\forall x A_i(x), \Gamma_{i+2,s} \Rightarrow \Delta, \Gamma_{i+1,s+1}|\Gamma_{i+2,s+1}};$$

$$(F_{\exists,\mathrm{rec}})\ \frac{\Delta, \Gamma_{i,s}|A_i(x), \Gamma_{i+2,s} \Rightarrow \Delta, \Gamma_{i,s}, A_i(x)|\Gamma_{i+2,s}}{\Delta, \Gamma_{i,s}|\exists x A_i(x), \Gamma_{i+2,s} \Rightarrow \Delta, \Gamma_{i+1,s+1}|\Gamma_{i+2,s+1}};$$

where

$$\begin{array}{l}\Gamma_{i+1,s+1} = \Gamma_{i,s} \cup \{A_i\},\\ A_i = p|\neg p|A_{i1} \wedge A_{i2}|A_{i1} \vee A_{i2}|\forall x A_i(x)|\exists x A_i(x),\end{array}$$

and

- **Axiom:**

$$(F^{\neg,\mathrm{rec}})\ \frac{\Delta, \Gamma_{i,s} \vdash_s \neg p}{\Delta, \Gamma_{i,s}|p, \Gamma_{i+2,s} \Rightarrow \Delta, \Gamma_{i,s+1}|\Gamma'_{i+2,s+1}};$$

- **Deduction rules:**

$$(F_1^{\wedge,\mathrm{rec}})\ \frac{\Delta, \Gamma_{i,s}|A_{i1}, \Gamma_{i+1,s} \Rightarrow \Delta, \Gamma_{i,s}|\Gamma'_{i+2,s}}{\Delta, \Gamma_{i,s}|A_{i1} \wedge A_{i2}, \Gamma_{i+2,s} \Rightarrow \Delta, \Gamma_{i+1,s+1}|\Gamma'_{i+2,s+1}};$$

$$(F_2^{\wedge,\mathrm{rec}})\ \frac{\begin{array}{l}\Delta, \Gamma_{i,s}|A_{i1}, \Gamma_{i+2,s} \Rightarrow \Delta, \Gamma_{i,s}, A_{i1}|\Gamma_{i+2,s}\\ \Delta, \Gamma_{i,s}, A_{i1}|A_{i2}, \Gamma_{i+2,s} \Rightarrow \Delta, \Gamma_{i,s}, A_{i1}|\Gamma'_{i+2,s}\end{array}}{\Delta, \Gamma_{i,s}|A_{i1} \wedge A_{i2}, \Gamma_{i+2,s} \Rightarrow \Delta, \Gamma_{i+1,s+1}|\Gamma'_{i+2,s}};$$

$$(F^{\vee,\mathrm{rec}})\ \frac{\begin{array}{l}\Delta, \Gamma_{i,s}|A_{i1}, \Gamma_{i+2,s} \Rightarrow \Delta, \Gamma_{i,s}|\Gamma'_{i+2,s}\\ \Delta, \Gamma_{i,s}|A_{i2}, \Gamma_{i+2,s} \Rightarrow \Delta, \Gamma_{i,s}|\Gamma'_{i+2,s}\end{array}}{\Delta, \Gamma_{i,s}|A_{i1} \vee A_{i2}, \Gamma_{i+2,s} \Rightarrow \Delta, \Gamma_{i+1,s+1}|\Gamma'_{i+2,s+1}};$$

$$(F^{\forall,\mathrm{rec}})\ \frac{\Delta, \Gamma_{i,s}|A_i(t), \Gamma_{i+1,s} \Rightarrow \Delta, \Gamma_{i,s}|\Gamma'_{i+2,s}}{\Delta, \Gamma_{i,s}|\forall x A_i(x), \Gamma_{i+2,s} \Rightarrow \Delta, \Gamma_{i+1,s+1}|\Gamma'_{i+2,s+1}};$$

$$(F^{\exists,\mathrm{rec}})\ \frac{\Delta, \Gamma_{i,s}|A_i(x), \Gamma_{i+1,s} \Rightarrow \Delta, \Gamma_{i,s}|\Gamma'_{i+2,s}}{\Delta, \Gamma_{i,s}|\exists x A_i(x), \Gamma_{i+2,s} \Rightarrow \Delta, \Gamma_{i+1,s+1}|\Gamma'_{i+2,s+1}};$$

where $\Delta, \Gamma_{i,s}$ is s-consistent; and

$$\Gamma_{i+1,s+1} = \begin{cases} \Gamma_{i,s} \cup \{A_i\} & \text{if } \Gamma_{i,s} \nvdash_s \neg A_i \\ \Gamma_{i,s} & \text{otherwise.} \end{cases}$$

By soundness and completeness theorem of R-calculus **F**, for any $i \in \omega$, there is a stage s_i such that

$$\Gamma_{i,s_i} = \lim_{s\to\infty} \Gamma_{i,s},$$

and $\Delta|\Gamma \Rightarrow \Sigma$ is provable in $\mathbf{F}^{\mathrm{rec}}$, where

$$\Sigma = \Delta \cup \bigcup_i \Gamma_{i,s_i}.$$

Definition 9.3.10 $\Delta|\Gamma \Rightarrow \Delta, \Theta$ is provable in $\mathbf{F}^{\mathrm{rec}}$, denoted by $\vdash_{\mathbf{F}^{\mathrm{rec}}} \Delta|\Gamma \Rightarrow \Delta, \Theta$, if there is a sequence $\Delta_1|\Gamma_1 \Rightarrow \Delta_2|\Gamma_2, ..., \Delta_{n-1}|\Gamma_{n-1} \Rightarrow \Delta_n|\Gamma_n, ...$ such that $\Delta_1|\Gamma_1 = \Delta|\Gamma$, $\Delta \cup \Theta = \lim_{n\to\infty} \Delta_n$, and for each $j < n$, $\Delta_j|\Gamma_j \Rightarrow \Delta_{j+1}|\Gamma_{j+1}$ is an axiom or is deduced from the previous statements by the deduction rules in $\mathbf{F}^{\mathrm{rec}}$.

Therefore, we have

Theorem 9.3.11 *Given theories* Γ, Δ *and* Θ,

$$\vdash_{\mathbf{F}^{\mathrm{rec}}} \Delta|\Gamma \Rightarrow \Sigma;$$

and conversely, if $\vdash_{\mathbf{F}^{\mathrm{rec}}} \Delta|\Gamma \Rightarrow \Delta, \Theta$ *then* $\Theta \cup \Delta = \Sigma$. □

9.4 Default Logic and Priority Method

Similarly we will consider default logic and the finite injury priority method. The difference between R-calculus and default logic with the priority method is that

• in R-calculus given a formula $A \in \Gamma$, either A is put in Θ eventually or never be put in Θ anymore; and A being put in Θ only is injured by some formula B before A in an ordering of Γ; and

• in default logic, a default $A \rightsquigarrow B$ inactive at some stage may become active after some default with lower priority puts some element in an extension. To consider such a case, the priority method may become the infinite injury priority method.

9.4.1 Construction of an Extension Without Injury

Let $D = \{\delta_0, \delta_1, ...\}$ be a set of defaults. We will construct in stages a sequence $\{\Theta_i : i \in \omega\}$ of theories such that $\Theta_0 = \Delta$, and $\Theta = \bigcup_i \Theta_i$ is a pseudo-extension of (Δ, D).

It suffices to meet for each e the following requirements:

$$P_e : \Theta \vdash A_e \& \Theta \nvdash \neg B_e \Rightarrow \Theta \vdash B_e,$$
$$N_e : \Theta_e \text{ is consistent},$$

where $\delta_e = A_e \rightsquigarrow B_e$.

Define

$$\Theta_s \upharpoonright e = \{B_{e'} \in \Theta_s : e' < e\}.$$

The priority ranking of requirements are

$$P_0, N_0, P_1, N_1, ..., P_e, N_e, ...$$

A requirement P_e requires attention at stage $s + 1$ if $\Theta_s \vdash_s A_e$, $\Theta_s \lceil e \nvdash_s \neg B_e$ and $\Theta_s \lceil e \nvdash_s B_e$,

A requirement P_e is satisfied at stage $s + 1$ if $\Theta_s \vdash_s A_e$ and $\Theta_s \nvdash_s \neg B_e$ imply $\Theta_s \vdash_s B_e$.

The construction:

```
Stage s = 0. Define Θ0 = Δ.
Stage s + 1. Find the least e ≤ s such that Pe requires attention.
Set Θs+1 = Θs ∪ {Be}. We say that Pe receives attention.
Define
```

$$\Theta = \lim_{s \to \infty} \Theta_s.$$

```
This ends the construction.
```

Lemma 9.4.1 *For each e, if $\Theta \vdash A_e$ and $\Theta \nvdash B_e$ then there is a stage s_e at which P_e is satisfied.*

Proof Assume that $\Theta \vdash A_e$ and $\Theta \nvdash \neg B_e$. There is a stage s_e such that P_e requires attention at stage $s \geq s_e$, $\Theta_{s_e+1} \vdash_{s_e+1} B_e$, and P_e is satisfied, and for any $t \geq s_e$, P_e never require attention. That is, P_e is eventually satisfied. □

Lemma 9.4.2 *Θ is an extension of (Δ, D).*

Proof By Lemma 9.4.1, each positive requirement P_e is satisfied. Θ is an extension of (Δ, D), because for any $\delta = A \rightsquigarrow B \in D$, if $\Theta \vdash A$ and $\Theta \nvdash \neg B$ then there is a stage s such that each P_e with higher priority than δ is satisfied eventually, $\Theta \vdash_s A$, and P_δ receives attention at stage $s + 1$. That is, $B \in \Theta_{s+1}$, and for any $t \geq s$, $B \in \Theta_{t+1}$, i.e., $B \in \Theta$. □

9.4.2 Construction of a Strong Extension with Finite Injury Priority Method

We first consider an example.

Example 9.4.3 Let $\Delta = \{p, r\}$ and $D = \{\frac{s:q}{q}, \frac{p:\neg q}{\neg q}, \frac{r:s}{s}\}$. Assume that

$$\frac{s:q}{q} \prec \frac{p:\neg q}{\neg q} \prec \frac{r:s}{s},$$

that is, $q \prec \neg q \prec s$. Then, traditionally we have

$$\begin{aligned} & p, r|\frac{s:q}{q}, \frac{p:\neg q}{\neg q}*, \frac{r:s}{s} \\ \Rightarrow\ & p, r, \neg q|\frac{s:q}{q}, \frac{p:\neg q}{\neg q}, \frac{r:s}{s}* \\ \Rightarrow\ & p, r, \neg q, s|\frac{s:q}{q}*, \frac{p:\neg q}{\neg q}, \frac{r:s}{s}; \end{aligned}$$

and $\{p, r, \neg q, s\}$ is a pseudo-extension, where $*$ marks the default active at the current stage.

We hope to have

$$\begin{aligned} & p, r|\frac{s:q}{q}, \frac{p:\neg q}{\neg q}*, \frac{r:s}{s} \\ \Rightarrow\ & p, r, \neg q|\frac{s:q}{q}, \frac{p:\neg q}{\neg q}, \frac{r:s}{s}* \\ \Rightarrow\ & p, r, \neg q, s|\frac{s:q}{q}*, \frac{p:\neg q}{\neg q}, \frac{r:s}{s} \\ \Rightarrow\ & p, r, q, s|\frac{s:q}{q}*, \frac{p:\neg q}{\neg q}, \frac{r:s}{s}. \end{aligned}$$

and $\{p, r, \neg q, s\}$ is a pseudo-extension. We think that under the ordering $\prec$, $\{p, r, q, s\}$ is better than $\{p, r, \neg q, s\}$, because $\{p, r, q, s\} \prec \{p, r, \neg q, s\}$.

Definition 9.4.4 Given an ordering $\prec$ on D, an extension S of default theory (Δ, D) is strong if S is an extension of (Δ, D) and for any other extension E of (Δ, D), there is an e such that $B_e \in S - E$, and for any $e' < e$, $B_{e'} \in E$ iff $B_{e'} \in S$.

Let $D = \{\delta_0, \delta_1, ...\}$. We will construct in stages a sequence $\{\Theta_i : i \in \omega\}$ of theories such that $\Theta_0 = \Delta$, and $\Theta = \bigcup_i \Theta_i$ is a strong extension of (Δ, D).

It suffices to meet for each e the following requirements:

$$\begin{aligned} & P_e : \Theta \vdash A_e \& \Theta \nvdash \neg B_e \Rightarrow \Theta \vdash B_e, \\ & N_e : \Theta_e \text{ is consistent,} \end{aligned}$$

where $\delta_e = A_e \rightsquigarrow B_e$.

The priority ranking of requirements are

$$P_0, N_0, P_1, N_1, ..., P_e, N_e, ...$$

A requirement P_e requires attention at stage $s+1$ if there are $e_1, ..., e_k > e$ such that

(i) $\Theta_s - \{B_{e_1}, ..., B_{e_k}\} \vdash_{s+1} A_e$, $\Theta_s - \{B_{e_1}, ..., B_{e_k}\} \nvdash_{s+1} \neg B_e$ and $\Theta_s - \{B_{e_1}, ..., B_{e_k}\} \nvdash_{s+1} B_e$, and

(ii) for each $k' \leq k$, $\Theta_s \cup \{B_e\} - \{B_{e_1}, ..., B_{e_k}\} \vdash_{s+1} \neg B_{e_{k'}}$.

A requirement P_e is satisfied at stage $s+1$ if $\Theta_s \vdash_s A_e$, $\Theta_s \nvdash_s \neg B_e$ and $\Theta_s \vdash_s B_e$.

The construction:

Stage $s = 0$. Define $\Theta_0 = \Delta$.

Stage $s+1$. Find the least $e \leq s$ such that P_e requires attention. Define $\Theta_{s+1} = (\Theta_s \cup \{B_e\}) - \{B_{e_1}, ..., B_{e_k}\}$, and we say that P_e receives attention.

Define

$$\Theta = \lim_{s\to\infty} \Theta_s.$$

This ends the construction.

We say that P_e is injured at stage $s+1$ if $B_e \in \Theta_s - \Theta_{s+1}$.

Define the injury set of P_e:

$$I_e = \{s : \exists i(B_i \in \Theta_s - \Theta_{s+1} \& A_e \in \Theta_{s+1} - \Theta_s)\}.$$

Lemma 9.4.5 *I_e is finite.*

Proof By definition of requiring attention, for any s, if $s \in I_e$ then there is an $i < e$ such that $B_i \in \Theta_s - \Theta_{s+1}$. □

Lemma 9.4.6 *For each e, there is a stage s_e such that for any $s \geq s_e$, if P_e requires attention at $s+1$ then P_e is satisfied eventually.*

Proof Assume that $\Theta \vdash A_e$ and $\Theta \nvdash \neg B_e$. By Lemma 9.4.5, there is a stage s_e such that $P_{e'}$ for no $e' < e$ requires attention after s_e. Then, P_e requires attention at stage $s \geq s_e$ such that $\Theta \vdash_s A_e$, and P_e is satisfied, and for any $t \geq s$, P_e never require attention. That is, P_e is eventually satisfied. □

Lemma 9.4.7 *Θ is an extension of (Δ, D).*

Proof By Lemma 9.4.6, each positive requirement P_e is satisfied. Θ is an extension of (Δ, D), because for any $\delta = A \rightsquigarrow B \in D$, if $\Theta \vdash A$ and $\Theta \nvdash \neg B$ then there is a stage s such that each P_e with higher priority than δ is satisfied eventually, $\Theta \vdash_s A$, and P_δ receives attention at stage $s+1$. That is, $B \in \Theta_{s+1}$, and for any $t \geq s$, $B \in \Theta_{t+1}$, i.e., $B \in \Theta$. □

Lemma 9.4.8 *Θ has the highest priority, that is, for any pseudo-extension E of (Δ, D), $\Theta \preceq E$, that is, there is a formula A such that $E[\prec A] = \Theta[\prec A]$ and $A \prec B$ for any $B \in E - E[\prec A]$, where $E[\prec A] = \{B \in E : B \prec A]$.*

Proof By the satisfaction of requirements in the construction. □

References

1. R.M. Friedberg, Two recursively enumerable sets of incomparable degrees of unsolvability. Proc. Natl. Acad. Sci. **43**, 236–238 (1957)
2. M.L. Ginsberg (ed.), *Readings in Nonmonotonic Reasoning* (Morgan Kaufmann, San Francisco, 1987)
3. D.S. Hochbaum (ed.), *Approximation Algorithms for NP-Hard Problems* (PWS Publishing Company, 1997)
4. J. Horty, Defaults with priorities. J. Philos. Log. **36**, 367–413 (2007)
5. W. Li, R-calculus: an inference system for belief revision. Comput. J. **50**, 378–390 (2007)
6. W. Li, Y. Sui, The R-calculus and the finite injury priority method. J Comput. **12**, 127–134 (2017)
7. A.A. Muchnik, On the separability of recursively enumerable sets (in Russian). Dokl. Akad. Nauk SSSR, N.S. **109**, 29–32 (1956)
8. R. Reiter, A logic for default reasoning. Artif. Intell. **13**, 81–132 (1980)
9. H. Rogers, *Theory of Recursive Functions and Effective Computability* (The MIT Press, Cambridge, 1967)
10. R.I. Soare, *Recursively Enumerable Sets and Degrees, a Study of Computable Functions and Computably Generated Sets* (Springer, Berlin, 1987)

Chapter 10
An Application to Default Logic

Given a default theory [1, 2, 6] (Δ, D), where Δ is a theory in propositional logic and D is a set of defaults, we will define three kinds of the minimal change in default logic: a theory Θ is a $\subseteq$-/$\preceq$-/$\vdash_{\preceq}$-minimal change of D by Δ, where Θ is a $\subseteq$-minimal change of D by Δ if and only if $\mathrm{Th}(\Theta \cup \Delta)$ is an extension of (Δ, D).

$\preceq$-minimal change has the actual importance. For example, let (Δ, D) be a default theory, where Δ says that *someone is a human* (denoted by p) *and this man has no arms* (denoted by $\neg q$); and D contains one default $\frac{p : q \wedge r}{q \wedge r}$, which says that *a man defaultly has arms and legs*, where r denotes that *this man has legs*. By default logic, (Δ, D) has an extension $\{p\}$, that is, we know that this man is a human. Practically, we hope to deduce that this man is a human (p) and this man has legs. Formally,

$$(*) \qquad \text{from } p, \neg q, \frac{p : q \wedge r}{q \wedge r}, \text{ defaultly infer } r.$$

I.e.,*if a man has no arms then defaultly this man has legs.*

Also, $\vdash_{\preceq}$-minimal change has the actual importance. For example, let (Δ, D) be a default theory, where $\Delta = \{\neg p, \neg r\}$ says that *someone has not the left arm* (denoted by $\neg p$), *and this man has no legs* (denoted by $\neg r$); and D contains one default

$$\frac{: p \wedge (r \vee q)}{p \wedge (r \vee q)},$$

which says that *a man defaultly has the left arm and either the right arm or legs*, where q denotes that *this man has the right arm*. By default logic, (Δ, D) has an extension $\{\neg p, \neg r\}$. We hope to deduce that this man has the right arm [3–5, 7, 8]. Formally,

W. Li and Y. Sui, *R-CALCULUS: A Logic of Belief Revision*, Perspectives in Formal Induction, Revision and Evolution,
https://doi.org/10.1007/978-981-16-2944-0_10

$$(**) \qquad \text{from } \neg p, \neg r, \frac{: p \wedge (r \vee q)}{p \wedge (r \vee q)}, \text{ defaultly infer } q.$$

Given a default theory (Δ, D), we will given three calculi $\mathbf{S}^{\mathrm{D}}$, $\mathbf{T}^{\mathrm{D}}$ and $\mathbf{U}^{\mathrm{D}}$ and prove that $\mathbf{S}^{\mathrm{D}}$, $\mathbf{T}^{\mathrm{D}}$ and $\mathbf{U}^{\mathrm{D}}$ are sound and complete with respect to $\subseteq$-minimal change, $\preceq$-minimal change and $\vdash_{\preceq}$-minimal change, respectively. Hence, if Θ is a $\subseteq$-/$\preceq$-/$\vdash_{\preceq}$-minimal change of D by Δ, then $\Theta \cup \Delta$ can be taken as some kinds of extensions of (Δ, D).

10.1 Default Logic and Subset-Minimal Change

Given a default theory (Δ, D), Θ is a $\subseteq$-minimal change of D by Δ, denoted by $\models_{\mathrm{S}} \Delta|D \Rightarrow \Delta, \Theta$, if Θ is minimal such that

(i) Θ is consistent with Δ;
(ii) for any formula $B \in \Theta$, there is a default $A \rightsquigarrow B \in D$ such that $\Delta, \Theta \vdash A$; and
(iii) for any $A \rightsquigarrow B \in D$, either $B \in \Theta$, or $\Delta, \Theta \nvdash A$ or $\Delta \cup \Theta \cup \{B\}$ is inconsistent.

In this section, we will give a Gentzen deduction system $\mathbf{S}^{\mathrm{D}}$ such that for any default theory (Δ, D) and a theory Θ, $\Delta|D \Rightarrow \Delta, \Theta$ is provable in $\mathbf{S}^{\mathrm{D}}$ if and only if Θ is a $\subseteq$-minimal change of D by Δ.

10.1.1 Deduction System $\mathbf{S}^{\mathrm{D}}$ for a Default

Deduction system $\mathbf{S}^{\mathrm{D}}$ for a default $C \rightsquigarrow A$ consists of the following axioms and deduction rules:

- **Axioms:**

$$(S^{\mathbf{A}}) \ \frac{\begin{array}{l}\Delta \vdash C \\ \Delta \nvdash \neg l\end{array}}{\Delta|C \rightsquigarrow l \Rightarrow \Delta, l} \qquad (S_{\mathbf{A}}) \ \frac{\begin{array}{l}\Delta \vdash C \\ \Delta \vdash \neg l\end{array}}{\Delta|C \rightsquigarrow l \Rightarrow \Delta}$$

- **Deduction rules**:

$$(S^{\wedge}) \ \frac{\begin{array}{l}\Delta|C \rightsquigarrow A_1 \Rightarrow \Delta, A_1 \\ \Delta, A_1|C \rightsquigarrow A_2 \Rightarrow \Delta, A_1, A_2\end{array}}{\Delta|C \rightsquigarrow A_1 \wedge A_2 \Rightarrow \Delta, A_1 \wedge A_2|}$$

$$(S^1_{\wedge}) \ \frac{\Delta|C \rightsquigarrow A_1 \Rightarrow \Delta}{\Delta|C \rightsquigarrow A_1 \wedge A_2 \Rightarrow \Delta}$$

$$(S^2_{\wedge}) \ \frac{\Delta, A_1|C \rightsquigarrow A_2 \Rightarrow \Delta, A_1}{\Delta|C \rightsquigarrow A_1 \wedge A_2 \Rightarrow \Delta}$$

$$(S^{\vee}_1) \ \frac{\Delta|C \rightsquigarrow A_1 \Rightarrow \Delta, A_1}{\Delta|C \rightsquigarrow A_1 \vee A_2 \Rightarrow \Delta, A_1 \vee A_2}$$

$$(S^{\vee}_2) \ \frac{\Delta|C \rightsquigarrow A_2 \Rightarrow \Delta, A_2}{\Delta|C \rightsquigarrow A_1 \vee A_2 \Rightarrow \Delta, A_1 \vee A_2}$$

$$(S_{\vee}) \ \frac{\Delta|C \rightsquigarrow A_1 \Rightarrow \Delta \quad \Delta|C \rightsquigarrow A_2 \Rightarrow \Delta}{\Delta|C \rightsquigarrow A_1 \vee A_2 \Rightarrow \Delta}$$

where the rules of the left-hand side are to put formula A into Θ, and the ones of the right-hand side are not to.

Definition 10.1.1 $\Delta|C \leadsto A \Rightarrow \Delta, B$ is provable in $\mathbf{S}^{\mathrm{D}}$, denoted by $\vdash_{\mathbf{S}^{\mathrm{D}}} \Delta|C \leadsto A \Rightarrow \Delta, B$, if there is a sequence $S_1, ..., S_m$ of statements such that

$$\begin{aligned} S_1 &= \Delta|C \leadsto A_1 \Rightarrow \Delta|C \leadsto A'_1, \\ &\cdots \\ S_m &= \Delta|C \leadsto A_m \Rightarrow \Delta, A'_m; \\ A_1 &= A, \\ A'_m &= B; \end{aligned}$$

and for each $i < m$, S_{i+1} is an axiom or is deduced from the previous statements by a deduction rule in $\mathbf{S}^{\mathrm{D}}$.

Example 10.1.2 Let (Δ, D) be a default theory, where

$$\begin{aligned} \Delta &= \{\neg p, \neg r\}, \\ D &= \{\top \leadsto p \wedge (r \vee q)\}, \end{aligned}$$

where p says that someone has the left arm, r says that this man has legs, and the default says that the man defaultly has the left arm and either the right arm or legs. By default logic, (Δ, D) has an extension $\{\neg p, \neg r\}$. We have the following deduction:

$$\begin{aligned} &\neg p, \neg r|\top \leadsto p \Rightarrow \neg p, \neg r \\ &\neg p, \neg r|\top \leadsto p \wedge (r \vee q) \Rightarrow \neg p, \neg r. \end{aligned}$$

That is, $\emptyset$ is a $\subseteq$-minimal change of $\top \leadsto p \wedge (r \vee q)$ by $\neg p, \neg r$, i.e.,

$$\vdash_{\mathbf{S}^{\mathrm{D}}} \neg p, \neg r|\top \leadsto p \wedge (r \vee q) \Rightarrow \neg p, \neg r.$$

Theorem 10.1.3 (Completeness theorem) *For any consistent formula set Δ and a default $C \leadsto A$ such that $\Delta \vdash C$, if $\Delta \cup \{A\}$ is consistent then $\Delta|C \leadsto A \Rightarrow \Delta, A$ is provable in $\mathbf{S}^{\mathrm{D}}$, i.e.,*

$$\models_{\mathbf{S}^{\mathrm{D}}} \Delta|C \leadsto A \Rightarrow \Delta, A \textit{ implies } \vdash_{\mathbf{S}^{\mathrm{D}}} \Delta|C \leadsto A \Rightarrow \Delta, A;$$

and if $\Delta \cup \{A\}$ is inconsistent then $\Delta|C \leadsto A \Rightarrow \Delta$ is provable in $\mathbf{S}^{\mathrm{D}}$, i.e.,

$$\models_{\mathbf{S}^{\mathrm{D}}} \Delta|C \leadsto A \Rightarrow \Delta \textit{ implies } \vdash_{\mathbf{S}^{\mathrm{D}}} \Delta|C \leadsto A \Rightarrow \Delta.$$

Proof We prove the theorem by induction on the structure of formula A.

Assume that $\Delta \cup \{A\}$ is consistent.

If $A = l$ then $\Delta \nvdash \neg A$, and by $(S^{\mathbf{A}})$, $\Delta|C \leadsto A \Rightarrow \Delta, l$ is provable;

If $A = A_1 \wedge A_2$ then $\Delta \cup \{A_1\}$ and $\Delta \cup \{A_1, A_2\}$ are consistent, and by induction assumption, $\Delta|C \rightsquigarrow A_1 \Rightarrow \Delta, A_1$ is provable and $\Delta, A_1|C \rightsquigarrow A_2 \Rightarrow \Delta, A_1, A_2$ is provable, and by $(S^{\wedge})$, $\Delta|C \rightsquigarrow A_1 \wedge A_2 \Rightarrow \Delta, A_1 \wedge A_2$ is provable;

If $A = A_1 \vee A_2$ then either $\Delta \cup \{A_1\}$ or $\Delta \cup \{A_2\}$ is consistent, and by induction assumption, either $\Delta|C \rightsquigarrow A_1 \Rightarrow \Delta, A_1$ or $\Delta|C \rightsquigarrow A_2 \Rightarrow \Delta, A_2$ are provable, and by $(S_1^{\vee})$ or $(S_2^{\vee})$, $\Delta|C \rightsquigarrow A_1 \vee A_2 \Rightarrow \Delta, A_1 \vee A_2$ is provable.

Assume that $\Delta \cup \{A\}$ is inconsistent.

If $A = l$ then $\Delta \vdash \neg A$, and by $(S_{\mathbf{A}})$, $\Delta|C \rightsquigarrow A \Rightarrow \Delta$ is provable;

If $A = A_1 \wedge A_2$ then either $\Delta \cup \{A_1\}$ or $\Delta \cup \{A_1, A_2\}$ is inconsistent, and by induction assumption, either $\Delta|A_1 \Rightarrow \Delta$ is provable, or $\Delta, A_1|C \rightsquigarrow A_2 \Rightarrow \Delta, A_1|$ is provable, and by $(S_{\wedge}^1)$ and $(S_{\wedge}^2)$, $\Delta|C \rightsquigarrow A_1 \wedge A_2 \Rightarrow \Delta$ is provable;

If $A = A_1 \vee A_2$ then $\Delta \cup \{A_1\}$ and $\Delta \cup \{A_2\}$ are inconsistent, and by induction assumption, $\Delta|C \rightsquigarrow A_1 \Rightarrow \Delta$ and $\Delta|C \rightsquigarrow A_2 \Rightarrow \Delta$ are provable, and by $(S_{\vee})$, $\Delta|C \rightsquigarrow A_1 \vee A_2 \Rightarrow \Delta$ is provable. □

Theorem 10.1.4 (Soundness theorem) *For any consistent formula set* Δ *and a default* $C \rightsquigarrow A$ *such that* $\Delta \vdash C$, *if* $\Delta|C \rightsquigarrow A \Rightarrow \Delta, A$ *is provable in* $\mathbf{S}^{\mathrm{D}}$ *then* $\Delta \cup \{A\}$ *is consistent, i.e.,*

$$\vdash_{\mathbf{S}^{\mathrm{D}}} \Delta|C \rightsquigarrow A \Rightarrow \Delta, A \text{ implies } \models_{\mathbf{S}^{\mathrm{D}}} \Delta|C \rightsquigarrow A \Rightarrow \Delta, A;$$

and if $\Delta|C \rightsquigarrow A \Rightarrow \Delta$ *is provable in* $\mathbf{S}^{\mathrm{D}}$ *then* $\Delta \cup \{A\}$ *is inconsistent, i.e.,*

$$\vdash_{\mathbf{S}^{\mathrm{D}}} \Delta|C \rightsquigarrow A \Rightarrow \Delta \text{ implies } \models_{\mathbf{S}^{\mathrm{D}}} \Delta|C \rightsquigarrow A \Rightarrow \Delta.$$

Proof Given a theory Δ and a default $C \rightsquigarrow A$, assume that $\Delta \vdash C$. We prove the theorem by induction on the structure of formula A.

Assume that $\Delta|C \rightsquigarrow A \Rightarrow \Delta, A$ is $\mathbf{S}^{\mathrm{D}}$-provable.

If $A = l$ then $\Delta \nvdash \neg A$, and $\Delta \cup \{A\}$ is consistent;

If $A = A_1 \wedge A_2$ then $\Delta|C \rightsquigarrow A_1 \Rightarrow \Delta, A_1$ and $\Delta, A_1|C \rightsquigarrow A_2 \Rightarrow \Delta, A_1, A_2$ are $\mathbf{S}^{\mathrm{D}}$-provable. By induction assumption, $\Delta \cup \{A_1\}$ and $\Delta \cup \{A_1, A_2\}$ are consistent, and so is $\Delta \cup \{A_1 \wedge A_2\}$;

If $A = A_1 \vee A_2$ then either $\Delta|C \rightsquigarrow A_1 \Rightarrow \Delta, A_1$ or $\Delta|C \rightsquigarrow A_2 \Rightarrow \Delta, A_2$ is $\mathbf{S}^{\mathrm{D}}$-provable. By induction assumption, either $\Delta \cup \{A_1\}$ or $\Delta \cup \{A_2\}$ is consistent, and so is $\Delta \cup \{A_1 \vee A_2\}$.

Assume that $\Delta|A \Rightarrow \Delta$ is $\mathbf{S}^{\mathrm{D}}$-provable.

If $A = l$ then $\Delta \vdash \neg A$, and $\Delta \cup \{A\}$ is inconsistent;

If $A = A_1 \wedge A_2$ then either $\Delta|C \rightsquigarrow A_1 \Rightarrow \Delta$ or $\Delta, A_1|C \rightsquigarrow A_2 \Rightarrow \Delta, A_1$ is $\mathbf{S}^{\mathrm{D}}$-provable, and by induction assumption, either $\Delta \cup \{A_1\}$ or $\Delta \cup \{A_1, A_2\}$ is inconsistent. Hence, $\Delta \cup \{A_1 \wedge A_2\}$ is inconsistent;

If $A = A_1 \vee A_2$ then $\Delta|C \rightsquigarrow A_1 \Rightarrow \Delta$ and $\Delta|C \rightsquigarrow A_2 \Rightarrow \Delta$ are $\mathbf{S}^{\mathrm{D}}$-provable, and by induction assumption, $\Delta \cup \{A_1\}$ and $\Delta \cup \{A_2\}$ are inconsistent. Hence, $\Delta \cup \{A_1 \vee A_2\}$ is inconsistent. □

We have the following

Proposition 10.1.5 *There is no default $C \rightsquigarrow A$ such that*

$$\vdash_{\mathbf{S}^{\mathrm{D}}} \Delta | C \rightsquigarrow A \Rightarrow \Delta$$

and

$$\vdash_{\mathbf{S}^{\mathrm{D}}} \Delta | C \rightsquigarrow A \Rightarrow \Delta, A.$$

□

10.1.2 Deduction System $\mathbf{S}^{\mathrm{D}}$ *for a Set of Defaults*

Let $D = (C_1 \rightsquigarrow A_1, ..., C_n \rightsquigarrow A_n)$, i.e., set $\{C_1 \rightsquigarrow A_1, ..., C_n \rightsquigarrow A_n\}$ with an ordering $<$ such that $C_1 \rightsquigarrow A_1 < C_2 \rightsquigarrow A_2 < \cdots < C_n \rightsquigarrow A_n$.

Deduction system $\mathbf{S}^{\mathrm{D}}$ for a set D of defaults consists of the following axioms and deduction rules: assume that $D = D_1 \cup \{C \rightsquigarrow A\} \cup D_2$, and for each $C' \rightsquigarrow A' \in D_1, \Delta \nvdash C'$; and $\Delta \vdash C$.

- **Axioms:**

$$(S^{\mathbf{A}})\ \frac{\Delta \vdash C \quad \Delta \nvdash \neg l}{\Delta | D_1, C \rightsquigarrow l, D_2 \Rightarrow \Delta, l | D_1, D_2}$$

$$(S_{\mathbf{A}})\ \frac{\Delta \vdash C \quad \Delta \vdash \neg l}{\Delta | D_1, C \rightsquigarrow l, D_2 \Rightarrow \Delta | D_1, D_2}$$

- **Deduction rules**:

$$(S^{\wedge})\ \frac{\Delta | D_1, C \rightsquigarrow A_1, D_2 \Rightarrow \Delta, A_1 | D_1, D_2 \quad \Delta, A_1 | D_1, C \rightsquigarrow A_2, D_2 \Rightarrow \Delta, A_1, A_2 | D_1, D_2}{\Delta | D_1, C \rightsquigarrow A_1 \wedge A_2, D_2 \Rightarrow \Delta, A_1 \wedge A_2 | D_1, D_2}$$

$$(S^1_{\wedge})\ \frac{\Delta | D_1, C \rightsquigarrow A_1, D_2 \Rightarrow \Delta | D_1, D_2}{\Delta | D_1, C \rightsquigarrow A_1 \wedge A_2, D_2 \Rightarrow \Delta | D_1, D_2}$$

$$(S^2_{\wedge})\ \frac{\Delta, A_1 | D_1, C \rightsquigarrow A_2, D_2 \Rightarrow \Delta, A_1 | D_1, D_2}{\Delta | D_1, C \rightsquigarrow A_1 \wedge A_2, D_2 \Rightarrow \Delta | D_1, D_2}$$

$$(S^{\vee}_1)\ \frac{\Delta | D_1, C \rightsquigarrow A_1, D_2 \Rightarrow \Delta, A_1 | D_1, D_2}{\Delta | D_1, C \rightsquigarrow A_1 \vee A_2, D_2 \Rightarrow \Delta, A_1 \vee A_2 | D_1, D_2}$$

$$(S^{\vee}_2)\ \frac{\Delta | D_1, C \rightsquigarrow A_2, D_2 \Rightarrow \Delta, A_2 | D_1, D_2}{\Delta | D_1, C \rightsquigarrow A_1 \vee A_2, D_2 \Rightarrow \Delta, A_1 \vee A_2 | D_1, D_2}$$

$$(S_{\vee})\ \frac{\Delta | D_1, C \rightsquigarrow A_1, D_2 \Rightarrow \Delta | D_1, D_2 \quad \Delta | D_1, C \rightsquigarrow A_2, D_2 \Rightarrow \Delta | D_1, D_2}{\Delta | D_1, C \rightsquigarrow A_1 \vee A_2, D_2 \Rightarrow \Delta | D_1, D_2}$$

Definition 10.1.6 $\Delta|D \Rightarrow \Delta, \Theta$ is provable in $\mathbf{S}^{\mathrm{D}}$, denoted by $\vdash_{\mathbf{S}^{\mathrm{D}}} \Delta|D \Rightarrow \Delta, \Theta$, if there is a sequence S_1, ..., S_m of statements such that

$$S_1 = \Delta|D \Rightarrow \Delta_1|D_1,$$
$$\ldots$$
$$S_m = \Delta_{m-1}|D_{m-1} \Rightarrow \Delta_m|D_m = \Delta, \Theta$$

and for each $i < m$, S_{i+1} is an axiom or is deduced from the previous statements by a deduction rule in $\mathbf{S}^{\mathrm{D}}$.

Theorem 10.1.7 (Soundness theorem) *For any consistent formula sets Θ, Δ and any finite default set D, if $\Delta|D \Rightarrow \Delta, \Theta$ is provable in $\mathbf{S}^{\mathrm{D}}$ then Θ is a $\subseteq$-minimal change of D by Δ. That is,*

$$\vdash_{\mathbf{S}^{\mathrm{D}}} \Delta|D \Rightarrow \Delta, \Theta \textit{ implies } \models_{\mathbf{S}^{\mathrm{D}}} \Delta|D \Rightarrow \Delta, \Theta.$$

Proof We prove the theorem by induction on n.

Assume that $\Delta|D \Rightarrow \Delta, \Theta$ is provable in $\mathbf{S}^{\mathrm{D}}$.

Let $n = 1$. Then, either $\Theta = A_1$ or $\Theta = \lambda$. If $\Theta = A_1$ then $\Delta \cup \{A_1\}$ is consistent, and Θ is a $\subseteq$-minimal change of $C \rightsquigarrow A_1$ by Δ; otherwise, $\Delta \cup \{A_1\}$ is inconsistent, and $\Theta = \lambda$ is a $\subseteq$-minimal change of A_1 by Δ.

Assume that the theorem holds for n, that is, if $\Delta|D \Rightarrow \Delta, \Theta$ then Θ is a $\subseteq$-minimal change of D by Δ.

Let $D' = (D, A_{n+1}) = (A_1, ..., A_{n+1})$. Then, if $\Delta|D' \Rightarrow \Delta, \Theta'$ is provable then $\Delta|D \Rightarrow \Delta, \Theta$ and $\Delta, \Theta|C_{n+1} \rightsquigarrow A_{n+1} \Rightarrow \Delta, \Theta'$ are provable. By the case $n = 1$ and the induction assumption, Θ' is a $\subseteq$-minimal change of A_{n+1} by $\Delta \cup \Theta$, and Θ is a $\subseteq$-minimal change of D by Δ, therefore, Θ' is a $\subseteq$-minimal change of D' by Δ. □

Theorem 10.1.8 (Completeness Theorem) *For any consistent formula sets Θ, Δ and any finite default set D, if Θ is a $\subseteq$-minimal change of D by Δ then $\Delta|D \Rightarrow \Delta, \Theta$ is provable in $\mathbf{S}^{\mathrm{D}}$. That is,*

$$\models_{\mathbf{S}^{\mathrm{D}}} \Delta|D \Rightarrow \Delta, \Theta \textit{ implies } \vdash_{\mathbf{S}^{\mathrm{D}}} \Delta|D \Rightarrow \Delta, \Theta.$$

Proof Assume that Θ is a $\subseteq$-minimal change of D by Δ. Then, there is an ordering $<$ of D such that $D = (C_1 \rightsquigarrow A_1, C_2 \rightsquigarrow A_2, ..., C_n \rightsquigarrow A_n)$, where $C_1 \rightsquigarrow A_1 < C_2 \rightsquigarrow A_2 < \cdots < C_n \rightsquigarrow A_n$, and Θ is a maximal subset of D such that $\Delta \cup \Theta$ is consistent.

We prove the theorem by induction on n.

Let $n = 1$. By the last theorem, if $\Theta = \{A_1\}$ then $\Delta|C_1 \rightsquigarrow A_1 \Rightarrow \Delta, A_1$ is provable; and if $\Theta = \emptyset$ then $\Delta|C_1 \rightsquigarrow A_1 \Rightarrow \Delta, \Delta$ is provable.

Assume that the theorem holds for n, that is, if Θ is a $\subseteq$-minimal change of D by Δ then $\Delta|D \Rightarrow \Delta, \Theta$ is provable.

Let $D' = (D, A_{n+1}) = (A_1, ..., A_{n+1})$ and Θ' is a $\subseteq$-minimal change of D' by Δ. Then, Θ' is a $\subseteq$-minimal change of $C_{n+1} \rightsquigarrow A_{n+1}$ by $\Delta \cup \Theta$, and $\Delta, \Theta | C_{n+1} \rightsquigarrow A_{n+1} \Rightarrow \Delta, \Theta'$ is provable. By the induction assumption, $\Delta | D \Rightarrow \Delta, \Theta$ is provable and so is $\Delta | D' \Rightarrow \Delta, \Theta | C_{n+1} \rightsquigarrow A_{n+1}$, and hence, $\Delta | D' \Rightarrow \Delta, \Theta'$ is provable in $\mathbf{S}^{\mathrm{D}}$. □

Theorem 10.1.9 *For any consistent formula sets* Θ, Δ *and any finite default set* D, *if* Θ *is a* $\subseteq$*-minimal change of* D *by* Δ *then* $\mathrm{Th}(\Theta \cup \Delta)$ *is an extension of* (Δ, D). □

Hence, if Θ is a $\subseteq$-minimal change of D by Δ then Θ is a pseudo-extension of default theory (Δ, D).

10.2 Default Logic and Pseudo-subformula-minimal Change

Definition 10.2.1 Given a default theory (Δ, D), a theory Θ is a $\preceq$-minimal change of D by Δ, denoted by $\models_{\mathrm{T}} \Delta | D \Rightarrow \Delta, \Theta$, if Θ is minimal such that

(i) $\Theta \cup \Delta$ is consistent;
(ii) $\Theta \preceq D$, that is, for each formula $B \in \Theta$, there is a default $C \rightsquigarrow A \in D$ such that $\Delta \cup \Theta \vdash C$ and $B \preceq A$, and
(iii) for any theory Ξ with $\Theta \prec \Xi \preceq \Gamma$, $\Xi \cup \Delta$ is inconsistent, where $\Gamma = \{A : C \rightsquigarrow A \in D, \Delta \cup \Theta \vdash C\}$.

In this section, we will give a Gentzen deduction system $\mathbf{T}^{\mathrm{D}}$ such that for any default theory (Δ, D) and theory Θ, $\Delta | D \Rightarrow \Theta, \Delta$ is provable in $\mathbf{T}^{\mathrm{D}}$ if and only if Θ is a $\subseteq$-minimal change of D by Δ.

10.2.1 Deduction System $\mathbf{T}^{\mathrm{D}}$ *for a Default*

Deduction system $\mathbf{T}^{\mathrm{D}}$ for a default $C \rightsquigarrow A$ consists of the following axioms and deduction rules:

- **Axioms:**

$$(T^{\mathrm{A}}) \frac{\begin{array}{l}\Delta \vdash C \\ \Delta \nvdash \neg A\end{array}}{\Delta | C \rightsquigarrow A \Rightarrow \Delta, A} \qquad (T_{\mathrm{A}}) \frac{\begin{array}{l}\Delta \vdash C \\ \Delta \vdash \neg l\end{array}}{\Delta | C \rightsquigarrow l \Rightarrow \Delta, \lambda}$$

• **Deduction rules**:

$$(T^{\wedge})\ \frac{\Delta|C \rightsquigarrow A_1 \Rightarrow \Delta, B_1}{\Delta|C \rightsquigarrow A_1 \wedge A_2 \Rightarrow \Delta, B_1|C \rightsquigarrow A_2}$$

$$(T_1^{\vee})\ \frac{\Delta|C \rightsquigarrow A_1 \Rightarrow \Delta, B_1 \neq \lambda}{\Delta|C \rightsquigarrow A_1 \vee A_2 \Rightarrow \Delta, C_1 \vee A_2}$$

$$(T_2^{\vee})\ \frac{\begin{array}{l}\Delta|C \rightsquigarrow A_1 \Rightarrow \Delta, \lambda\\ \Delta|C \rightsquigarrow A_2 \Rightarrow \Delta, B_2\\ B_2 \neq \lambda\end{array}}{\Delta|C \rightsquigarrow A_1 \vee A_2 \Rightarrow \Delta, A_1 \vee B_2}$$

$$(T_3^{\vee})\ \frac{\Delta|C \rightsquigarrow A_1 \Rightarrow \Delta, \lambda\ \ \Delta|C \rightsquigarrow A_2 \Rightarrow \Delta, \lambda}{\Delta|C \rightsquigarrow A_1 \vee A_2 \Rightarrow \Delta, \lambda}$$

We assume that if B is consistent then

$$\lambda \vee B \equiv B \vee \lambda \equiv B; \ \ \lambda \wedge B \equiv B \wedge \lambda \equiv B; \ \ \Delta, \lambda \equiv \Delta$$

and if B is inconsistent then

$$\lambda \vee B \equiv B \vee \lambda \equiv \lambda; \ \ \lambda \wedge B \equiv B \wedge \lambda \equiv \lambda$$

Definition 10.2.2 $\Delta|C \rightsquigarrow A \Rightarrow \Delta, B$ is provable in $\mathbf{T}^{\mathrm{D}}$, denoted by $\vdash_{\mathbf{T}^{\mathrm{D}}} \Delta|C \rightsquigarrow A \Rightarrow \Delta, B$, if there is a sequence $S_1, ..., S_m$ of statements such that

$$\begin{array}{l} S_1 = \Delta|C_1 \rightsquigarrow A_1 \Rightarrow \Delta|C_1' \rightsquigarrow A_1', \\ \ldots \\ S_m = \Delta|C_m \rightsquigarrow A_m \Rightarrow \Delta, A_m'; \\ A_1 = A, \\ A_m' = B; \end{array}$$

and for each $i < m$, S_{i+1} is an axiom or is deduced from the previous statements by a deduction rule in $\mathbf{T}^{\mathrm{D}}$.

For Example 10.1.2, we have different result in $\mathbf{T}^{\mathrm{D}}$.

Example 10.2.3 Let (Δ, D) be a default theory, where

$$\begin{array}{l} \Delta = \{\neg p, \neg r\}, \\ D = \{\top \rightsquigarrow p \wedge (r \vee q)\}, \end{array}$$

In $\mathbf{T}^{\mathrm{D}}$, we have the following deduction:

$$\begin{array}{l} \neg p, \neg r|\top \rightsquigarrow p \Rightarrow \neg p, \neg r \\ \neg p, \neg r, \lambda|\top \rightsquigarrow r \Rightarrow \neg p, \neg r \\ \neg p, \neg r, \lambda|\top \rightsquigarrow q \Rightarrow \neg p, \neg r, q \\ \neg p, \neg r, \lambda|\top \rightsquigarrow r \vee q \Rightarrow \neg p, \neg r, r \vee q \\ \neg p, \neg r|\top \rightsquigarrow p \wedge (r \vee q) \Rightarrow \neg p, \neg r, r \vee q. \end{array}$$

That is, $\{r \vee q\}$ is a $\preceq$-minimal change of $\top \rightsquigarrow p \wedge (r \vee q)$ by $\neg p, \neg r$, i.e.,

$$\vdash_{\mathbf{T}^{\mathrm{D}}} \neg p, \neg r | \top \rightsquigarrow p \wedge (r \vee q) \Rightarrow \neg p, \neg r, r \vee q.$$

That means that the man has either legs or the right arm. □

Theorem 10.2.4 *For any consistent formula set Δ and default $C \rightsquigarrow A$ such that $\Delta \vdash C$, there is a formula B such that $B \preceq A$ and $\Delta | C \rightsquigarrow A \Rightarrow \Delta, B$ is provable.*

Proof We prove the theorem by induction on the structure of formula A.

Case $A = l$. Then, if $\Delta \vdash \neg l$ then let $B = \lambda$; if $\Delta \nvdash \neg l$ then let $B = l$. Then, by $(T^{\mathbf{A}})$ and $(T_{\mathbf{A}})$, $\Delta | C \rightsquigarrow A \Rightarrow \Delta, B$ is provable.

Case $A = A_1 \wedge A_2$. Then, either $\Delta \vdash \neg A_1$ or $\Delta, \neg A_1 \vdash \neg A_2$. By induction assumption, there are formulas $B_1 \preceq A_1, B_2 \preceq A_2$ such that either $\Delta | C \rightsquigarrow A_1 \Rightarrow \Delta, B_1$ or $\Delta, B_1 | C \rightsquigarrow A_2 \Rightarrow \Delta, B_1, B_2$ is provable in $\mathbf{T}^{\mathrm{D}}$. Hence, so is $\Delta | C \rightsquigarrow A_1 \wedge A_2 \Rightarrow \Delta, B_1 \wedge B_2$.

Case $A = A_1 \vee A_2$. Then, by induction assumption, there are formulas $B_1 \preceq A_1$ and $B_2 \preceq A_2$ such that $\Delta | C \rightsquigarrow A_1 \Rightarrow \Delta, B_1$ and $\Delta | C \rightsquigarrow A_2 \Rightarrow \Delta, B_2$ are provable. If $B_1 \neq \lambda$ then by $(T_1^{\vee})$, $\vdash_{\mathbf{T}^{\mathrm{D}}} \Delta | C \rightsquigarrow A_1 \vee A_2 \Rightarrow \Delta, B_1 \vee A_2$; if $B_1 = \lambda$ and $B_2 \neq \lambda$ then by $(T_2^{\vee})$, $\vdash_{\mathbf{T}^{\mathrm{D}}} \Delta | C \rightsquigarrow A_1 \vee A_2 \Rightarrow A_1 \vee B_2$; if $B_1 = B_2 = \lambda$ then by $(T_3^{\vee})$, $\vdash_{\mathbf{T}^{\mathrm{D}}} \Delta | C \rightsquigarrow A_1 \vee A_2 \Rightarrow \Delta$. Let

$$B = \begin{cases} B_1 \vee A_2 \text{ if } B_1 \neq \lambda \\ A_1 \vee B_2 \text{ if } B_1 = \lambda \neq B_2 \\ \lambda \qquad\quad \text{otherwise,} \end{cases}$$

and $\vdash_{\mathbf{T}^{\mathrm{D}}} \Delta | C \rightsquigarrow A_1 \vee A_2 \Rightarrow \Delta, B$. □

Lemma 10.2.5 *If Θ is a $\preceq$-minimal change of D by Δ and Θ' is a $\preceq$-minimal change of $C_{n+1} \rightsquigarrow A_{n+1}$ by $\Delta \cup \Theta$ then Θ' is a $\preceq$-minimal change of $D \cup \{C_{n+1} \rightsquigarrow A_{n+1}\}$ by Δ.*

Proof Assume that Θ is a $\preceq$-minimal change of D by Δ and Θ' is a $\preceq$-minimal change of $C_{n+1} \rightsquigarrow A_{n+1}$ by $\Delta \cup \Theta$. Then, by the definition, Θ' is a $\preceq$-minimal change of $D \cup \{C_{n+1} \rightsquigarrow A_{n+1}\}$ by Δ. □

Lemma 10.2.6 *If B_1 is a $\preceq$-minimal change of $C \rightsquigarrow A_1$ by Δ and B_2 is a $\preceq$-minimal change of $C \rightsquigarrow A_2$ by $\Delta \cup \{B_1\}$ then $B_1 \wedge B_2$ is a $\preceq$-minimal change of $C \rightsquigarrow A_1 \wedge A_2$ by Δ.*

Proof Assume that B_1 is a $\preceq$-minimal changes of $C \rightsquigarrow A_1$ by Δ and B_2 of $C \rightsquigarrow A_2$ by $\Delta \cup \{C_1\}$.

Then, $C_1 \wedge C_2 \preceq A_1 \wedge A_2$, and $\Delta \cup \{C_1 \wedge C_2\}$ is consistent.

For any E with $B_1 \wedge B_2 \prec E \preceq A_1 \wedge A_2$, there are E_1, E_2 such that $E = E_1 \wedge E_2$, and

$$\begin{aligned} B_1 &\preceq E_1 \preceq A_1, \\ B_2 &\preceq E_2 \preceq A_2. \end{aligned}$$

Then, if $B_1 \prec E_1$ then $\Delta \cup \{E_1\}$ is inconsistent and so is $\Delta \cup \{E_1 \wedge E_2\}$; if $B_2 \prec E_2$ then $\Delta \cup \{E_2\}$ is inconsistent and so is $\Delta \cup \{E_1 \wedge E_2\}$. □

Lemma 10.2.7 *If B_1, B_2 are $\preceq$-minimal changes of $C \rightsquigarrow A_1$ and $C \rightsquigarrow A_2$ by Δ, respectively, then B' is a $\preceq$-minimal change of $C \rightsquigarrow A_1 \vee A_2$ by Δ, where*

$$B' = \begin{cases} B_1 \vee A_2 \ \textit{if } B_1 \neq \lambda \\ A_1 \vee B_2 \ \textit{if } B_1 = \lambda \textit{ and } B_2 \neq \lambda \\ \lambda \qquad\qquad \textit{if } B_1 = B_2 = \lambda \end{cases}$$

Proof It is clear that $B' \preceq A_1 \vee A_2$, and $\Delta \cup \{B'\}$ is consistent.

For any E with $B' \prec E \preceq A_1 \vee A_2$, there are E_1, E_2 such that

$$\begin{aligned} B &= B_1 \vee B_2, \\ B_1' &\preceq E_1 \preceq A_1, \\ B_2' &\preceq E_2 \preceq A_2, \end{aligned}$$

and either $B_1' \prec E_1$ or $B_2' \prec E_2$, where B_1' is either B_1 or A_1 and B_2' is either B_2 or A_2.

If $B_1' \prec E_1$ then $B_1' = B_1$, $B_2' = A_2$, and by induction assumption, $\Delta \cup \{E_1\}$ is inconsistent and so is $\Delta \cup \{E_1 \vee A_2\}$; and if $B_2' \prec E_2$ then $B_2' = B_2$, $B_1' = A_1$, and by induction assumption, $\Delta \cup \{E_2\}$ is inconsistent and so is $\Delta \cup \{A_1 \vee E_2\}$. □

Theorem 10.2.8 *For any formula set Δ, default $C \rightsquigarrow A$ and formula B such that $\Delta \vdash C$, if $\Delta | C \rightsquigarrow A \Rightarrow \Delta, B$ is provable then B is a $\preceq$-minimal change of $C \rightsquigarrow A$ by Δ. That is,*

$$\vdash_{\mathbf{T}^{\mathrm{D}}} \Delta | C \rightsquigarrow A \Rightarrow \Delta, B \textit{ implies } \models_{\mathbf{T}^{\mathrm{D}}} \Delta | C \rightsquigarrow A \Rightarrow \Delta, B.$$

Proof We prove the theorem by induction on the structure of A. Assume that $\Delta \vdash C$.

Case $A = l$, a literal. Either $\Delta | C \rightsquigarrow l \Rightarrow \Delta, l$ or $\Delta | C \rightsquigarrow l \Rightarrow \Delta$ is provable. That is, either $B = l$ or $B = \lambda$, which is a $\preceq$-minimal change of $C \rightsquigarrow A$ by Δ.

Case $A = A_1 \wedge A_2$. Then, if $\Delta | C \rightsquigarrow A_1 \Rightarrow \Delta, B_1$ is provable then B_1 is a $\preceq$-minimal change of $C \rightsquigarrow A_1$ by Δ, and if $\Delta, B_1 | C \rightsquigarrow A_2 \Rightarrow \Delta, B_1, B_2$ is provable then B_2 is a $\preceq$-minimal change of $C \rightsquigarrow A_2$ by $\Delta \cup \{B_1\}$. By Lemma 10.2.6, $B_1 \wedge B_2$ is a $\preceq$-minimal change of $C \rightsquigarrow A_1 \wedge A_2$ by Δ.

Case $A = A_1 \vee A_2$. Then $\Delta \cup \{A_1\}$ and $\Delta \cup \{A_2\}$ are inconsistent, and there are B_1', B_2' such that $\Delta | C \rightsquigarrow A_1 \Rightarrow \Delta, B_1'$ and $\Delta | C \rightsquigarrow A_2 \Rightarrow \Delta, B_2'$ are provable, and by induction assumption, B_1' and B_2' are $\preceq$-minimal changes of $C \rightsquigarrow A_1$ and $C \rightsquigarrow A_2$ by Δ, respectively. Then, $\Delta | C \rightsquigarrow A_1 \vee A_2 \Rightarrow \Delta, B_1' \vee B_2'$, and by Lemma 10.2.7, $B_1' \vee B_2'$ is a $\preceq$-minimal change of $C \rightsquigarrow A_1 \vee A_2$ by Δ. □

Theorem 10.2.9 *For any formula sets Δ, default $C \rightsquigarrow A$ and formula B, if B is a $\preceq$-minimal change of $C \rightsquigarrow A$ by Δ then there is a formula A' such that $A \simeq A'$ and $\Delta | C \rightsquigarrow A' \Rightarrow B$ is $\mathbf{T}^{\mathrm{D}}$-provable. That is,*

$$\models_{\mathbf{T}^{\mathrm{D}}} \Delta|C \rightsquigarrow A \Rightarrow, \Delta, B \textit{ implies } \vdash_{\mathbf{T}^{\mathrm{D}}} \Delta|C \rightsquigarrow A' \Rightarrow, \Delta, B.$$

Proof Let $B \preceq A$ be a $\preceq$-minimal change of $C \rightsquigarrow A$ by Δ.

Case $A = l$. Then $B = \lambda$ (if Δ, A is inconsistent) or $B = A$ (if Δ, A is consistent), and $\Delta|C \rightsquigarrow A \Rightarrow \Delta, B$ is provable.

Case $A = A_1 \wedge A_2$. Then there are B_1, B_2 such that $B = B_1 \wedge B_2$, and B_1 and B_2 are $\preceq$-minimal changes of $C \rightsquigarrow A_1$ and $C \rightsquigarrow A_2$ by Δ and Δ, B_1, respectively. Hence, $B_1 \wedge B_2$ is a $\preceq$-minimal change of $C \rightsquigarrow A_1 \wedge A_2$ by Δ. By induction assumption, $\Delta|C \rightsquigarrow A_1 \Rightarrow \Delta, B_1$ and $\Delta, B_1|C \rightsquigarrow A_2 \Rightarrow \Delta, B_1, B_2$ are provable, and so is $\Delta|C \rightsquigarrow A_1 \wedge A_2 \Rightarrow \Delta, B_1 \wedge B_2$.

Case $A = A_1 \vee A_2$. Then there are B_1 and B_2 such that $B = B'$, and B_1 and B_2 are $\preceq$-minimal changes of $C \rightsquigarrow A_1$ and $C \rightsquigarrow A_2$ by Δ, respectively, where

$$B' = \begin{cases} B_1 \vee A_2 \text{ if } B_1 \neq \lambda \\ A_1 \vee B_2 \text{ if } B_1 = \lambda \text{ and } B_2 \neq \lambda \\ \lambda \qquad\quad \text{if } B_1 = B_2 = \lambda \end{cases}$$

Then, B' is a $\preceq$-minimal change of $C \rightsquigarrow A_1 \vee A_2$ by Δ. By induction assumption, either (i) $\Delta|C \rightsquigarrow A_1 \Rightarrow \Delta, B_1$ or (ii) $\Delta|C \rightsquigarrow A_2 \Rightarrow \Delta, B_2$, or (iii) $\Delta|C \rightsquigarrow A_1 \Rightarrow \Delta$ and $\Delta|C \rightsquigarrow A_2 \Rightarrow \Delta$ are provable, and so is $\Delta|C \rightsquigarrow A_1 \vee A_2 \Rightarrow \Delta, B'$, where if $B_1 \neq \lambda$ then $\Delta|C \rightsquigarrow A_1 \vee A_2 \Rightarrow \Delta, B_1 \vee A_2$ is provable; if $B_1 = \lambda$ and $B_2 \neq \lambda$ then $\Delta|C \rightsquigarrow A_1 \vee A_2 \Rightarrow \Delta, A_1 \vee B_2$ is provable; and if $B_1 = \lambda$ and $B_2 = \lambda$ then $\Delta|C \rightsquigarrow A_1 \vee A_2 \Rightarrow \Delta$ is provable. □

10.2.2 Deduction System $\mathbf{T}^{\mathrm{D}}$ for a Set of Defaults

Let $D = (C_1 \rightsquigarrow A_1, ..., C_n \rightsquigarrow A_n)$.

Deduction system $\mathbf{T}^{\mathrm{D}}$ for a set D of defaults consists of the following axioms and deduction rules: assume that $D = D_1 \cup \{C \rightsquigarrow A\} \cup D_2$, and for each $C' \rightsquigarrow A' \in D_1, \Delta \nvdash C'$; and $\Delta \vdash C$.

- **Axioms:**

$$(T^{\mathbf{A}})\ \frac{\begin{array}{l}\Delta \vdash C \\ \Delta \nvdash \neg A\end{array}}{\Delta|D_1, C \rightsquigarrow A, D_2 \Rightarrow \Delta, A|D_1, D_2}$$

$$(T_{\mathbf{A}})\ \frac{\begin{array}{l}\Delta \vdash C \\ \Delta \vdash \neg l\end{array}}{\Delta|D_1, C \rightsquigarrow l, D_2 \Rightarrow \Delta|D_1, D_2}$$

• **Deduction rules**:

$$(T^{\wedge})\ \frac{\Delta|D_1, C \rightsquigarrow A_1, D_2 \Rightarrow \Delta, B_1|D_1, D_2}{\Delta|D_1, C \rightsquigarrow A_1 \wedge A_2, D_2 \Rightarrow \Delta, B_1|D_1, C \rightsquigarrow A_2, D_2}$$

$$(T_1^{\vee})\ \frac{\Delta|D_1, C \rightsquigarrow A_1, D_2 \Rightarrow \Delta, B_1|D_1, D_2 \quad B_1 \neq \lambda}{\Delta|D_1, C \rightsquigarrow A_1 \vee A_2, D_2 \Rightarrow \Delta, B_1 \vee A_2|D_1, D_2}$$

$$(T_2^{\vee})\ \frac{\begin{array}{l}\Delta|D_1, C \rightsquigarrow A_1, D_2 \Rightarrow \Delta|D_1, D_2\\ \Delta|D_1, C \rightsquigarrow A_2, D_2 \Rightarrow \Delta, B_2|D_1, D_2\\ B_2 \neq \lambda\end{array}}{\Delta|D_1, C \rightsquigarrow A_1 \vee A_2, D_2 \Rightarrow \Delta, A_1 \vee B_2|D_1, D_2}$$

$$(T_3^{\vee})\ \frac{\Delta|D_1, C \rightsquigarrow A_1, D_2 \Rightarrow \Delta|D_1, D_2 \quad \Delta|D_1, C \rightsquigarrow A_2, D_2 \Rightarrow \Delta|D_1, D_2}{\Delta|D_1, C \rightsquigarrow A_1 \vee A_2, D_1 \Rightarrow \Delta|D_1, D_2}$$

Theorem 10.2.10 (Soundness Theorem) *For any formula sets* Θ, Δ *and any finite default set* D*, if* $\Delta|D \Rightarrow \Delta, \Theta$ *is provable in* $\mathbf{T}^{\mathrm{D}}$ *then* Θ *is a* $\preceq$*-minimal change of* D *by* Δ*. That is,*

$$\vdash_{\mathbf{T}^{\mathrm{D}}} \Delta|D \Rightarrow \Delta, \Theta \textit{ implies } \models_{\mathbf{T}^{\mathrm{D}}} \Delta|D \Rightarrow \Delta, \Theta.$$

Proof We prove the theorem by induction on n.

Assume that $\Delta|D \Rightarrow \Delta, \Theta$ is provable in $\mathbf{T}^{\mathrm{D}}$.

Let $n = 1$. By Theorem 10.1.3, $\Theta = C$ for some B is a $\preceq$-minimal change of $C \rightsquigarrow A$ by Δ.

Assume that the theorem holds for n, that is, if $\Delta|D \Rightarrow \Delta, \Theta$ is provable then Θ is a $\preceq$-minimal change of D by Δ, where $D = (C_1 \rightsquigarrow A_1, ..., C_n \rightsquigarrow A_n)$.

Let $D' = (D, C_{n+1} \rightsquigarrow A_{n+1}) = (C_1 \rightsquigarrow A_1, ..., C_{n+1} \rightsquigarrow A_{n+1})$. Then, if $\Delta|D' \Rightarrow \Delta, \Theta'$ is provable then there is a Θ such that $\Delta|D \Rightarrow \Delta, \Theta$ and $\Delta, \Theta|C_{n+1} \rightsquigarrow A_{n+1} \Rightarrow \Delta, \Theta'$ are provable. By the case $n = 1$ and the induction assumption, Θ' is a $\preceq$-minimal change of $C_{n+1} \rightsquigarrow A_{n+1}$ by $\Delta \cup \Theta$, and Θ is a $\preceq$-minimal change of D by Δ, therefore, Θ' is a $\preceq$-minimal change of D' by Δ. □

Theorem 10.2.11 (Completeness Theorem) *For any formula sets* Θ, Δ *and any finite default set* D*, if* Θ *is a* $\preceq$*-minimal change of* D *by* Δ *then* $\Delta|D \Rightarrow \Delta, \Theta$ *is provable in* $\mathbf{T}^{\mathrm{D}}$*. That is,*

$$\models_{\mathbf{T}^{\mathrm{D}}} \Delta|D \Rightarrow \Delta, \Theta \textit{ implies } \vdash_{\mathbf{T}^{\mathrm{D}}} \Delta|D \Rightarrow \Delta, \Theta.$$

Proof Assume that Θ is a $\preceq$-minimal change of D by Δ. Then, there is an ordering $<$ of D such that $D = (C_1 \rightsquigarrow A_1, C_2 \rightsquigarrow A_2, ..., C_n \rightsquigarrow A_n)$, where $C_1 \rightsquigarrow A_1 < C_2 \rightsquigarrow A_2 < \cdots < C_n \rightsquigarrow A_n$, and Θ is a $\preceq$-minimal change of D by Δ.

We prove the theorem by induction on n.

Let $n = 1$. By Theorem 10.2.9, there is Θ such that $\Delta|C_1 \rightsquigarrow A_1 \Rightarrow \Delta, \Theta$ is provable.

Assume that the theorem holds for n, that is, if Θ is a $\preceq$-minimal change of D by Δ then $\Delta|D \Rightarrow \Delta, \Theta$ is provable.

Let $D' = (D, C_{n+1} \rightsquigarrow A_{n+1}) = (C_1 \rightsquigarrow A_1, ..., C_{n+1} \rightsquigarrow A_{n+1})$ and Θ' is a $\preceq$-minimal change of D' by Δ. Then, Θ' is a $\preceq$-minimal change of $C_{n+1} \rightsquigarrow A_{n+1}$ by Θ, and $\Theta|C_{n+1} \rightsquigarrow A_{n+1} \Rightarrow \Delta, \Theta'$ is provable. By the induction assumption, $\Delta|D \Rightarrow \Delta, \Theta$ is provable and $\Delta|D' \Rightarrow \Delta, \Theta|C_{n+1} \rightsquigarrow A_{n+1}$, and hence, $\Delta|D' \Rightarrow \Delta, \Theta'$ is provable. □

10.3 Default Logic and Deduction-Based Minimal Change

In this section we assume that formula A is in conjunctive normal form.

Definition 10.3.1 Given a default theory (Δ, D) and a theory Θ, Θ is a $\vdash_{\preceq}$-minimal change of D by Δ, denoted by $\models_{\mathbf{U^D}} \Delta|D \Rightarrow \Delta, \Theta$, if

(i) $\Theta \cup \Delta$ is consistent;
(ii) $\Theta \preceq \Gamma$, and
(ii) for any theory Ξ with $\Gamma \succeq \Xi \succ \Theta$, either $\Delta, \Xi \vdash \Theta$ and $\Delta, \Theta \vdash \Xi$, or $\Xi \cup \Delta$ is inconsistent, where $\Gamma = \{A : C \rightsquigarrow A \in D, \Delta \cup \Theta \vdash C\}$.

10.3.1 Deduction System $\mathbf{U^D}$ *for a Default*

Deduction system $\mathbf{U^D}$ for a default $C \rightsquigarrow A$ consists of the following axioms and deduction rules:

- **Axioms**:

$$(U^{\mathbf{A}})\ \frac{\Delta \vdash C \quad \Delta \nvdash \neg l}{\Delta|C \rightsquigarrow l \Rightarrow \Delta, l} \qquad (U_{\mathbf{A}})\ \frac{\Delta \vdash C \quad \Delta \vdash \neg l}{\Delta|C \rightsquigarrow l \Rightarrow \Delta, \lambda}$$

- **Deduction rules**:

$$(U^{\wedge})\ \frac{\Delta|C \rightsquigarrow A_1 \Rightarrow \Delta, B_1}{\Delta|C \rightsquigarrow A_1 \wedge A_2 \Rightarrow \Delta, B_1|C \rightsquigarrow A_2}$$

$$(U^{\vee})\ \frac{\Delta|C \rightsquigarrow A_1 \Rightarrow \Delta, B_1 \quad \Delta|C \rightsquigarrow A_2 \Rightarrow \Delta, B_2}{\Delta|C \rightsquigarrow A_1 \vee A_2 \Rightarrow \Delta, B_1 \vee B_2}$$

where if B is consistent then

$$\lambda \vee B \equiv B \vee \lambda \equiv B$$
$$\lambda \wedge B \equiv B \wedge \lambda \equiv B$$
$$\Delta, \lambda \equiv \Delta$$

and if B is inconsistent then

$$\lambda \vee B \equiv B \vee \lambda \equiv \lambda$$
$$\lambda \wedge B \equiv B \wedge \lambda \equiv \lambda$$

Definition 10.3.2 $\Delta|C \rightsquigarrow A \Rightarrow \Delta, B$ is provable in $\mathbf{U}^{\mathrm{D}}$, denoted by $\vdash_{\mathbf{U}^{\mathrm{D}}} \Delta|C \rightsquigarrow A \Rightarrow \Delta, B$, if there is a sequence $S_1, ..., S_m$ of statements such that

$$\begin{aligned}
&S_1 = \Delta|C_1 \rightsquigarrow A_1 \Rightarrow \Delta|C_1' \rightsquigarrow A_1',\\
&\ldots\\
&S_m = \Delta|C_m \rightsquigarrow A_m \Rightarrow \Delta, A_m';\\
&A_1 = A,\\
&A_m' = B;
\end{aligned}$$

and for each $i < m$, S_{i+1} is an axiom or is deduced from the previous statements by a deduction rule in $\mathbf{U}^{\mathrm{D}}$.

For Example 10.1.2, we have different result in $\mathbf{U}^{\mathrm{D}}$.

Example 10.3.3 Let (Δ, D) be a default theory, where

$$\begin{aligned}
&\Delta = \{\neg p, \neg r\},\\
&D = \{\top \rightsquigarrow p \wedge (r \vee q)\},
\end{aligned}$$

In $\mathbf{U}^{\mathrm{D}}$, we have the following deduction:

$$\begin{aligned}
&\neg p, \neg r|\top \rightsquigarrow p \Rightarrow \neg p, \neg r\\
&\neg p, \neg r, \lambda|\top \rightsquigarrow r \Rightarrow \neg p, \neg r, \lambda\\
&\neg p, \neg r, \lambda|\top \rightsquigarrow q \Rightarrow \neg p, \neg r, q\\
&\neg p, \neg r, \lambda|\top \rightsquigarrow r \vee q \Rightarrow \neg p, \neg r, \lambda \vee q\\
&\neg p, \neg r|\top \rightsquigarrow p \wedge (r \vee q) \Rightarrow \neg p, \neg r, q.
\end{aligned}$$

That is, $\{q\}$ is a $\vdash_{\preceq}$-minimal change of $\top \rightsquigarrow p \wedge (r \vee q)$ by $\neg p, \neg r$, i.e.,

$$\vdash_{\mathbf{U}^{\mathrm{D}}} \neg p, \neg r|\top \rightsquigarrow p \wedge (r \vee q) \Rightarrow \neg p, \neg r, q.$$

That means that the man has the right arm. □

Theorem 10.3.4 *For any consistent set Δ of formulas and default $C \rightsquigarrow A$, where A is in conjunctive normal form, such that $\Delta \vdash C$, there is a formula B such that*
(1) $\Delta|C \rightsquigarrow A \Rightarrow \Delta, B$ is provable in $\mathbf{U}^{\mathrm{D}}$;
(2) $B \preceq A$, and
(3) $\Delta \cup \{B\}$ is consistent, and for any E with $B \prec E \preceq A$, either $\Delta, B \vdash E$ and $\Delta, E \vdash B$, or $\Delta \cup \{E\}$ is inconsistent.

Proof By induction on the structure of A we prove the theorem.

Case $A = l$, a literal. Then by the assumption, if Δ, l is consistent then $\vdash_{\mathbf{U}^{\mathrm{D}}} \Delta|C \rightsquigarrow l \Rightarrow \Delta, l$ and let $B = l$; and if Δ, l is inconsistent, then $\Delta \vdash \neg l$, and by $(U_{\mathbf{A}})$, $\vdash_{\mathbf{U}^{\mathrm{D}}} \Delta|C \rightsquigarrow l \Rightarrow \Delta$, and let $B = \lambda$. B satisfies (2) and (3).

Case $A = A_1 \wedge A_2$. Then by induction assumption, there are B_1, B_2 such that

$$\vdash_{\mathbf{U^D}} \Delta|C \rightsquigarrow A_1 \Rightarrow \Delta, B_1;$$
$$\vdash_{\mathbf{U^D}} \Delta, B_1|C \rightsquigarrow A_2 \Rightarrow \Delta, B_1, B_2.$$

By $(U^{\wedge})$, $\vdash_{\mathbf{U^D}} \Delta|C \rightsquigarrow A_1 \wedge A_2 \Rightarrow \Delta, B_1 \wedge B_2$, and let $B = B_1 \wedge B_2$. B satisfies (3), because for any E with $B \prec E \preceq A_1 \wedge A_2$, there are formulas E_1, E_2 such that

$$B_1 \preceq E_1;\ B_2 \preceq E_2.$$

By the induction assumption, either

$$\Delta, B_1 \vdash E_1;\ \Delta, E_1 \vdash B_1;$$
$$\Delta, B_2 \vdash E_2;\ \Delta, E_2 \vdash B_2;$$

or $\Delta \cup \{E_1\}$ or $\Delta \cup \{E_2\}$ are inconsistent. Therefore, we have

$$\Delta, B_1 \wedge B_2 \vdash E_1 \wedge E_2;$$
$$\Delta, E_1 \wedge E_2 \vdash B_1 \wedge B_2,$$

or $\Delta \cup \{E_1 \wedge E_2\}$ is inconsistent.

Case $A = A_1 \vee A_2$. By the induction assumption, there are B_1, B_2 such that

$$\vdash_{\mathbf{U^D}} \Delta|C \rightsquigarrow A_1 \Rightarrow \Delta, B_1$$
$$\vdash_{\mathbf{U^D}} \Delta|C \rightsquigarrow A_2 \Rightarrow \Delta, B_2.$$

By $(U^{\vee})$, $\vdash_{\mathbf{U^D}} \Delta|C \rightsquigarrow A_1 \vee A_2 \Rightarrow \Delta, B_1 \vee B_2$. To prove that $B_1 \vee B_2$ satisfies (3), for any E with $B_1 \vee B_2 \prec E \preceq A_1 \vee A_2$, there are formulas E_1, E_2 such that $B_1 \preceq E_1 \preceq A_1$, $B_2 \preceq E_2 \preceq A_2$, and either

$$\Delta, B_1 \vdash E_1;\ \Delta, E_1 \vdash B_1;$$
$$\Delta, B_2 \vdash E_2;\ \Delta, E_2 \vdash B_2;$$

or $\Delta \cup \{E_1\}$ is inconsistent and $\Delta \cup \{E_2\}$ is inconsistent. Therefore, either

$$\Delta, B_1 \vee B_2 \vdash E_1 \vee E_2;$$
$$\Delta, E_1 \vee E_2 \vdash B_1 \vee B_2.$$

□

Theorem 10.3.5 *Assume that* $\Delta|C \rightsquigarrow A \Rightarrow \Delta, B$ *is provable in* $\mathbf{U^D}$. *If* A *is consistent with* Δ *then* $\Delta, A \vdash B$ *and* $\Delta, B \vdash A$.

Proof Assume that $\Delta|C \rightsquigarrow A \Rightarrow \Delta, B$ is provable in $\mathbf{U^D}$. We prove the theorem by induction on the structure of A.

Case $A = l$. By $(U^{\mathbf{A}})$, $B = l$, and

$$\Delta, l \vdash l.$$

Case $A = A_1 \wedge A_2$. By induction assumption, there are formulas B_1 and B_2 such that

$$\vdash_{\mathbf{U}^{\mathrm{D}}} \Delta|C \rightsquigarrow A_1 \Rightarrow \Delta, B_1,$$
$$\vdash_{\mathbf{U}^{\mathrm{D}}} \Delta|C \rightsquigarrow A_2 \Rightarrow \Delta, B_2,$$

and $B = B_1 \wedge B_2$. Because A is consistent with Δ iff A_1 and A_2 are consistent with Δ and $\Delta \cup \{B_1\}$, respectively, we have

$$\Delta, A_1 \vdash B_1; \Delta, B_1 \vdash A_1;$$
$$\Delta, B_1, A_2 \vdash B_2; \Delta, B_1, B_2 \vdash A_2;$$

and hence,

$$\Delta, A_1 \wedge A_2 \vdash B_1 \wedge B_2;$$
$$\Delta, B_1 \wedge B_2 \vdash A_1 \wedge A_2.$$

That is,

$$\Delta, A \vdash B; \Delta, B \vdash A.$$

Case $A = A_1 \vee A_2$. By induction assumption, there are formulas B_1 and B_2 such that

$$\vdash_{\mathbf{U}^{\mathrm{D}}} \Delta|C \rightsquigarrow A_1 \Rightarrow \Delta, B_1,$$
$$\vdash_{\mathbf{U}^{\mathrm{D}}} \Delta|C \rightsquigarrow A_2 \Rightarrow \Delta, B_2,$$

and $B = B_1 \vee B_2$. Because A is consistent with Δ iff either A_1 is consistent with Δ or A_2 is consistent with Δ, we have

$$\Delta, A_1 \vdash B_1; \Delta, B_1 \vdash A_1;$$
$$\Delta, A_2 \vdash B_2; \Delta, B_2 \vdash A_2;$$

and hence,

$$\Delta, A_1 \vee A_2 \vdash B_1 \vee B_2;$$
$$\Delta, B_1 \vee B_2 \vdash A_1 \vee A_2.$$

That is,

$$\Delta, A \vdash B; \Delta, B \vdash A.$$

Notice that we assume that A is in conjunctive normal form. $A = A_1 \vee A_2$ is taken as a set of literals. □

Theorem 10.3.6 *Assume that* $\Delta|C \rightsquigarrow A \Rightarrow \Delta, B$ *is provable in* $\mathbf{U}^{\mathrm{D}}$. *Then,* B *is a* $\vdash_{\preceq}$*-minimal change of* $C \rightsquigarrow A$ *by* Δ.

Proof By the definition and the last theorem. □

Theorem 10.3.7 $\mathbf{U}^{\mathrm{D}}$ *is sound and complete with respect to* $\vdash_{\preceq}$*-minimal change.* □

10.3.2 Deduction System $\mathbf{U}^{\mathrm{D}}$ for a Set of Defaults

Let $D = (C_1 \rightsquigarrow A_1, ..., C_n \rightsquigarrow A_n)$.

Deduction system $\mathbf{U}^{\mathrm{D}}$ for $\vdash_{\preceq}$-minimal change consists of the following axioms and deduction rules: assume that $D = D_1 \cup \{C \rightsquigarrow A\} \cup D_2$, and for each $C' \rightsquigarrow A' \in D_1, \Delta \nvdash C'$; and $\Delta \vdash C$.

- **Axioms:**

$$(U^{\mathbf{A}})\ \frac{\Delta \vdash C \quad \Delta \nvdash \neg l}{\Delta | D_1, C \rightsquigarrow l, D_2 \Rightarrow \Delta, l | D_1, D_2}$$

$$(U_{\mathbf{A}})\ \frac{\Delta \vdash C \quad \Delta \vdash \neg l}{\Delta | D_1, l, D_2 \Rightarrow \Delta | D_1, D_2}$$

- **Deduction rules**:

$$(U^{\wedge})\ \frac{\Delta | D_1, C \rightsquigarrow A_1, D_2 \Rightarrow \Delta, B_1 | D_1, D_2}{\Delta | D_1, C \rightsquigarrow A_1 \wedge A_2, D_2 \Rightarrow \Delta, B_1 | D_1, C \rightsquigarrow A_2, D_2}$$

$$(U^{\vee})\ \frac{\Delta | D_1, C \rightsquigarrow A_1, D_2 \Rightarrow \Delta, B_1 | D_1, D_2 \quad \Delta | D_1, C \rightsquigarrow A_2, D_2 \Rightarrow \Delta, B_2 | D_1, D_2}{\Delta | D_1, C \rightsquigarrow A_1 \vee A_2, D_2 \Rightarrow \Delta, B_1 \vee B_2 | D_1, D_2}$$

Definition 10.3.8 $\Delta | D \Rightarrow \Delta, \Theta$ is provable in $\mathbf{U}^{\mathrm{D}}$, denoted by $\vdash_{\mathbf{U}^{\mathrm{D}}} \Delta | D \Rightarrow \Delta, \Theta$, if there is a sequence $S_1, ..., S_m$ of statements such that

$$\begin{aligned} S_1 &= \Delta | D \Rightarrow \Delta_1 | D_1, \\ &\dots \\ S_m &= \Delta_{m-1} | D_{m-1} \Rightarrow \Delta_m | D_m = \Delta, \Theta \end{aligned}$$

and for each $i < m$, S_{i+1} is an axiom or is deduced from the previous statements by a deduction rule in $\mathbf{U}^{\mathrm{D}}$.

Theorem 10.3.9 (Soundness Theorem) *For any consistent formula sets Θ, Δ and any finite default set D, if $\Delta | D \Rightarrow \Delta, \Theta$ is provable in $\mathbf{U}^{\mathrm{D}}$ then Θ is a $\vdash_{\preceq}$-minimal change of D by Δ. That is,*

$$\vdash_{\mathbf{U}^{\mathrm{D}}} \Delta | D \Rightarrow \Delta, \Theta \textit{ implies } \models_{\mathbf{U}^{\mathrm{D}}} \Delta | D \Rightarrow \Delta, \Theta.$$

□

Theorem 10.3.10 (Completeness theorem) *For any consistent formula sets Θ, Δ and any finite default set D, if Θ is a $\vdash_{\preceq}$-minimal change of D by Δ then $\Delta | D \Rightarrow \Delta, \Theta$ is provable in $\mathbf{U}^{\mathrm{D}}$. That is,*

$$\models_{\mathbf{U}^{\mathrm{D}}} \Delta | D \Rightarrow \Delta, \Theta \textit{ implies } \vdash_{\mathbf{U}^{\mathrm{D}}} \Delta | D \Rightarrow \Delta, \Theta.$$

References

1. M.L. Ginsberg (ed.), *Readings in Nonmonotonic Reasoning* (Morgan Kaufmann, San Francisco, 1987)
2. J. Horty, Defaults with Priorities. J. Philos. Log. **36**, 367–413 (2007)
3. J. Horty, Some direct theories of nonmonotonic inheritance, in *Handbook of Logic in Artificial Intelligence and Logic Programming*, vol. 3: Nonmonotonic Reasoning and Uncertain Reasoning, ed. by D.M. Gabbay, C.J. Hogger, J.A. Robinson (Oxford University Press, Oxford, 1994), pp. 111–187
4. D.J. Lehmann, Another Perspective on Default Reasoning. Ann. Math. Artif. Intell. **15**, 61–82 (1995)
5. D. Nute, Defeasible logics, in *Handbook of Logic in Artificial Intelligence and Logic Programming*, vol. 3 (Oxford University Press, Oxford, 1994), pp. 353–395
6. R. Reiter, A logic for default reasoning. Artif. Intell. **13**, 81–132 (1980)
7. Y. Shoham, A Semantical Approach to Nonmonotonic Logics, in *Readings in Non-Monotonic Reasoning*. ed. by M.L. Ginsberg (Morgan Kaufmann, Los Altos, 1987), pp. 227–249
8. R. Stalnaker, A note on non-monotonic modal logic. Artif. Intell. **64**, 183–196 (1993)

Chapter 11
An Application to Semantic Networks

In Chap. 4, we gave R-calculi $\mathbf{S}^{\mathrm{DL}}$, $\mathbf{T}^{\mathrm{DL}}$, $\mathbf{U}^{\mathrm{DL}}$ for description logic, where $\mathbf{S}^{\mathrm{DL}}$, $\mathbf{T}^{\mathrm{DL}}$, $\mathbf{U}^{\mathrm{DL}}$ are sound and complete with respect to $\subseteq$-minimal change, $\preceq$-minimal change and $\vdash_{\preceq}$-minimal change, respectively, where we assume that theories does not contain statements of form $C \not\sqsubseteq D$.

By the following rules: for any concepts C_1, C_2, D_1, D_2,

$$\begin{array}{l} C_1 \sqcap C_2 \sqsubseteq D \Leftarrow C_1 \sqsubseteq D \sqcup C_2 \sqsubseteq D \\ C_1 \sqcup C_2 \sqsubseteq D \equiv C_1 \sqsubseteq D \sqcap C_2 \sqsubseteq D \\ C \sqsubseteq D_1 \sqcap D_2 \equiv C \sqsubseteq D_1 \sqcap C \sqsubseteq D_2 \\ C \sqsubseteq D_1 \sqcup D_2 \Leftarrow C \sqsubseteq D_1 \sqcup C \sqsubseteq D_2; \end{array}$$

and

$$\begin{array}{l} C_1 \sqcap C_2 \not\sqsubseteq D \Rightarrow C_1 \not\sqsubseteq D \wedge C_2 \not\sqsubseteq D \\ C_1 \sqcup C_2 \not\sqsubseteq D \equiv C_1 \not\sqsubseteq D \vee C_2 \not\sqsubseteq D \\ C \not\sqsubseteq D_1 \sqcap D_2 \equiv C \not\sqsubseteq D_1 \vee C \not\sqsubseteq D_2 \\ C \not\sqsubseteq D_1 \sqcup D_2 \Rightarrow C \not\sqsubseteq D_1 \wedge C \not\sqsubseteq D_2. \end{array}$$

we will give R-calculi $\mathbf{S}^{\mathrm{SN}}$, $\mathbf{T}^{\mathrm{SN}}$, $\mathbf{U}^{\mathrm{SN}}$ for the semantic networks [1–10].

11.1 Semantic Networks

Semantic networks are similar to logic programs in the correspondence given in the following.

W. Li and Y. Sui, *R-CALCULUS: A Logic of Belief Revision*, Perspectives in Formal Induction, Revision and Evolution,
https://doi.org/10.1007/978-981-16-2944-0_11

- literals correspond to statements $C \sqsubseteq D$;
- clauses $l_1, ..., l_m \leftarrow l'_1, ..., l'_n$ correspond to sequents $C_{11} \sqsubseteq D_{11}, ..., C_{1n} \sqsubseteq D_{1n} \Rightarrow C_{21} \sqsubseteq D_{21}, ..., C_{2m} \sqsubseteq D_{2m}$,

where $l_1, ..., l_m, l'_1, ..., l'_n$ are literals and

$$C_{11}, D_{11}, ..., C_{1n}, D_{1n}, C_{21}, D_{21}, C_{2m}, D_{2m}$$

are concepts.

This section gives basic definitions and a deduction system $\mathbf{G}_4$ for semantic networks, based on which R-calculi $\mathbf{S}^{\mathrm{SN}}, \mathbf{T}^{\mathrm{SN}}, \mathbf{U}^{\mathrm{SN}}$ for semantic networks will be built.

11.1.1 Basic Definitions

The logical language of $\mathcal{AL}^{\mathrm{SN}}$ contains the following symbols:

- atomic concept symbols: $A_0, A_1, ...$;
- concept constructors: $\neg, \sqcup, \sqcap$; and
- concept connectives: $\sqsubseteq, \not\sqsubseteq$.

Concepts are defined as follows:

$$C ::= A|\neg A|C_1 \sqcap C_2|C_1 \sqcup C_2.$$

Statements are defined as follows:

$$\delta ::= C \sqsubseteq D|C \not\sqsubseteq D,$$

where C, D are concepts.

Let U be a universe, and I an interpretation such that for each concept symbol $A, I(A) \subseteq U$.

I interprets concept C as follows:

$$C^I = \begin{cases} I(A) & \text{if } C = A \\ U - I(A) & \text{if } C = \neg A \\ C_1^I \cap C_2^I & \text{if } C = C_1 \sqcap C_2 \\ C_1^I \cup C_2^I & \text{if } C = C_1 \sqcup C_2. \end{cases}$$

A statement δ is satisfied in I, denoted by $I \models \delta$, if

$$\begin{cases} I(C \sqsubseteq D)(a) = 1 \text{ if } \delta = C \sqsubseteq D \\ I(C \not\sqsubseteq D)(a) = 1 \text{ if } \delta = C \not\sqsubseteq D, \end{cases}$$

where

$$\begin{array}{l} I(C \sqsubseteq D)(a) = 1 \text{ if } a \in C^I \text{ implies } a \in D^I; \\ I(C \not\sqsubseteq D)(a) = 1 \text{ if } a \in C^I \text{ and } a \notin D^I. \end{array}$$

Given two sets Γ, Δ of statements, define $\Gamma \Rightarrow \Delta$ a sequent, and I satisfies $\Gamma \Rightarrow \Delta$, denoted by $I \models_{G_4} \Gamma \Rightarrow \Delta$, if for each element $a \in U$, if

(1) *for each* $C \sqsubseteq D \in \Gamma$, $C^I(a)$ *implies* $D^I(a)$; *and for each* $C \not\sqsubseteq D \in \Gamma$, $C^I(a)$ *does not imply* $D^I(a)$,

implies

(2) *for some* $C \sqsubseteq D \in \Delta$, $C^I(a)$ *implies* $D^I(a)$, *or for some* $C \not\sqsubseteq D \in \Delta$, $C^I(a)$ *does not imply* $D^I(a)$,

where $C^I(a)$ does not imply $D^I(a)$ if $C^I(a) = 1$ and $D^I(a) = 0$.

A sequent $\Gamma \Rightarrow \Delta$ is valid, denoted by $\models_{G_4} \Gamma \Rightarrow \Delta$, if for any interpretation I, $I \models_{G_4} \Gamma \Rightarrow \Delta$.

A statement $B_1 \sqsubseteq B_2$ is called a literal statement if $B_1, B_2 ::= A|\neg A$.

Proposition 11.1.1 *For any concepts* C_1, C_2, D_1, D_2,

$$\begin{aligned}
C_1 \sqcap C_2 \sqsubseteq D &\Leftarrow C_1 \sqsubseteq D \sqcup C_2 \sqsubseteq D \\
C_1 \sqcup C_2 \sqsubseteq D &\equiv C_1 \sqsubseteq D \sqcap C_2 \sqsubseteq D \\
C \sqsubseteq D_1 \sqcap D_2 &\equiv C \sqsubseteq D_1 \sqcap C \sqsubseteq D_2 \\
C \sqsubseteq D_1 \sqcup D_2 &\Leftarrow C \sqsubseteq D_1 \sqcup C \sqsubseteq D_2;
\end{aligned}$$

and

$$\begin{aligned}
C_1 \sqcap C_2 \not\sqsubseteq D &\Rightarrow C_1 \not\sqsubseteq D \wedge C_2 \not\sqsubseteq D \\
C_1 \sqcup C_2 \not\sqsubseteq D &\equiv C_1 \not\sqsubseteq D \vee C_2 \not\sqsubseteq D \\
C \not\sqsubseteq D_1 \sqcap D_2 &\equiv C \not\sqsubseteq D_1 \vee C \not\sqsubseteq D_2 \\
C \not\sqsubseteq D_1 \sqcup D_2 &\Rightarrow C \not\sqsubseteq D_1 \wedge C \not\sqsubseteq D_2.
\end{aligned}$$

□

These statements are corresponding to those in propositional logic:

$$\begin{aligned}
A_1 \wedge A_2 \to B &\Leftarrow A_1 \to B \vee A_2 \to B \\
A_1 \vee A_2 \to B &\equiv A_1 \to B \wedge A_2 \to B \\
A \to B_1 \wedge B_2 &\equiv A \to B_1 \wedge A \to B_2 \\
A \to B_1 \vee B_2 &\Leftarrow A \to B_1 \vee A \to B_2;
\end{aligned}$$

and

$$\begin{aligned}
A_1 \wedge A_2 \not\to B &\Rightarrow A_1 \not\to B \wedge A_2 \not\to B \\
A_1 \vee A_2 \not\to B &\equiv A_1 \not\to B \vee A_2 \not\to B \\
A \not\to B_1 \wedge B_2 &\equiv A \not\to B_1 \vee A \not\to B_2 \\
A \not\to B_1 \vee B_2 &\Rightarrow A \not\to B_1 \wedge A \not\to B_2;
\end{aligned}$$

where A, B are formulas in propositional logic.

Proposition 11.1.2 *Let Δ, Γ be sets of literal statements such that*

$$\Gamma = \{B_{11} \sqsubseteq B_{12}, ..., B_{n1} \sqsubseteq B_{n2}; B'_{11} \not\sqsubseteq B'_{12}, ..., B'_{n'1} \not\sqsubseteq B'_{n'2}\}$$
$$\Delta = \{B''_{11} \sqsubseteq B''_{12}, ..., B''_{m1} \sqsubseteq B''_{m2}; B'''_{11} \not\sqsubseteq B'''_{12}, ..., B'''_{m'1} \not\sqsubseteq B'''_{m'2}\}.$$

Then, $\models_{\mathbf{G}_4} \Gamma \Rightarrow \Delta$ if and only if for any $f : \{1, ..., n\} \to \{1, 2\}$ and $g : \{1, ..., m'\} \to \{1, 2\}$ either

(i) $\mathbf{E}B(B, \neg B \in \{\neg^{f(1)} B_{1f(1)}, ..., \neg^{f(n)} B_{nf(n)}; B'_{11}, \neg B'_{12}, ..., B'_{n'1}, \neg B'_{n'2}\})$, *or*
(ii) $\mathbf{E}B(B, \neg B \in \{B''_{11}, \neg B''_{12}, ..., B''_{m1}, \neg B''_{m2}; \neg^{g(1)} B'''_{1g(1)}, ..., \neg^{g(m')} B'''_{m'f(m')}\})$, *or*
(iii) $\sigma_f(\Gamma) \cap \tau_f(\Delta) \neq \emptyset$,

where

$$\sigma_f(\Gamma) = \{\neg^{f(1)} B_{1f(1)}, ..., \neg^{f(n)} B_{nf(n)}; B'_{11}, \neg B'_{12}, ..., B'_{n'1}, \neg B'_{n'2}\}$$
$$\tau_f(\Delta) = \{B''_{11}, \neg B''_{12}, ..., B''_{m1}, \neg B''_{m2}; \neg^{g(1)} B'''_{1g(1)}, ..., \neg^{g(m')} B'''_{m'f(m')}\},$$

Proof Because $\Gamma \Rightarrow \Delta$

$$\begin{aligned}
\text{iff } & B_{11} \sqsubseteq B_{12}, ..., B_{n1} \sqsubseteq B_{n2}; B'_{11} \not\sqsubseteq B'_{12}, ..., B'_{n'1} \not\sqsubseteq B'_{n'2} \\
& \Rightarrow B''_{11} \sqsubseteq B''_{12}, ..., B''_{m1} \sqsubseteq B''_{m2}; B'''_{11} \not\sqsubseteq B'''_{12}, ..., B'''_{m'1} \not\sqsubseteq B'''_{m'2} \\
\text{iff } & \neg B_{11} \sqcup B_{12}, ..., \neg B_{n1} \sqcup B_{n2}; B'_{11} \sqcap \neg B'_{12}, ..., B'_{n'1} \sqcap \neg B'_{n'2} \\
& \Rightarrow \neg B''_{11} \sqcup B''_{12}, ..., \neg B''_{m1} \sqcup B''_{m2}; B'''_{11} \sqcap \neg B'''_{12}, ..., B'''_{m'1} \sqcap \neg B'''_{m'2} \\
\text{iff } & \neg B_{11} \sqcup B_{12}, ..., \neg B_{n1} \sqcup B_{n2}; B'_{11}, \neg B'_{12}, ..., B'_{n'1}, \neg B'_{n'2} \\
& \Rightarrow \neg B''_{11}, B''_{12}, ..., \neg B''_{m1}, B''_{m2}; B'''_{11} \sqcap \neg B'''_{12}, ..., B'''_{m'1} \sqcap \neg B'''_{m'2}.
\end{aligned}$$

Assume that (i),(ii),(iii) do not hold, we define an interpretation I such that for any element $a \in U$,

$$I \models_{\mathbf{G}_4} \Gamma(a) \& I \not\models_{\mathbf{G}_4} \Delta(a).$$

Define an interpretation I such that for each $i \leq n$, $j \leq m$ and any element $a \in U$,

$$\begin{array}{ll}
I(\neg^{f(i)} B_{if(i)})(a) = 1; & a \in (\neg^{f(i)} B_{if(i)})^I \\
I(B'_{i1})(a) = 1; & a \in (B'_{i1})^I \\
I(\neg B'_{i2})(a) = 1; & a \in (\neg B'_{i2})^I \\
I(B''_{j1})(a) = 0, & a \notin (B''_{j1})^I \\
I(\neg B''_{j2})(a) = 0, & a \notin (\neg B''_{j2})^I \\
I((\neg^{g(j)} B'''_{jg(j)})(a) = 0, & a \notin (\neg^{g(j)} B'''_{jg(j)})^I
\end{array}$$

i.e.,

$$
\begin{array}{ll}
I(\neg^{f(i)} B_{if(i)})(a) = 1; & a \in (\neg^{f(i)} B_{if(i)})^I \\
I(B'_{i1})(a) = 1; & a \in (B'_{i1})^I \\
I(B'_{i2})(a) = 0; & a \notin (B'_{i2})^I \\
I(B''_{j1})(a) = 0, & a \notin (B''_{j1})^I \\
I(B''_{j2})(a) = 1, & a \in (B''_{j2})^I \\
I((\neg^{g(j)} B'''_{jg(j)})(a) = 0, \ a \notin (\neg^{g(j)} B'''_{jg(j)})^I &
\end{array}
$$

By (∗), I is well-defined and $I \models_{\mathbf{G}_4} \Gamma(a)$ and $I \not\models_{\mathbf{G}_4} \Delta(a)$. □

11.1.2 Deduction System $\mathbf{G}_4$ for Semantic Networks

Gentzen deduction system $\mathbf{G}_4$ consists of the following axioms and deduction rules:

- **Axioms**:

$$
(\mathbf{A}^{\sqsubseteq}) \quad \frac{\bigwedge_{f:\{1,\ldots,n\}\to\{1,2\}}^{g:\{1,\ldots,m'\}\to\{1,2\}} \begin{array}{l} \mathbf{E}B(B, \neg B \in \sigma_f(\Gamma)) \\ \vee\, \mathbf{E}B(B, \neg B \in \tau_g(\Delta)) \\ \vee\, \sigma_f(\Gamma) \cap \tau_f(\Delta) \neq \emptyset \end{array}}{\Gamma \Rightarrow \Delta},
$$

where Δ, Γ are sets of literal statements, and if

$$
\begin{array}{l}
\Gamma = \{B_{11} \sqsubseteq B_{12}, \ldots, B_{n1} \sqsubseteq B_{n2}; B'_{11} \not\sqsubseteq B'_{12}, \ldots, B'_{n'1} \not\sqsubseteq B'_{n'2}\} \\
\Delta = \{B''_{11} \sqsubseteq B''_{12}, \ldots, B''_{m1} \sqsubseteq B''_{m2}; B'''_{11} \not\sqsubseteq B'''_{12}, \ldots, B'''_{m'1} \not\sqsubseteq B'''_{m'2}\}
\end{array}
$$

then

$$
\begin{array}{l}
\sigma_f(\Gamma) = \{\neg^{f(1)} B_{1f(1)}, \ldots, \neg^{f(n)} B_{nf(n)}; B'_{11}, \neg B'_{12}, \ldots, B'_{n'1}, \neg B'_{n'2}\} \\
\tau_g(\Delta) = \{B''_{11}, \neg B''_{12}, \ldots, B''_{m1}, \neg B''_{m2}; \neg^{g(1)} B'''_{1g(1)}, \ldots, \neg^{g(m')} B'''_{m'f(m')}\},
\end{array}
$$

where $\neg^1_f = \neg, \neg^2_f = \lambda$; and $\neg^1_g = \lambda$ and $\neg^2_g = \neg$.

- **Deduction rules**:

$$
(+\sqcap^{LL}) \ \frac{\begin{array}{c}\Gamma, C_1 \sqsubseteq D \Rightarrow \Delta \\ \Gamma, C_2 \sqsubseteq D \Rightarrow \Delta\end{array}}{\Gamma, C_1 \sqcap C_2 \sqsubseteq D \Rightarrow \Delta} \qquad (+\sqcap^{RL}_1) \ \frac{\Gamma \Rightarrow C_1 \sqsubseteq D, \Delta}{\Gamma \Rightarrow C_1 \sqcap C_2 \sqsubseteq D, \Delta}
$$

$$
(+\sqcap^{RL}_2) \ \frac{\Gamma \Rightarrow C_2 \sqsubseteq D, \Delta}{\Gamma \Rightarrow C_1 \sqcap C_2 \sqsubseteq D, \Delta}
$$

$$
(+\sqcap^{LR}_1) \ \frac{\Gamma, C \sqsubseteq D_1 \Rightarrow \Delta}{\Gamma, C \sqsubseteq D_1 \sqcap D_2, \Delta} \qquad (+\sqcap^{RR}) \ \frac{\begin{array}{c}\Gamma \Rightarrow C \sqsubseteq D_1, \Delta \\ \Gamma \Rightarrow C \sqsubseteq D_2, \Delta\end{array}}{\Gamma \Rightarrow C \sqsubseteq D_1 \sqcap D_2, \Delta}
$$

$$
(+\sqcap^{LR}_2) \ \frac{\Gamma, C \sqsubseteq D_2 \Rightarrow \Delta}{\Gamma, C \sqsubseteq D_1 \sqcap D_2 \Rightarrow \Delta}
$$

$$
(-\sqcap^{LL}_1) \ \frac{\Gamma, C_1 \not\sqsubseteq D \Rightarrow \Delta}{\Gamma, C_1 \sqcap C_2 \not\sqsubseteq D \Rightarrow \Delta} \qquad (-\sqcap^{RL}) \ \frac{\begin{array}{c}\Gamma \Rightarrow C_1 \not\sqsubseteq D, \Delta \\ \Gamma \Rightarrow C_2 \not\sqsubseteq D, \Delta\end{array}}{\Gamma \Rightarrow C_1 \sqcap C_2 \not\sqsubseteq D, \Delta}
$$

$$
(-\sqcap^{LL}_2) \ \frac{\Gamma, C_2 \not\sqsubseteq D \Rightarrow \Delta}{\Gamma, C_1 \sqcap C_2 \not\sqsubseteq D \Rightarrow \Delta}
$$

$$
(-\sqcap^{LR}) \ \frac{\begin{array}{c}\Gamma, C \not\sqsubseteq D_1 \Rightarrow \Delta \\ \Gamma, C \not\sqsubseteq D_2 \Rightarrow \Delta\end{array}}{\Gamma, C \not\sqsubseteq D_1 \sqcap D_2 \Rightarrow \Delta} \qquad (-\sqcap^{RR}_1) \ \frac{\Gamma \Rightarrow C \not\sqsubseteq D_1, \Delta}{\Gamma \Rightarrow C \not\sqsubseteq D_1 \sqcap D_2, \Delta}
$$

$$
(-\sqcap^{RR}_2) \ \frac{\Gamma \Rightarrow C \not\sqsubseteq D_2, \Delta}{\Gamma \Rightarrow C \not\sqsubseteq D_1 \sqcap D_2, \Delta}
$$

and

$$(+\sqcup_1^{LL})\ \frac{\Gamma, C_1 \sqsubseteq D \Rightarrow \Delta}{\Gamma, C_1 \sqcup C_2 \sqsubseteq D \Rightarrow \Delta} \qquad (+\sqcup^{RL})\ \frac{\Gamma \Rightarrow C_1 \sqsubseteq D, \Delta \quad \Gamma \Rightarrow C_2 \sqsubseteq D, \Delta}{\Gamma \Rightarrow C_1 \sqcup C_2 \sqsubseteq D, \Delta}$$

$$(+\sqcup_2^{LL})\ \frac{\Gamma, C_2 \sqsubseteq D \Rightarrow \Delta}{\Gamma, C_1 \sqcup C_2 \sqsubseteq D \Rightarrow \Delta}$$

$$(+\sqcup^{LR})\ \frac{\Gamma, C \sqsubseteq D_1 \Rightarrow \Delta \quad \Gamma, C \sqsubseteq D_2 \Rightarrow \Delta}{\Gamma, C \sqsubseteq D_1 \sqcup D_2 \Rightarrow \Delta} \qquad (+\sqcup_1^{RR})\ \frac{\Gamma \Rightarrow C \sqsubseteq D_1, \Delta}{\Gamma \Rightarrow C \sqsubseteq D_1 \sqcup D_2, \Delta}$$

$$(+\sqcup_2^{RR})\ \frac{\Gamma \Rightarrow C \sqsubseteq D_2, \Delta}{\Gamma \Rightarrow C \sqsubseteq D_1 \sqcup D_2, \Delta}$$

$$(-\sqcup^{LL})\ \frac{\Gamma, C_1 \not\sqsubseteq D \Rightarrow \Delta \quad \Gamma, C_2 \not\sqsubseteq D \Rightarrow \Delta}{\Gamma, C_1 \sqcup C_2 \not\sqsubseteq D \Rightarrow \Delta} \qquad (-\sqcup_1^{RL})\ \frac{\Gamma \Rightarrow C_1 \not\sqsubseteq D, \Delta}{\Gamma \Rightarrow C_1 \sqcup C_2 \not\sqsubseteq D, \Delta}$$

$$(-\sqcup_2^{RL})\ \frac{\Gamma \Rightarrow C_2 \not\sqsubseteq D, \Delta}{\Gamma \Rightarrow C_1 \sqcup C_2 \not\sqsubseteq D, \Delta}$$

$$(-\sqcup_1^{LR})\ \frac{\Gamma, C \not\sqsubseteq D_1 \Rightarrow \Delta}{\Gamma, C \not\sqsubseteq D_1 \sqcup D_2 \Rightarrow \Delta} \qquad (-\sqcup^{RR})\ \frac{\Gamma \Rightarrow C \not\sqsubseteq D_1, \Delta \quad \Gamma \Rightarrow C \not\sqsubseteq D_2, \Delta}{\Gamma \Rightarrow C \not\sqsubseteq D_1 \sqcup D_2, \Delta}$$

$$(-\sqcup_2^{LR})\ \frac{\Gamma, C \not\sqsubseteq D_2 \Rightarrow \Delta}{\Gamma, C \not\sqsubseteq D_1 \sqcup D_2 \Rightarrow \Delta}$$

Definition 11.1.3 A sequent $\Gamma \Rightarrow \Delta$ is provable, denoted by $\vdash_{\mathbf{G}_4} \Gamma \Rightarrow \Delta$ if there is a sequence $\Gamma_1 \Rightarrow \Delta_1, ..., \Gamma_n \Rightarrow \Delta_n$ such that $\Gamma_n \Rightarrow \Delta_n = \Gamma \Rightarrow \Delta$, and for each $1 \leq i \leq n, \Gamma_i \Rightarrow \Delta_i$ is an axiom or deduced from the previous sequents by one of the deduction rules in $\mathbf{G}_4$.

11.1.3 Soundness and Completeness Theorem

Theorem 11.1.4 (Soundness theorem) *For any sequent* $\Gamma \Rightarrow \Delta$,

$$\vdash_{\mathbf{G}_4} \Gamma \Rightarrow \Delta \textit{ implies } \models_{\mathbf{G}_4} \Gamma \Rightarrow \Delta.$$

Proof We prove that each axiom is valid and each deduction rule preserves the validity.

To verify the validity of the axiom, assume that Γ and Δ satisfy the conditions in axiom $(\mathbf{A}^{\sqsubseteq})$, by Proposition 11.1.2, for any interpretation $I, I \models \Gamma \Rightarrow \Delta$.

To verify that $(+\sqcap^{LL})$ preserves the validity, assume that for any interpretation I and element $a \in U$,

$$I(\Gamma, C_1 \sqsubseteq D)(a) = 1 \text{ implies } I(\Delta)(a) = 1;$$
$$I(\Gamma, C_2 \sqsubseteq D)(a) = 1 \text{ implies } I(\Delta)(a) = 1.$$

For any interpretation I and element $a \in U$, assume that $I(\Gamma, C_1 \sqcap C_2 \sqsubseteq D)(a) = 1$. Then, $I(C_1 \sqcap C_2)(a) = 1$ implies $I(D)(a) = 1$. If $I(C_1)(a) = 0$ or $I(C_2)(a) = 0$ then $I(C_1 \sqsubseteq D)(a) = 1$ or $I(C_2 \sqsubseteq D)(a) = 1$, and by assumption, $I(\Delta)(a) = 1$; if $I(C_1)(a) = 1$ and $I(C_2)(a) = 1$ then $I(D)(a) = 1$, i.e., $I(C_1 \sqsubseteq D)(a) = 1, I(C_2 \sqsubseteq D)(a) = 1$, and by assumption, $I(\Delta)(a) = 1$.

To verify that $(+\sqcap_1^{RL})$ preserves the validity, assume that for any interpretation I and element $a \in U$,

$$I(\Gamma)(a) = 1 \text{ implies } I(C_1 \sqsubseteq D, \Delta)(a) = 1.$$

For any interpretation I and element $a \in U$, assume that $I(\Gamma)(a) = 1$. If $I(\Delta)(a) = 1$ then $I(C_1 \sqcap C_2 \sqsubseteq D, \Delta)(a) = 1$; otherwise, by the assumption, $I(C_1 \sqsubseteq D)(a) = 1$, and if $I(C_1 \sqcap C_2) = 0$ then $I(C_1 \sqcap C_2 \sqsubseteq D)(a) = 1$, and so $I(C_1 \sqcap C_2 \sqsubseteq D, \Delta)(a) = 1$; if $I(C_1 \sqcap C_2)(a) = 1$ then $I(C_1)(a) = 1$ and by assumption of $I(C_1 \sqsubseteq D)(a) = 1, I(D)(a) = 1$, i.e., $I(C_1 \sqcap C_2 \sqsubseteq D)(a) = 1$, and $I(C_1 \sqcap C_2 \sqsubseteq D, \Delta)(a) = 1$.

To verify that $(+\sqcap_1^{LR})$ preserves the validity, assume that for any interpretation I and element $a \in U$,

$$\mathrm{I}(\Gamma, C \sqsubseteq D_1)(a) = 1 \text{ implies } I(\Delta)(a) = 1.$$

For any interpretation I and element $a \in U$, assume that $I(\Gamma, C \sqsubseteq D_1 \sqcap D_2)(a) = 1$. Then, $I(C \sqsubseteq D_1 \sqcap D_2)(a) = 1$. If $I(C) = 0$ then $I(C \sqsubseteq D_1)(a) = 1$, and by the assumption, $I(\Delta)(a) = 1$; if $I(C)(a) = 1$ then $I(D_1 \sqcap D_2)(a) = 1, I(D_1)(a) = 1$, and by the assumption, $I(\Delta)(a) = 1$.

To verify that $(+\sqcap^{RR})$ preserves the validity, assume that for any interpretation I and element $a \in U$,

$$I(\Gamma)(a) = 1 \text{ implies } I(C \sqsubseteq D_1, \Delta)(a) = 1;$$
$$I(\Gamma)(a) = 1 \text{ implies } I(C \sqsubseteq D_2, \Delta)(a) = 1.$$

For any very interpretation I and element $a \in U$, assume that $I(\Gamma)(a) = 1$. Then, $I(C \sqsubseteq D_1, \Delta)(a) = 1$ and $I(C \sqsubseteq D_2, \Delta)(a) = 1$. If $I(\Delta)(a) = 1$ then $I(C \sqsubseteq D_1 \sqcap D_2, \Delta)(a) = 1$; if $I(\Delta) = 0$ then $I(C \sqsubseteq D_1)(a) = 1$ and $I(C \sqsubseteq D_2)(a) = 1$. If $I(C) = 0$ then $I(C \sqsubseteq D_1 \sqcap D_2)(a) = 1$, and so $I(C \sqsubseteq D_1 \sqcap D_2, \Delta)(a) = 1$; otherwise, $I(D_1)(a) = 1, I(D_2)(a) = 1$, i.e., $I(C \sqsubseteq D_1 \sqcap D_2)(a) = 1$, and so $I(C \sqsubseteq D_1 \sqcap D_2, \Delta)(a) = 1$.

Similar for other cases.

To verify that $(-\sqcap^{LL})$ preserves the validity, assume that for any interpretation I and any element $a \in U$,

$$\mathrm{I}(\Gamma, C_1 \not\sqsubseteq D)(a) = 1 \text{ implies } I(\Delta)(a) = 1.$$

For any very interpretation I and any element $a \in U$, assume that $I(\Gamma, C_1 \sqcap C_2 \not\sqsubseteq D)(a) = 1$. Then, $I(C_1 \sqcap C_2)(a) = 1$ and $I(D)(a) = 0$, which imply that $I(C_1)(a) = 1$ and $I(D)(a) = 0$. By the induction assumption, $I(\Delta)(a) = 1$.

To verify that $(-\sqcap_1^{RL})$ preserves the validity, assume that for any interpretation I and any element $a \in U$,

$$I(\Gamma)(a) = 1 \text{ implies } I(C_1 \not\sqsubseteq D, \Delta)(a) = 1;$$
$$I(\Gamma)(a) = 1 \text{ implies } I(C_2 \not\sqsubseteq D, \Delta)(a) = 1.$$

For any interpretation I and any element $a \in U$, assume that $I(\Gamma)(a) = 1$. If $I(\Delta)(a) = 1$ then $I(C_1 \sqcap C_2 \not\sqsubseteq D)(a) = 1$ and $I(C_1 \sqcap C_2 \not\sqsubseteq D, \Delta)(a) = 1$; otherwise, by the assumption, $I(C_1 \not\sqsubseteq D)(a) = 1$ and $I(C_2 \not\sqsubseteq D)(a) = 1$, i.e., $I(C_1)(a) = I(C_2)(a) = 1$ and $I(D)(a) = 0$, i.e., $I(C_1 \sqcap C_2 \not\sqsubseteq D)(a) = 1$, i.e., $I(C_1 \sqcap C_2 \not\sqsubseteq D, \Delta)(a) = 1$.

To verify that $(-\sqcap^{LR})$ preserves the validity, assume that for any interpretation I and any element $a \in U$,

$$I(\Gamma, C \not\sqsubseteq D_1)(a) = 1 \text{ implies } I(\Delta)(a) = 1;$$
$$I(\Gamma, C \not\sqsubseteq D_1)(a) = 1 \text{ implies } I(\Delta)(a) = 1.$$

For any interpretation I and any element $a \in U$, assume that $I(\Gamma, C \not\sqsubseteq D_1 \sqcap D_2)(a) = 1$. Then, $I(C \not\sqsubseteq D_1 \sqcap D_2)(a) = 1$, i.e., $I(C)(a) = 1$ and $I(D_1 \sqcap D_2)(a) = 0$, i.e., $I(C)(a) = 1$ and either $I(D_1)(a) = 0$ or $I(D_2)(a) = 0$. Let $i = 1, 2$. If $I(D_i)(a) = 0$ then $I(C \not\sqsubseteq D_i)(a) = 1$, and by induction assumption, $I(\Delta)(a) = 1$.

To verify that $(-\sqcap_1^{RR})$ preserves the validity, assume that for any interpretation I and any element $a \in U$,

$$I(\Gamma)(a) = 1 \text{ implies } I(C \not\sqsubseteq D_1, \Delta)(a) = 1.$$

For any interpretation I and any element $a \in U$, assume that $I(\Gamma)(a) = 1$. Then, $I(C \not\sqsubseteq D_1, \Delta)(a) = 1$, and if $I(\Delta)(a) = 1$ then $I(C \not\sqsubseteq D_1 \sqcap D_2, \Delta)(a) = 1$; if $I(C \not\sqsubseteq D_1)(a) = 1$ then $I(C)(a) = 1$ and $I(D_1)(a) = 0$, which imply that $I(C)(a) = 1$ and $I(D_1 \sqcap D_2)(a) = 0$, i.e., $I(C \not\sqsubseteq D_1 \sqcap D_2)(a) = 1$, and hence, $I(C \not\sqsubseteq D_1 \sqcap D_2, \Delta)(a) = 1$.

Similar for other cases. □

Theorem 11.1.5 (Completeness theorem) *For any sequent* $\Gamma \Rightarrow \Delta$,

$$\models_{\mathbf{G}_4} \Gamma \Rightarrow \Delta \textit{ implies } \vdash_{\mathbf{G}_4} \Gamma \Rightarrow \Delta.$$

Proof. Given a sequent $\Gamma \Rightarrow \Delta$, we construct a tree T as follows:

• the root of T is $\Gamma \Rightarrow \Delta$;

• if $\Gamma' \Rightarrow \Delta'$ is a node such that Γ', Δ' are sets of literal statements then the node is a leaf;

• if $\Gamma' \Rightarrow \Delta'$ is a node of T which is not a leaf then $\Gamma' \Rightarrow \Delta'$ has the direct child nodes

$$\left\{\begin{array}{ll}
\left\{\begin{array}{l}\Gamma_1, C_1 \sqsubseteq D \Rightarrow \Delta_1 \\ \Gamma_1, C_2 \sqsubseteq D \Rightarrow \Delta_1\end{array}\right. & \text{if } \Gamma' \Rightarrow \Delta' = \Gamma_1, C_1 \sqcap C_2 \sqsubseteq D \Rightarrow \Delta_1 \\
\left[\begin{array}{l}\Gamma_1 \Rightarrow C_1 \sqsubseteq D, \Delta_1 \\ \Gamma_1 \Rightarrow, C_2 \sqsubseteq D, \Delta_1\end{array}\right. & \text{if } \Gamma' \Rightarrow \Delta' = \Gamma_1 \Rightarrow C_1 \sqcap C_2 \sqsubseteq D, \Delta_1 \\
\left[\begin{array}{l}\Gamma_1, C \sqsubseteq D_1 \Rightarrow \Delta_1 \\ \Gamma_1, C \sqsubseteq D_2 \Rightarrow \Delta_1\end{array}\right. & \text{if } \Gamma' \Rightarrow \Delta' = \Gamma_1, C \sqsubseteq D_1 \sqcap D_2 \Rightarrow \Delta_1 \\
\left\{\begin{array}{l}\Gamma_1 \Rightarrow C \sqsubseteq D_1, \Delta_1 \\ \Gamma_1 \Rightarrow C \sqsubseteq D_2, \Delta_1\end{array}\right. & \text{if } \Gamma' \Rightarrow \Delta' = \Gamma_1 \Rightarrow C \sqsubseteq D_1 \sqcap D_2, \Delta_1
\end{array}\right.$$

and

$$\left\{\begin{array}{ll}
\left[\begin{array}{l}\Gamma_1, C_1 \not\sqsubseteq D \Rightarrow \Delta_1 \\ \Gamma_1, C_2 \not\sqsubseteq D \Rightarrow \Delta_1\end{array}\right. & \text{if } \Gamma' \Rightarrow \Delta' = \Gamma_1, C_1 \sqcap C_2 \not\sqsubseteq D \Rightarrow \Delta_1 \\
\left\{\begin{array}{l}\Gamma_1 \Rightarrow C_1 \not\sqsubseteq D, \Delta_1 \\ \Gamma_1 \Rightarrow C_2 \not\sqsubseteq D, \Delta_1\end{array}\right. & \text{if } \Gamma' \Rightarrow \Delta' = \Gamma_1 \Rightarrow C_1 \sqcap C_2 \not\sqsubseteq D, \Delta_1 \\
\left\{\begin{array}{l}\Gamma_1, C \not\sqsubseteq D_1 \Rightarrow \Delta_1 \\ \Gamma_1, C \not\sqsubseteq D_2 \Rightarrow \Delta_1\end{array}\right. & \text{if } \Gamma' \Rightarrow \Delta' = \Gamma_1, C \not\sqsubseteq D_1 \sqcap D_2 \Rightarrow \Delta_1 \\
\left[\begin{array}{l}\Gamma_1 \Rightarrow C \not\sqsubseteq D_1, \Delta_1 \\ \Gamma_1 \Rightarrow C \not\sqsubseteq D_2, \Delta_1\end{array}\right. & \text{if } \Gamma' \Rightarrow \Delta' = \Gamma_1 \Rightarrow C \not\sqsubseteq D_1 \sqcap D_2, \Delta_1
\end{array}\right.$$

and

$$\left\{\begin{array}{ll}
\left[\begin{array}{l}\Gamma_1, C_1 \sqsubseteq D \Rightarrow \Delta_1 \\ \Gamma_1, C_2 \sqsubseteq D \Rightarrow \Delta_1\end{array}\right. & \text{if } \Gamma' \Rightarrow \Delta' = \Gamma_1, C_1 \sqcup C_2 \sqsubseteq D \Rightarrow \Delta_1 \\
\left\{\begin{array}{l}\Gamma_1 \Rightarrow C_1 \sqsubseteq D, \Delta \\ \Gamma_1 \Rightarrow C_2 \sqsubseteq D, \Delta_1\end{array}\right. & \text{if } \Gamma' \Rightarrow \Delta' = \Gamma_1 \Rightarrow C_1 \sqcup C_2 \sqsubseteq D, \Delta_1 \\
\left\{\begin{array}{l}\Gamma_1, C \sqsubseteq D_1 \Rightarrow \Delta \\ \Gamma_1, C \sqsubseteq D_2 \Rightarrow \Delta_1\end{array}\right. & \text{if } \Gamma' \Rightarrow \Delta' = \Gamma_1, C \sqsubseteq D_1 \sqcup D_2 \Rightarrow \Delta_1 \\
\left[\begin{array}{l}\Gamma_1 \Rightarrow C \sqsubseteq D_1, \Delta_1 \\ \Gamma_1 \Rightarrow C \sqsubseteq D_2, \Delta_1\end{array}\right. & \text{if } \Gamma' \Rightarrow \Delta' = \Gamma_1 \Rightarrow C \sqsubseteq D_1 \sqcup D_2, \Delta_1
\end{array}\right.$$

and

$$\left\{\begin{array}{ll}
\left\{\begin{array}{l}\Gamma_1, C_1 \not\sqsubseteq D \Rightarrow \Delta_1 \\ \Gamma_1, C_2 \not\sqsubseteq D \Rightarrow \Delta_1\end{array}\right. & \text{if } \Gamma' \Rightarrow \Delta' = \Gamma_1, C_1 \sqcup C_2 \not\sqsubseteq D \Rightarrow \Delta_1 \\
\left[\begin{array}{l}\Gamma_1 \Rightarrow C_1 \not\sqsubseteq D, \Delta \\ \Gamma_1 \Rightarrow C_2 \not\sqsubseteq D, \Delta_1\end{array}\right. & \text{if } \Gamma' \Rightarrow \Delta' = \Gamma_1 \Rightarrow C_1 \sqcup C_2 \not\sqsubseteq D, \Delta_1 \\
\left[\begin{array}{l}\Gamma_1, C \not\sqsubseteq D_1 \Rightarrow \Delta \\ \Gamma_1, C \not\sqsubseteq D_2 \Rightarrow \Delta_1\end{array}\right. & \text{if } \Gamma' \Rightarrow \Delta' = \Gamma_1, C \not\sqsubseteq D_1 \sqcup D_2 \Rightarrow \Delta_1 \\
\left\{\begin{array}{l}\Gamma_1 \Rightarrow C \not\sqsubseteq D_1, \Delta_1 \\ \Gamma_1 \Rightarrow C \not\sqsubseteq D_2, \Delta_1\end{array}\right. & \text{if } \Gamma' \Rightarrow \Delta' = \Gamma_1 \Rightarrow C \not\sqsubseteq D_1 \sqcup D_2, \Delta_1
\end{array}\right.$$

where $\left[\begin{array}{l}\delta_1 \\ \delta_2\end{array}\right.$ represents that δ_1, δ_2 are at a same subling; and $\left\{\begin{array}{l}\delta_1 \\ \delta_2\end{array}\right.$ represents that δ_1, δ_2 are at different direct sublings.

A sequent $\Gamma \Rightarrow \Delta$ is provable in $\mathbf{G}_4$, denoted by $\vdash_{\mathbf{G}_4} \Gamma \Rightarrow \Delta$, if there is a proof tree T of $\Gamma \Rightarrow \Delta$ such that each leaf is an axiom.

Lemma 11.1.6 *If there is a branch $\alpha \subseteq T$ such that the leaf of α is not an axiom in $\mathbf{G}_4$ then there is an interpretation I such that $I \not\models \Gamma' \Rightarrow \Delta'$ for each $\Gamma' \Rightarrow \Delta' \in \alpha$.*

Proof Let $\Gamma' \Rightarrow \Delta'$ be the leaf of α. By Proposition 11.1.2, there is an interpretation I such that $I \models \Gamma'$ and $I \not\models \Delta'$.

Fix any $\gamma \in \alpha$, and assume that $I \not\models \gamma$.

Case $(+\sqcap^{LL})$. If γ is generated from $\beta \in \alpha$ by $(\sqcap^{LL})$ then there are concepts $C_1^1, C_2^1, ..., C_1^m, C_2^m, D^1, ..., D^m$ and $\beta \in \{\beta_1, ..., \beta_n\}$ such that $\beta \in \alpha$, and

$$\begin{array}{l}\gamma = \Gamma'', C_{f(1)}^1 \sqsubseteq D^1, ..., C_{f(m)}^m \sqsubseteq D^m \Rightarrow \Delta', \\ \beta = \Gamma'', C_1^1 \sqcap C_2^1 \sqsubseteq D^1, ..., C_1^m \sqcap C_2^m \sqsubseteq D^m \Rightarrow \Delta',\end{array}$$

where $f : \{1, ..., m\} \in \{1, 2\}$ is a function. By induction assumption,

$$I \models \Gamma''(a), (C_1^1 \sqcap C_2^1 \sqsubseteq D^1)(a), ..., (C_1^m \sqcap C_2^m \sqsubseteq D^m)(a).$$

Then

$$I \models \Gamma''(a), (C_{f(1)}^1 \sqsubseteq D^1)(a), ..., (C_{f(m)}^m \sqsubseteq D^m)(a),$$

and by induction assumption, $I \not\models \Delta'(a)$.

Case $(+\sqcap^{RL})$. If γ is generated from $\beta \in \alpha$ by $(\sqcap^R)$ then there are concepts $C^1, ..., C^m, D_1^1, D_2^1, ..., D_1^m, D_2^m$ such that

$$\begin{array}{l}\gamma = \Gamma'' \Rightarrow C^1 \sqsubseteq D_1^1, C^1 \sqsubseteq D_2^1, ..., C^m \sqsubseteq D_1^m, C^m \sqsubseteq D_2^m, \Delta'', \\ \beta = \Gamma'' \Rightarrow C^1 \sqsubseteq D_1^1 \sqcap D_2^1, ..., C^m \sqsubseteq D_1^m \sqcap D_2^m, \Delta''.\end{array}$$

By the induction assumption, $I \models \Gamma''(a)$, and

$$I \not\models (C^1 \sqsubseteq D_1^1)(a), (C^1 \sqsubseteq D_2^1)(a), ..., (C^m \sqsubseteq D_1^m)(a), (C^m \sqsubseteq D_2^m)(a), \Delta''(a).$$

Therefore,

$$I \not\models (C^1 \sqsubseteq D_1^1 \sqcap D_2^1)(a), ..., (C^m \sqsubseteq D_1^m \sqcap D_2^m)(a), \Delta''(a).$$

Case $(-\sqcap^{LL})$. If γ is generated from $\beta \in \alpha$ by $(\sqcap^{LL})$ then there are concepts $C_1^1, C_2^1, ..., C_1^m, C_2^m, D^1, ..., D^m$ and $\beta \in \{\beta_1, ..., \beta_n\}$ such that $\beta \in \alpha$, and

$$\begin{array}{l}\gamma = \Gamma'', C_{f(1)}^1 \not\sqsubseteq D^1, ..., C_{f(m)} \not\sqsubseteq D^m \Rightarrow \Delta'', \\ \beta = \Gamma'', C_1^1 \sqcap C_2^1 \not\sqsubseteq D^1, ..., C_m^1 \sqcap C_m^2 \not\sqsubseteq D^m \Rightarrow \Delta'',\end{array}$$

where $f : \{1, ..., m\} \in \{1, 2\}$ is a function. By induction assumption,

$$I \models \Gamma''(a), (C_1^1 \sqcap C_2^1 \not\sqsubseteq D^1)(a), ..., (C_m^1 \sqcap C_m^2 \not\sqsubseteq D^m)(a).$$

Then $I \models \Gamma''(a), (C^1_{f(1)} \not\sqsubseteq D^1)(a), ..., (C_{f(m)} \not\sqsubseteq D^m)(a)$, and by induction assumption, $I \not\models \Delta''(a)$.

Case $(-\sqcap^{RL})$. If γ is generated from $\beta \in \alpha$ by $(\sqcap^R)$ then there are concepts $C^1, ..., C^m, D^1_1, D^1_2, ..., D^m_1, D^m_2$ such that

$$\begin{aligned} \gamma &= \Gamma'' \Rightarrow C^1 \not\sqsubseteq D^1_1, C^1 \not\sqsubseteq D^1_2, ..., C^m \not\sqsubseteq D^m_1, C^m \not\sqsubseteq D^m_2, \Delta'', \\ \beta &= \Gamma'' \Rightarrow C^1 \not\sqsubseteq D^1_1 \sqcap D^1_2, ..., C^m \not\sqsubseteq D^m_1 \sqcap D^m_2, \Delta''. \end{aligned}$$

By the induction assumption, $I \models \Gamma''(a)$, and

$$I \not\models (C^1 \not\sqsubseteq D^1_1)(a), (C^1 \not\sqsubseteq D^1_2)(a), ..., (C^m \not\sqsubseteq D^m_1)(a), (C^m \not\sqsubseteq D^m_2)(a), \Delta''(a).$$

Therefore,

$$I \not\models (C^1 \not\sqsubseteq D^1_1 \sqcap D^1_2)(a), ..., (C^m \not\sqsubseteq D^m_1 \sqcap D^m_2)(a), \Delta''(a).$$

□

Lemma 11.1.7 *If for each branch $\alpha \subseteq T$, the leaf of α is an axiom in* $\mathbf{G}_4$ *then the tree is a proof tree of* $\Gamma \Rightarrow \Delta$.

Proof By the construction of the tree.

This completes the proof. □

Therefore, Γ is consistent with $C \sqsubseteq D$ $(C \not\sqsubseteq D)$ iff there is an interpretation I and an element $a \in U$ such that

$$I \models \Gamma(a)$$

and

$$I \models (C \sqsubseteq D)(a),$$

i.e., $a \in C^I$ implies $a \in D^I$ $(I \models (C \not\sqsubseteq D)(a)$, i.e., $a \in C^I$ and $a \notin D^I)$.

11.2 R-Calculus for ⊆-Minimal Change

Given theories Γ and Δ, Θ is a $\subseteq$-minimal change of Γ by Δ, denoted by $\models_{\mathbf{S}^{\mathrm{SN}}} \Delta|\Gamma \Rightarrow \Delta, \Theta$, if Θ is minimal such that (i) $\Theta \subseteq \Gamma$ is consistent with Δ, and (ii) for any statement $\theta \in \Gamma - \Theta$, $\Theta \cup \Delta \cup \{\theta\}$ is inconsistent.

In this section, we will give an R-calculus $\mathbf{S}^{\mathrm{SN}}$ such that for any theories Γ, Δ and Θ, $\Delta|\Gamma \Rightarrow \Delta, \Theta$ is provable in $\mathbf{S}^{\mathrm{SN}}$ if and only if Θ is a $\subseteq$-minimal change of Γ by Δ.

11.2.1 R-Calculus $\mathbf{S}^{\mathrm{SN}}$ for a Statement

R-calculus $\mathbf{S}^{\mathrm{SN}}$ for a statement $C \sqsubseteq D$ consists of the following axioms and deduction rules:

- **Axioms**:

$$(S^{\mathbf{A}})\frac{\Delta \nvdash B \not\sqsubseteq B'}{\Delta|B \sqsubseteq B' \Rightarrow \Delta, B \sqsubseteq B'} \quad (S_{\mathbf{A}})\frac{\Delta \vdash B \not\sqsubseteq B'}{\Delta|B \sqsubseteq B' \Rightarrow \Delta}$$

$$(S^{\neg})\frac{\Delta \nvdash B \sqsubseteq B'}{\Delta|B \not\sqsubseteq B' \Rightarrow \Delta, B \not\sqsubseteq B'} \quad (S_{\neg})\frac{\Delta \vdash B \sqsubseteq B'}{\Delta|B \not\sqsubseteq B' \Rightarrow \Delta}$$

- **Deduction rules for** $C \sqsubseteq D$:

$$({}^{\sqcap}S_1)\ \frac{\Delta|C_1 \sqsubseteq D \Rightarrow \Delta, C_1 \sqsubseteq D}{\Delta|C_1 \sqcap C_2 \sqsubseteq D \Rightarrow \Delta, C_1 \sqcap C_2 \sqsubseteq D}$$

$$({}^{\sqcap}S_2)\ \frac{\begin{array}{l}\Delta|C_1 \sqsubseteq D \Rightarrow \Delta\\ \Delta|C_2 \sqsubseteq D \Rightarrow \Delta, C_2 \sqsubseteq D\end{array}}{\Delta|C_1 \sqcap C_2 \sqsubseteq D \Rightarrow \Delta, C_1 \sqcap C_2 \sqsubseteq D}$$

$$({}_{\sqcap}S)\ \frac{\begin{array}{l}\Delta|C_1 \sqsubseteq D \Rightarrow \Delta\\ \Delta|C_2 \sqsubseteq D \Rightarrow \Delta\end{array}}{\Delta|C_1 \sqcap C_2 \sqsubseteq D \Rightarrow \Delta}$$

$$({}^{\sqcup}S)\ \frac{\begin{array}{l}\Delta|C_1 \sqsubseteq D \Rightarrow \Delta, C_1 \sqsubseteq D\\ \Delta, C_1 \sqsubseteq D|C_2 \sqsubseteq D \Rightarrow \Delta, C_1 \sqsubseteq D, C_2 \sqsubseteq D\end{array}}{\Delta|C_1 \sqcup C_2 \sqsubseteq D \Rightarrow \Delta, C_1 \sqcup C_2 \sqsubseteq D}$$

$$({}_{\sqcup}S_1)\ \frac{\Delta|C_1 \sqsubseteq D \Rightarrow \Delta}{\Delta|C_1 \sqcup C_2 \sqsubseteq D \Rightarrow \Delta}$$

$$({}_{\sqcup}S_2)\ \frac{\Delta, C_1 \sqsubseteq D|C_2 \sqsubseteq D \Rightarrow \Delta, C_1 \sqsubseteq D}{\Delta|C_1 \sqcup C_2 \sqsubseteq D \Rightarrow \Delta}$$

and

$$(S^{\sqcap})\ \frac{\begin{array}{l}\Delta|C \sqsubseteq D_1 \Rightarrow \Delta, C \sqsubseteq D_1|\\ \Delta, C \sqsubseteq D_1|C \sqsubseteq D_2 \Rightarrow \Delta, C \sqsubseteq D_1, C \sqsubseteq D_2\end{array}}{\Delta|C \sqsubseteq D_1 \sqcap D_2 \Rightarrow \Delta, C \sqsubseteq D_1 \sqcap D_2|}$$

$$(S_{\sqcap}^1)\ \frac{\Delta|C \sqsubseteq D_1 \Rightarrow \Delta}{\Delta|C \sqsubseteq D_1 \sqcap D_2 \Rightarrow \Delta|}$$

$$(S_{\sqcap}^2)\ \frac{\Delta, C \sqsubseteq D_1|C \sqsubseteq D_2 \Rightarrow \Delta, C \sqsubseteq D_1}{\Delta|C \sqsubseteq D_1 \sqcap D_2 \Rightarrow \Delta}$$

$$(S_1^{\sqcup})\ \frac{\Delta|C \sqsubseteq D_1 \Rightarrow \Delta, C \sqsubseteq D_1}{\Delta|C \sqsubseteq D_1 \sqcup D_2 \Rightarrow \Delta, C \sqsubseteq D_1 \sqcup D_2}$$

$$(S_2^{\sqcup})\ \frac{\Delta|C \sqsubseteq D_2 \Rightarrow \Delta, C \sqsubseteq D_2}{\Delta|C \sqsubseteq D_1 \sqcup D_2 \Rightarrow \Delta, C \sqsubseteq D_1 \sqcup D_2}$$

$$(S_{\sqcup})\ \frac{\Delta|C \sqsubseteq D_1 \Rightarrow \Delta \quad \Delta|C \sqsubseteq D_2 \Rightarrow \Delta}{\Delta|C \sqsubseteq D_1 \sqcup D_2 \Rightarrow \Delta}$$

where the rules of the left-hand side are to put a statement into Θ, and the ones of the right-hand side are not to.

• **Deduction rules for** $C \not\sqsubseteq D$:

$$(-^{\sqcap}S)\ \frac{\begin{array}{l}\Delta|C_1 \not\sqsubseteq D \Rightarrow \Delta, C_1 \not\sqsubseteq D\\ \Delta, C_1 \not\sqsubseteq D|C_2 \not\sqsubseteq D \Rightarrow \Delta, C_1 \not\sqsubseteq D, C_2 \not\sqsubseteq D\end{array}}{\Delta|C_1 \sqcap C_2 \not\sqsubseteq D \Rightarrow \Delta, C_1 \sqcap C_2 \not\sqsubseteq D}$$

$$(-_{\sqcap}S_1)\ \frac{\Delta|C_1 \not\sqsubseteq D \Rightarrow \Delta}{\Delta|C_1 \sqcap C_2 \not\sqsubseteq D \Rightarrow \Delta}$$

$$(-_{\sqcap}S_2)\ \frac{\Delta, C_1 \not\sqsubseteq D|C_2 \not\sqsubseteq D \Rightarrow \Delta, C_1 \not\sqsubseteq D}{\Delta|C_1 \sqcap C_2 \not\sqsubseteq D \Rightarrow \Delta}$$

$$(-^{\sqcup}S_1)\ \frac{\Delta|C_1 \not\sqsubseteq D \Rightarrow \Delta, C_1 \not\sqsubseteq D}{\Delta|C_1 \sqcup C_2 \not\sqsubseteq D \Rightarrow \Delta, C_1 \sqcup C_2 \not\sqsubseteq D}$$

$$(-^{\sqcup}S_2)\ \frac{\begin{array}{l}\Delta|C_1 \not\sqsubseteq D \Rightarrow \Delta\\ \Delta|C_2 \not\sqsubseteq D \Rightarrow \Delta, C_2 \not\sqsubseteq D\end{array}}{\Delta|C_1 \sqcup C_2 \not\sqsubseteq D \Rightarrow \Delta, C_1 \sqcup C_2 \not\sqsubseteq D}$$

$$(-_{\sqcup}S)\ \frac{\Delta|C_1 \not\sqsubseteq D \Rightarrow \Delta \quad \Delta|C_2 \not\sqsubseteq D \Rightarrow \Delta}{\Delta|C_1 \sqcup C_2 \not\sqsubseteq D \Rightarrow \Delta}$$

and

$$(-S_1^{\sqcap})\ \frac{\Delta|C \not\sqsubseteq D_1 \Rightarrow \Delta, C \not\sqsubseteq D_1}{\Delta|C \not\sqsubseteq D_1 \sqcap D_2 \Rightarrow \Delta, C \not\sqsubseteq D_1 \sqcap D_2}$$

$$(-S_2^{\sqcap})\ \frac{\begin{array}{l}\Delta|C \not\sqsubseteq D_1 \Rightarrow \Delta\\ \Delta|C \not\sqsubseteq D_2 \Rightarrow \Delta, C \not\sqsubseteq D_2\end{array}}{\Delta|C \not\sqsubseteq D_1 \sqcap D_2 \Rightarrow \Delta, C \not\sqsubseteq D_1 \sqcap D_2}$$

$$(-S_{\sqcap})\ \frac{\begin{array}{l}\Delta|C \not\sqsubseteq D_1 \Rightarrow \Delta\\ \Delta|C \not\sqsubseteq D_2 \Rightarrow \Delta\end{array}}{\Delta|C \not\sqsubseteq D_1 \sqcap D_2 \Rightarrow \Delta}$$

$$(-S^{\sqcup})\ \frac{\begin{array}{l}\Delta|C \not\sqsubseteq D_1 \Rightarrow \Delta, C \not\sqsubseteq D_1\\ \Delta, C \not\sqsubseteq D_1|C \not\sqsubseteq D_2 \Rightarrow \Delta, C \not\sqsubseteq D_1, C \not\sqsubseteq D_2\end{array}}{\Delta|C \not\sqsubseteq D_1 \sqcup D_2 \Rightarrow \Delta, C \not\sqsubseteq D_1 \sqcup D_2}$$

$$(-S_{\sqcup}^1)\ \frac{\Delta|C \not\sqsubseteq D_1 \Rightarrow \Delta}{\Delta|C \not\sqsubseteq D_1 \sqcup D_2 \Rightarrow \Delta}$$

$$(-S_{\sqcup}^2)\ \frac{\begin{array}{l}\Delta|C \not\sqsubseteq D_1 \Rightarrow \Delta, C \not\sqsubseteq D_1\\ \Delta, C \not\sqsubseteq D_1|C \not\sqsubseteq D_2 \Rightarrow \Delta, C \not\sqsubseteq D_1\end{array}}{\Delta|C \not\sqsubseteq D_1 \sqcup D_2 \Rightarrow \Delta}$$

Definition 11.2.1 $\Delta|C \sqsubseteq D \Rightarrow \Delta, C' \sqsubseteq D'$ is provable in $\mathbf{S}^{\mathrm{SN}}$, denoted by $\vdash_{\mathbf{S}^{\mathrm{SN}}} \Delta|C \sqsubseteq D \Rightarrow \Delta, C' \sqsubseteq D'$, if there is a sequence $S_1, ..., S_m$ of statements such that

$$\begin{array}{l} S_1 = \Delta|C \sqsubseteq D \Rightarrow \Delta|C_1' \sqsubseteq D_1', \\ \cdots \\ S_m = \Delta|C_m' \sqsubseteq D_m \Rightarrow \Delta, C' \sqsubseteq D'; \end{array}$$

and for each $i < m$, S_{i+1} is an axiom or is deduced from the previous statements by a deduction rule in $\mathbf{S}^{\mathrm{SN}}$.

11.2.2 Soundness and Completeness Theorem

Theorem 11.2.2 (Completeness theorem 1) *For any consistent theory Δ and a statement $C \sqsubseteq D$, if $\Delta \cup \{C \sqsubseteq D\}$ is consistent then $\Delta|C \sqsubseteq D \Rightarrow \Delta, C \sqsubseteq D$ is provable in $\mathbf{S}^{\mathrm{SN}}$, i.e.,*

$$\models_{\mathbf{S}^{\mathrm{SN}}} \Delta|C \sqsubseteq D \Rightarrow \Delta, C \sqsubseteq D \text{ implies } \vdash_{\mathbf{S}^{\mathrm{SN}}} \Delta|C \sqsubseteq D \Rightarrow \Delta, C \sqsubseteq D;$$

and if $\Delta \cup \{C \sqsubseteq D\}$ is inconsistent then $\Delta|C \sqsubseteq D \Rightarrow \Delta$ is provable in $\mathbf{S}^{\mathrm{SN}}$, i.e.,

$$\models_{\mathbf{S}^{\mathrm{SN}}} \Delta|C \sqsubseteq D \Rightarrow \Delta \text{ implies } \vdash_{\mathbf{S}^{\mathrm{SN}}} \Delta|C \sqsubseteq D \Rightarrow \Delta.$$

Proof We prove the theorem by induction on the structure of concepts C and D.

Assume that $\Delta \cup \{C \sqsubseteq D\}$ is consistent.

If $C \sqsubseteq D = B_1 \sqsubseteq B_2$ then $\Delta \nvdash B_1 \not\sqsubseteq B_2$, and by $(S^{\mathbf{A}})$, $\Delta|C \sqsubseteq D \Rightarrow \Delta, C \sqsubseteq D$ is provable;

If $C = C_1 \sqcap C_2$ then either $\Delta \cup \{C_1 \sqsubseteq D\}$ or $\Delta \cup \{C_2 \sqsubseteq D\}$ are consistent, and by induction assumption, either

$$\vdash_{\mathbf{S}^{\mathrm{SN}}} \Delta|C_1 \sqsubseteq D \Rightarrow \Delta, C_1 \sqsubseteq D$$

or

$$\vdash_{\mathbf{S}^{\mathrm{SN}}} \Delta|C_2 \sqsubseteq D \Rightarrow \Delta, C_2 \sqsubseteq D;$$

and by $({}^{\sqcap}S_1)$ or $({}^{\sqcap}S_2)$, $\Delta|C_1 \sqcap C_2 \sqsubseteq D \Rightarrow \Delta, C_1 \sqcap C_2 \sqsubseteq D$ is provable;

If $C = C_1 \sqcup C_2$ then $\Delta \cup \{C_1 \sqsubseteq D\}$ and $\Delta \cup \{C_1 \sqsubseteq D, C_2 \sqsubseteq D\}$ are consistent, and by induction assumption,

$$\begin{array}{l} \vdash_{\mathbf{S}^{\mathrm{SN}}} \Delta|C_1 \sqsubseteq D \Rightarrow \Delta, C_1 \sqsubseteq D; \\ \vdash_{\mathbf{S}^{\mathrm{SN}}} \Delta, C_1 \sqsubseteq D|C_2 \sqsubseteq D \Rightarrow \Delta, C_1 \sqsubseteq D, C_2 \sqsubseteq D; \end{array}$$

and by $({}^{\sqcup}S)$, $\Delta|C_1 \sqcup C_2 \sqsubseteq D \Rightarrow \Delta, C_1 \sqcup C_2 \sqsubseteq D$ is provable.

If $D = D_1 \sqcap D_2$ then $\Delta \cup \{C \sqsubseteq D_1\}$ and $\Delta \cup \{C \sqsubseteq D_1, C \sqsubseteq D_2\}$ are consistent, and by induction assumption,

$$\begin{array}{l} \vdash_{\mathbf{S}^{\mathrm{SN}}} \Delta|C \sqsubseteq D_1 \Rightarrow \Delta, C \sqsubseteq D_1; \\ \vdash_{\mathbf{S}^{\mathrm{SN}}} \Delta, C \sqsubseteq D_1|C \sqsubseteq D_2 \Rightarrow \Delta, C \sqsubseteq D_1, C \sqsubseteq D_2; \end{array}$$

and by $(S^{\sqcap})$, $\Delta|C \sqsubseteq D_1 \sqcap D_2 \Rightarrow \Delta, C \sqsubseteq D_1 \sqcap D_2$ is provable;

If $D = D_1 \sqcup D_2$ then either $\Delta \cup \{C \sqsubseteq D_1\}$ or $\Delta \cup \{C \sqsubseteq D_2\}$ is consistent, and by induction assumption, either

$$\vdash_{\mathbf{S}^{\mathrm{SN}}} \Delta|C \sqsubseteq D_1 \Rightarrow \Delta, C \sqsubseteq D_1$$

or

$$\vdash_{\mathbf{S}^{\mathrm{SN}}} \Delta|C \sqsubseteq D_2 \Rightarrow \Delta, C \sqsubseteq D_2;$$

and by $(S_1^{\sqcup})$ or $(S_2^{\sqcup})$, $\Delta|C \sqsubseteq D_1 \sqcup D_2 \Rightarrow \Delta, C \sqsubseteq D_1 \sqcup D_2$ is provable.

Assume that $\Delta \cup \{C \sqsubseteq D\}$ is inconsistent.

If $C \sqsubseteq D = B_1 \sqsubseteq B_2$ then $\Delta \vdash B_1 \not\sqsubseteq B_2$, and by $(S_{\mathbf{A}})$, $\Delta|C \sqsubseteq D \Rightarrow \Delta$ is provable;

If $C = C_1 \sqcap C_2$ then $\Delta \cup \{C_1 \sqsubseteq D\}$ and $\Delta \cup \{C_2 \sqsubseteq D\}$ are inconsistent, and by induction assumption,

$$\vdash_{\mathbf{S}^{\mathrm{SN}}} \Delta|C_1 \sqsubseteq D \Rightarrow \Delta;$$
$$\vdash_{\mathbf{S}^{\mathrm{SN}}} \Delta|C_2 \sqsubseteq D \Rightarrow \Delta;$$

by $({}_{\sqcap}S)$, $\Delta|C_1 \sqcap C_2 \sqsubseteq D \Rightarrow \Delta$ is provable;

If $C = C_1 \sqcup C_2$ then either $\Delta \cup \{C_1 \sqsubseteq D\}$ or $\Delta \cup \{C_1 \sqsubseteq D, C_2 \sqsubseteq D\}$ are inconsistent, and by induction assumption, either

$$\vdash_{\mathbf{S}^{\mathrm{SN}}} \Delta|C_1 \sqsubseteq D \Rightarrow \Delta;$$

or

$$\vdash_{\mathbf{S}^{\mathrm{SN}}} \Delta, C_1 \sqsubseteq D|C_2 \sqsubseteq D \Rightarrow \Delta, C_1 \sqsubseteq D;$$

and by $(-_{\sqcup}S_1)$ or $(-_{\sqcup}S_2)$, $\Delta|C_1 \sqcup C_2 \sqsubseteq D \Rightarrow \Delta$ is provable.

If $D = D_1 \sqcap D_2$ then either $\Delta \cup \{C \sqsubseteq D_1\}$ or $\Delta \cup \{C \sqsubseteq D_1, C \sqsubseteq D_2\}$ is inconsistent, and by induction assumption, either

$$\vdash_{\mathbf{S}^{\mathrm{SN}}} \Delta|C \sqsubseteq D_1 \Rightarrow \Delta;$$

or

$$\vdash_{\mathbf{S}^{\mathrm{SN}}} \Delta, C \sqsubseteq D_1|C \sqsubseteq D_2 \Rightarrow \Delta, C \sqsubseteq D_1;$$

by $(S_{\sqcap}^1)$ or $(S_{\sqcap}^2)$, $\Delta|C \sqsubseteq D_1 \sqcap D_2 \Rightarrow \Delta$ is provable;

If $D = D_1 \sqcup D_2$ then $\Delta \cup \{C \sqsubseteq D_1\}$ and $\Delta \cup \{C \sqsubseteq D_2\}$ are inconsistent, and by induction assumption,

$$\vdash_{\mathbf{S}^{\mathrm{SN}}} \Delta|C_1 \sqsubseteq D \Rightarrow \Delta;$$
$$\vdash_{\mathbf{S}^{\mathrm{SN}}} \Delta|C_2 \sqsubseteq D \Rightarrow \Delta;$$

and by $(S_{\sqcup})$, $\Delta|C \sqsubseteq D_1 \sqcup D_2 \Rightarrow \Delta$ is provable. □

Similarly we have the following

Theorem 11.2.3 (Completeness theorem 2) *For any consistent theory* Δ *and a statement* $C \not\sqsubseteq D$, *if* $\Delta \cup \{C \not\sqsubseteq D\}$ *is consistent then* $\Delta|C \not\sqsubseteq D \Rightarrow \Delta, C \not\sqsubseteq D$ *is provable in* $\mathbf{S}^{\mathrm{SN}}$, *i.e.,*

$$\models_{\mathbf{S}^{\mathrm{SN}}} \Delta|C \not\sqsubseteq D \Rightarrow \Delta, C \not\sqsubseteq D \text{ implies } \vdash_{\mathbf{S}^{\mathrm{SN}}} \Delta|C \not\sqsubseteq D \Rightarrow \Delta, C \not\sqsubseteq D;$$

and if $\Delta \cup \{C \not\sqsubseteq D\}$ *is inconsistent then* $\Delta|C \not\sqsubseteq D \Rightarrow \Delta$ *is provable in* $\mathbf{S}^{\mathrm{SN}}$, *i.e.,*

$$\models_{\mathbf{S}^{\mathrm{SN}}} \Delta|C \not\sqsubseteq D \Rightarrow \Delta \text{ implies } \vdash_{\mathbf{S}^{\mathrm{SN}}} \Delta|C \not\sqsubseteq D \Rightarrow \Delta.$$

□

Theorem 11.2.4 (Soundness theorem 1) *For any consistent theory* Δ *and a statement* $C \sqsubseteq D$, *if* $\Delta|C \sqsubseteq D \Rightarrow \Delta, C \sqsubseteq D$ *is provable in* $\mathbf{S}^{\mathrm{SN}}$ *then* $\Delta \cup \{C \sqsubseteq D\}$ *is consistent, i.e.,*

$$\vdash_{\mathbf{S}^{\mathrm{SN}}} \Delta|C \sqsubseteq D \Rightarrow \Delta, C \sqsubseteq D \text{ implies } \models_{\mathbf{S}^{\mathrm{SN}}} \Delta|C \sqsubseteq D \Rightarrow \Delta, C \sqsubseteq D;$$

and if $\Delta|C \sqsubseteq D \Rightarrow \Delta$ *is provable in* $\mathbf{S}^{\mathrm{SN}}$ *then* $\Delta \cup \{C \sqsubseteq D\}$ *is inconsistent, i.e.,*

$$\vdash_{\mathbf{S}^{\mathrm{SN}}} \Delta|C \sqsubseteq D \Rightarrow \Delta \text{ implies } \models_{\mathbf{S}^{\mathrm{SN}}} \Delta|C \sqsubseteq D \Rightarrow \Delta.$$

Proof We prove the theorem by induction on the structure of concepts C and D. Assume that $\vdash_{\mathbf{S}^{\mathrm{SN}}} \Delta|C \sqsubseteq D \Rightarrow \Delta, C \sqsubseteq D$.

If $C \sqsubseteq D = B_1 \sqsubseteq B_2$ then if $\Delta|B_1 \sqsubseteq B_2 \Rightarrow \Delta, B_1 \sqsubseteq B_2$ then $\Delta \nvdash B_1 \not\sqsubseteq B_2$, and $\Delta \cup \{B_1 \sqsubseteq B_2\}$ is consistent;

If $C = C_1 \sqcap C_2$ then either

$$\vdash_{\mathbf{S}^{\mathrm{SN}}} \Delta|C_1 \sqsubseteq D \Rightarrow \Delta, C_1 \sqsubseteq D$$

or

$$\vdash_{\mathbf{S}^{\mathrm{SN}}} \Delta|C_2 \sqsubseteq D \Rightarrow \Delta, C_2 \sqsubseteq D;$$

and by induction assumption, $\Delta \cup \{C_1 \sqsubseteq D\}$ and $\Delta \cup \{C_2 \sqsubseteq D\}$ are consistent, and so is $\Delta \cup \{C_1 \sqcap C_2 \sqsubseteq D\}$;

If $C = C_1 \sqcup C_2$ then

$$\vdash_{\mathbf{S}^{\mathrm{SN}}} \Delta|C_1 \sqsubseteq D \Rightarrow \Delta, C_1 \sqsubseteq D;$$
$$\vdash_{\mathbf{S}^{\mathrm{SN}}} \Delta, C_1 \sqsubseteq D|C_2 \sqsubseteq D \Rightarrow \Delta, C_1 \sqsubseteq D, C_2 \sqsubseteq D;$$

and by induction assumption, $\Delta \cup \{C_1 \sqsubseteq D\}$ and $\Delta \cup \{C_1 \sqsubseteq D, C_2 \sqsubseteq D\}$ are consistent and so is $\Delta \cup \{C_1 \sqcup C_2 \sqsubseteq D\}$.

If $D = D_1 \sqcap D_2$ then,

$$\vdash_{\mathbf{S}^{\mathrm{SN}}} \Delta|C \sqsubseteq D_1 \Rightarrow \Delta, C \sqsubseteq D_1;$$
$$\vdash_{\mathbf{S}^{\mathrm{SN}}} \Delta, C \sqsubseteq D_1|C \sqsubseteq D_2 \Rightarrow \Delta, C \sqsubseteq D_1, C \sqsubseteq D_2;$$

and by induction assumption, $\Delta \cup \{C \sqsubseteq D_1\}$ and $\Delta \cup \{C \sqsubseteq D_1, C \sqsubseteq D_2\}$ are consistent, and so is $\Delta \cup \{C \sqsubseteq D_1 \sqcap D_2\}$;

If $D = D_1 \sqcup D_2$ then either

$$\vdash_{\mathbf{S}^{\mathrm{SN}}} \Delta | C \sqsubseteq D_1 \Rightarrow \Delta, C \sqsubseteq D_1$$

or

$$\vdash_{\mathbf{S}^{\mathrm{SN}}} \Delta | C \sqsubseteq D_2 \Rightarrow \Delta, C \sqsubseteq D_2;$$

and by induction assumption, either $\Delta \cup \{C \sqsubseteq D_1\}$ or $\Delta \cup \{C \sqsubseteq D_2\}$ is consistent, so is $\Delta \cup \{C \sqsubseteq D_1 \sqcup D_2\}$.

Assume that $\vdash_{\mathbf{S}^{\mathrm{SN}}} \Delta | C \sqsubseteq D \Rightarrow \Delta$.

If $C \sqsubseteq D = B_1 \sqsubseteq B_2$ then $\Delta \vdash B_1 \not\sqsubseteq B_2$, i.e., $\Delta \cup \{B_1 \sqsubseteq B_2\}$ is inconsistent;

If $C = C_1 \sqcap C_2$ then

$$\vdash_{\mathbf{S}^{\mathrm{SN}}} \Delta | C_1 \sqsubseteq D \Rightarrow \Delta;$$
$$\vdash_{\mathbf{S}^{\mathrm{SN}}} \Delta | C_2 \sqsubseteq D \Rightarrow \Delta;$$

and by induction assumption, $\Delta \cup \{C_1 \sqsubseteq D\}$ and $\Delta \cup \{C_2 \sqsubseteq D\}$ are inconsistent, and so is $\Delta \cup \{C_1 \sqcap C_2 \sqsubseteq D\}$;

If $C = C_1 \sqcup C_2$ then either

$$\vdash_{\mathbf{S}^{\mathrm{SN}}} \Delta | C_1 \sqsubseteq D \Rightarrow \Delta;$$

or

$$\vdash_{\mathbf{S}^{\mathrm{SN}}} \Delta, C_1 \sqsubseteq D | C_2 \sqsubseteq D \Rightarrow \Delta, C_1 \sqsubseteq D;$$

and by induction assumption, either $\Delta \cup \{C_1 \sqsubseteq D\}$ or $\Delta \cup \{C_1 \sqsubseteq D, C_2 \sqsubseteq D\}$ is inconsistent, so is $\Delta \cup \{C_1 \sqcup C_2 \sqsubseteq D\}$.

If $D = D_1 \sqcap D_2$ then either

$$\vdash_{\mathbf{S}^{\mathrm{SN}}} \Delta | C \sqsubseteq D_1 \Rightarrow \Delta;$$

or

$$\vdash_{\mathbf{S}^{\mathrm{SN}}} \Delta, C \sqsubseteq D_1 | C \sqsubseteq D_2 \Rightarrow \Delta, C \sqsubseteq D_1;$$

and by induction assumption, either $\Delta \cup \{C \sqsubseteq D_1\}$ or $\Delta \cup \{C \sqsubseteq D_1, C \sqsubseteq D_2\}$ is inconsistent, so is $\Delta \cup \{C \sqsubseteq D_1 \sqcap D_2\}$;

If $D = D_1 \sqcup D_2$ then ,

$$\vdash_{\mathbf{S}^{\mathrm{SN}}} \Delta | C_1 \sqsubseteq D \Rightarrow \Delta;$$
$$\vdash_{\mathbf{S}^{\mathrm{SN}}} \Delta | C_2 \sqsubseteq D \Rightarrow \Delta;$$

and by induction assumption, $\Delta \cup \{C_1 \sqsubseteq D\}$ and $\Delta \cup \{C_2 \sqsubseteq D\}$ are inconsistent, and so is $\Delta \cup \{C \sqsubseteq D_1 \sqcup D_2\}$. □

Similarly we have the following

Theorem 11.2.5 (Soundness theorem 2) *For any consistent theory* Δ *and a statement* $C \not\sqsubseteq D$, *if* $\Delta|C \not\sqsubseteq D \Rightarrow \Delta, C \not\sqsubseteq D$ *is provable in* $\mathbf{S}^{\rm SN}$ *then* $\Delta \cup \{C \not\sqsubseteq D\}$ *is consistent, i.e.,*

$$\vdash_{\mathbf{S}^{\rm SN}} \Delta|C \not\sqsubseteq D \Rightarrow \Delta, C \not\sqsubseteq D \textit{ implies } \models_{\mathbf{S}^{\rm SN}} \Delta|C \not\sqsubseteq D \Rightarrow \Delta, C \not\sqsubseteq D;$$

and if $\Delta|C \not\sqsubseteq D \Rightarrow \Delta$ *is provable in* $\mathbf{S}^{\rm SN}$ *then* $\Delta \cup \{C \not\sqsubseteq D\}$ *is inconsistent, i.e.,*

$$\vdash_{\mathbf{S}^{\rm SN}} \Delta|C \not\sqsubseteq D \Rightarrow \Delta \textit{ implies } \models_{\mathbf{S}^{\rm SN}} \Delta|C \not\sqsubseteq D \Rightarrow \Delta.$$

Hence, we have the following

Proposition 11.2.6 *There is no statement* $C \sqsubseteq D$ *such that*

$$\vdash_{\mathbf{S}^{\rm SN}} \Delta|C \sqsubseteq D \Rightarrow \Delta$$

and

$$\vdash_{\mathbf{S}^{\rm SN}} \Delta|C \sqsubseteq D \Rightarrow \Delta, C \sqsubseteq D.$$

□

11.2.3 Examples

In R-calculus $\mathbf{S}^{\rm SN}$, we have the following deductions.

Example 11.2.7 Let

$$\begin{array}{l}\Gamma = \{\mathbf{Fish} \not\sqsubseteq \mathbf{Mammal}, \mathbf{Mammal} \not\sqsubseteq \mathbf{Fish}, \mathbf{Whale} \sqsubseteq \mathbf{Fish}\},\\ \Delta = \{\mathbf{Whale} \sqsubseteq \mathbf{Mammal}\}.\end{array}$$

Then,

$$\begin{array}{l}\Delta|\mathbf{Fish} \not\sqsubseteq \mathbf{Mammal}, \mathbf{Mammal} \not\sqsubseteq \mathbf{Fish}, \mathbf{Whale} \sqsubseteq \mathbf{Fish}\\ \Rightarrow \Delta, \mathbf{Fish} \not\sqsubseteq \mathbf{Mammal}, \mathbf{Mammal} \not\sqsubseteq \mathbf{Fish}|\mathbf{Whale} \sqsubseteq \mathbf{Fish}\\ \Rightarrow \Delta, \mathbf{Fish} \not\sqsubseteq \mathbf{Mammal}, \mathbf{Mammal} \not\sqsubseteq \mathbf{Fish};\end{array}$$

and

$$\begin{array}{l}\Delta|\mathbf{Whale} \sqsubseteq \mathbf{Fish}, \mathbf{Fish} \not\sqsubseteq \mathbf{Mammal}, \mathbf{Mammal} \not\sqsubseteq \mathbf{Fish}\\ \Rightarrow \Delta, \mathbf{Whale} \sqsubseteq \mathbf{Fish}|\mathbf{Fish} \not\sqsubseteq \mathbf{Mammal}, \mathbf{Mammal} \not\sqsubseteq \mathbf{Fish}\\ \Rightarrow \Delta, \mathbf{Whale} \sqsubseteq \mathbf{Fish}.\end{array}$$

□

We extend the semantic networks to decompose the concepts C into three classes:
• the individual concept a which extent contains only one element a;

• the concepts C, and
• the properties ϕ which is a concept whose extent is the set of all the individuals satisfying the properties;
and correspondingly we decompose the subsumption relation $\sqsubseteq$ into three relations:
• the instantiation relation $a \in C$ between individual concepts and concepts;
• the subsumption relation $C \sqsubseteq D$ between concepts; and
• the satisfaction relation $C \mapsto \phi$ between concepts and properties, where $C \mapsto \phi$ means that the extent of C is a subset of the extent of ϕ.

Example 11.2.8 Let $\Gamma = \{\mathbf{Bird} \mapsto fly, \mathbf{Penguin} \sqsubseteq \mathbf{Bird}\}$ and $\Delta = \{\mathbf{Penguin} \mapsto \neg fly\}$.

$$\begin{aligned}&\mathbf{Penguin} \mapsto \neg fly|\mathbf{Bird} \mapsto fly, \mathbf{Penguin} \sqsubseteq \mathbf{Bird}\\ &\quad \Rightarrow \mathbf{Penguin} \mapsto \neg fly, \mathbf{Bird} \mapsto fly|\mathbf{Penguin} \sqsubseteq \mathbf{Bird}\\ &\quad \Rightarrow \mathbf{Penguin} \mapsto \neg fly, \mathbf{Bird} \mapsto fly, \mathbf{Penguin} \not\sqsubseteq \mathbf{Bird};\end{aligned}$$

and

$$\begin{aligned}&\mathbf{Penguin} \mapsto \neg fly|\mathbf{Penguin} \sqsubseteq \mathbf{Bird}, \mathbf{Bird} \mapsto fly\\ &\quad \Rightarrow \mathbf{Penguin} \mapsto \neg fly, \mathbf{Penguin} \sqsubseteq \mathbf{Bird}|\mathbf{Bird} \mapsto fly\\ &\quad \Rightarrow \mathbf{Penguin} \mapsto \neg fly, \mathbf{Penguin} \sqsubseteq \mathbf{Bird}.\end{aligned}$$

Remark. Actually, we have

$$\mathbf{Penguin} \mapsto \neg fly, \mathbf{Penguin} \sqsubseteq \mathbf{Bird}, \mathbf{Bird} \mapsto_d fly.$$

□

Example 11.2.9 Let $\Gamma = \{\mathbf{Swan} \mapsto white, white \rightsquigarrow \neg black, black \rightsquigarrow \neg white\}$ and $\Delta = \{\mathbf{Swan}(\text{tweety}), \mathbf{black}(\text{tweety})\}$.

$$\begin{aligned}&\Delta|\mathbf{Swan} \mapsto white, white \rightsquigarrow \neg black, black \rightsquigarrow \neg white\\ &\Rightarrow \mathbf{Swan}(\text{tweety}), \mathbf{black}(\text{tweety}), white \rightsquigarrow \neg black, black \rightsquigarrow \neg white;\end{aligned}$$

(actually we have that $\mathbf{Swan}(\text{tweety}), \mathbf{black}(\text{tweety}), \mathbf{Swan} \mapsto_d white, white \rightsquigarrow \neg black, black \rightsquigarrow \neg white$) and

$$\begin{aligned}&\Delta|\mathbf{Swan} \mapsto white, white \rightsquigarrow \neg black, black \rightsquigarrow \neg white\\ &\Rightarrow \mathbf{Swan}(\text{tweety}), \mathbf{black}(\text{tweety}), white \rightsquigarrow \neg black, black \rightsquigarrow \neg white.\end{aligned}$$

(Actually we have that $\mathbf{Swan}(\text{tweety}), \mathbf{black}(\text{tweety}), \mathbf{Swan} \mapsto white \sqcup black, white \rightsquigarrow \neg black, black \rightsquigarrow \neg white$.) □

11.3 R-Calculus for $\preceq$-Minimal Change

Definition 11.3.1 Given two theories Δ and Γ, a theory Θ is a $\preceq$-minimal change of Γ by Δ, denoted by $\models_{\mathbf{T}} \Delta|\Gamma \Rightarrow \Delta, \Theta$, if Θ is minimal such that

(i) $\Theta \cup \Delta$ is consistent;
(ii) $\Theta \preceq D$, that is, for each statement $C \sqsubseteq D \in \Theta$, there is a statement $C' \sqsubseteq D' \in \Gamma$ such that $C \preceq C'$ and $D \preceq D'$, and
(iii) for any theory Ξ with $\Theta \prec \Xi \preceq \Gamma$, $\Xi \cup \Delta$ is inconsistent.

In this section, we will give an an Gentzen-typed R-calculus $\mathbf{T}^{\mathrm{SN}}$ such that for any statement theories Δ, Γ and theory Θ, $\Delta|\Gamma \Rightarrow \Delta, \Theta$ is provable in $\mathbf{T}^{\mathrm{SN}}$ if and only if Θ is a $\preceq$-minimal change of Γ by Δ.

11.3.1 R-Calculus $\mathbf{T}^{\mathrm{SN}}$ *for a Statement*

R-calculus $\mathbf{T}^{\mathrm{SN}}$ for a statement $C \sqsubseteq D$ consists of the following axioms and deduction rules:

• **Axioms**:

$$(T^{\mathrm{A}})\frac{\Delta \nvdash B \not\sqsubseteq B'}{\Delta|B \sqsubseteq B' \Rightarrow \Delta, B \sqsubseteq B'} \quad (T_{\mathrm{A}})\frac{\Delta \vdash B \not\sqsubseteq B'}{\Delta|B \sqsubseteq B' \Rightarrow \Delta}$$

$$(T^{\neg})\frac{\Delta \nvdash B \sqsubseteq B'}{\Delta|B \not\sqsubseteq B' \Rightarrow \Delta, B \not\sqsubseteq B'} \quad (T_{\neg})\frac{\Delta \vdash B \sqsubseteq B'}{\Delta|B \not\sqsubseteq B' \Rightarrow \Delta}$$

• **Deduction rules for** $C \sqsubseteq D$:

$$({}^{\sqcap}T_1)\frac{\Delta|C_1 \sqsubseteq D \Rightarrow \Delta, E_1 \sqsubseteq D}{\Delta|C_1 \sqcap C_2 \sqsubseteq D \Rightarrow \Delta, E_1 \sqcap C_2 \sqsubseteq D}$$

$$({}^{\sqcap}T_2)\frac{\begin{array}{l}\Delta|C_1 \sqsubseteq D \Rightarrow \Delta\\ \Delta|C_2 \sqsubseteq D \Rightarrow \Delta, E_2 \sqsubseteq D\end{array}}{\Delta|C_1 \sqcap C_2 \sqsubseteq D \Rightarrow \Delta, C_1 \sqcap E_2 \sqsubseteq D}$$

$$({}_{\sqcap}T)\frac{\begin{array}{l}\Delta|C_1 \sqsubseteq D \Rightarrow \Delta\\ \Delta|C_2 \sqsubseteq D \Rightarrow \Delta\end{array}}{\Delta|C_1 \sqcap C_2 \sqsubseteq D \Rightarrow \Delta}$$

$$({}^{\sqcup}T)\frac{\Delta|C_1 \sqsubseteq D \Rightarrow \Delta, E_1 \sqsubseteq D}{\Delta|C_1 \sqcup C_2 \sqsubseteq D \Rightarrow \Delta, E_1 \sqsubseteq D|C_2 \sqsubseteq D}$$

and

$$(T^{\sqcap})\ \frac{\Delta|C \sqsubseteq D_1 \Rightarrow \Delta, C \sqsubseteq F_1|}{\Delta|C \sqsubseteq D_1 \sqcap D_2 \Rightarrow \Delta, C \sqsubseteq F_1|C \sqsubseteq D_2}$$

$$(T_1^{\sqcup})\ \frac{\Delta|C \sqsubseteq D_1 \Rightarrow \Delta, C \sqsubseteq F_1}{\Delta|C \sqsubseteq D_1 \sqcup D_2 \Rightarrow \Delta, C \sqsubseteq F_1 \sqcup D_2}$$

$$(T_2^{\sqcup})\ \frac{\Delta|C \sqsubseteq D_2 \Rightarrow \Delta, C \sqsubseteq F_2}{\Delta|C \sqsubseteq D_1 \sqcup D_2 \Rightarrow \Delta, C \sqsubseteq D_1 \sqcup F_2}$$

$$(T_{\sqcup})\ \frac{\Delta|C \sqsubseteq D_1 \Rightarrow \Delta \quad \Delta|C \sqsubseteq D_2 \Rightarrow \Delta}{\Delta|C \sqsubseteq D_1 \sqcup D_2 \Rightarrow \Delta}$$

- **Deduction rules for** $C \not\sqsubseteq D$:

$$(-^{\sqcap}T)\ \frac{\Delta|C_1 \not\sqsubseteq D \Rightarrow \Delta, E_1 \not\sqsubseteq D}{\Delta|C_1 \sqcap C_2 \not\sqsubseteq D \Rightarrow \Delta, E_1 \sqsubseteq D|C_2 \not\sqsubseteq D}$$

$$(-^{\sqcup}T_1)\ \frac{\Delta|C_1 \not\sqsubseteq D \Rightarrow \Delta, E_1 \not\sqsubseteq D}{\Delta|C_1 \sqcup C_2 \not\sqsubseteq D \Rightarrow \Delta, E_1 \sqcup C_2 \not\sqsubseteq D}$$

$$(-^{\sqcup}T_2)\ \frac{\begin{array}{l}\Delta|C_1 \not\sqsubseteq D \Rightarrow \Delta \\ \Delta|C_2 \not\sqsubseteq D \Rightarrow \Delta, E_2 \not\sqsubseteq D\end{array}}{\Delta|C_1 \sqcup C_2 \not\sqsubseteq D \Rightarrow \Delta, C_1 \sqcup E_2 \not\sqsubseteq D}$$

$$(-_{\sqcup}T)\ \frac{\Delta|C_1 \not\sqsubseteq D \Rightarrow \Delta \quad \Delta|C_2 \not\sqsubseteq D \Rightarrow \Delta}{\Delta|C_1 \sqcup C_2 \not\sqsubseteq D \Rightarrow \Delta}$$

and

$$(-T_1^{\sqcap})\ \frac{\Delta|C \not\sqsubseteq D_1 \Rightarrow \Delta, C \not\sqsubseteq F_1}{\Delta|C \not\sqsubseteq D_1 \sqcap D_2 \Rightarrow \Delta, C \not\sqsubseteq F_1 \sqcap D_2}$$

$$(-T_2^{\sqcap})\ \frac{\Delta|C \not\sqsubseteq D_1 \Rightarrow \Delta \quad \Delta|C \not\sqsubseteq D_2 \Rightarrow \Delta, C \not\sqsubseteq F_2}{\Delta|C \not\sqsubseteq D_1 \sqcap D_2 \Rightarrow \Delta, C \not\sqsubseteq D_1 \sqcap F_2}$$

$$(-T_{\sqcap})\ \frac{\Delta|C \not\sqsubseteq D_1 \Rightarrow \Delta \quad \Delta|C \not\sqsubseteq D_2 \Rightarrow \Delta}{\Delta|C \not\sqsubseteq D_1 \sqcap D_2 \Rightarrow \Delta}$$

$$(-T^{\sqcup})\ \frac{\Delta|C \not\sqsubseteq D_1 \Rightarrow \Delta, C \not\sqsubseteq F_1}{\Delta|C \not\sqsubseteq D_1 \sqcup D_2 \Rightarrow \Delta, C \not\sqsubseteq F_1|C \sqsubseteq D_2}$$

Definition 11.3.2 $\Delta|C \sqsubseteq D \Rightarrow \Delta, B$ is provable in $\mathbf{T}^{\mathrm{SN}}$, denoted by $\vdash_{\mathbf{T}^{\mathrm{SN}}} \Delta|C \sqsubseteq D \Rightarrow \Delta, B$, if there is a sequence $S_1, ..., S_m$ of statements such that

$$\begin{array}{l}
S_1 = \Delta|C_1 \sqsubseteq D_1 \Rightarrow \Delta|C_1' \sqsubseteq D_1', \\
\cdots \\
S_m = \Delta|C_m \sqsubseteq D_m \Rightarrow \Delta, A_m'; \\
A_1 = A, \\
A_m' = B;
\end{array}$$

and for each $i < m$, S_{i+1} is an axiom or is deduced from the previous statements by a deduction rule in $\mathbf{T}^{\mathrm{SN}}$.

11.3.2 Soundness and Completeness Theorem of $\mathbf{T}^{\mathrm{SN}}$

Theorem 11.3.3 *For any consistent theory Δ and statement $C \sqsubseteq D$, there is a statement $C' \sqsubseteq D'$ such that $C' \preceq C$ and $D' \preceq D$ and $\Delta|C \sqsubseteq D \Rightarrow \Delta, C' \sqsubseteq D'$ is provable.*

Proof We prove the theorem by induction on the structure of concepts C and D.

Case $C \sqsubseteq D = B \sqsubseteq B'$. Then, if $\Delta \vdash B \not\sqsubseteq B'$ then let $C' \sqsubseteq D' = \lambda$; if $\Delta \nvdash B \not\sqsubseteq B'$ then let $C' \sqsubseteq D' = B \sqsubseteq B'$. Then, by $(T^{\mathbf{A}})$ and $(T_{\mathbf{A}})$, $\Delta|C \sqsubseteq D \Rightarrow \Delta, C' \sqsubseteq D'$ is provable.

Case $C = C_1 \sqcap C_2$. Then, by induction assumption, there are statements $E_1 \preceq C_1$ and $E_2 \preceq C_2$ such that

$$\vdash_{\mathbf{T}^{\mathrm{SN}}} \Delta|C_1 \sqsubseteq D \Rightarrow \Delta, E_1 \sqsubseteq D;$$
$$\vdash_{\mathbf{T}^{\mathrm{SN}}} \Delta|C_2 \sqsubseteq D \Rightarrow \Delta, E_2 \sqsubseteq D.$$

If $E_1 \neq \lambda$ then by $(T_1^{\sqcup})$,

$$\vdash_{\mathbf{T}^{\mathrm{SN}}} \Delta|C_1 \sqcap C_2 \sqsubseteq D \Rightarrow \Delta, E_1 \sqcup C_2 \sqsubseteq D;$$

if $E_1 = \lambda$ and $E_2 \neq \lambda$ then by $(T_2^{\sqcup})$,

$$\vdash_{\mathbf{T}^{\mathrm{SN}}} \Delta|C_1 \sqcap C_2 \sqsubseteq D \Rightarrow \Delta, C_1 \sqcup E_2 \sqsubseteq D;$$

and if $B_1 = B_2 = \lambda$ then by $(T_{\sqcup})$,

$$\vdash_{\mathbf{T}^{\mathrm{SN}}} \Delta|C_1 \sqcap C_2 \sqsubseteq D \Rightarrow \Delta.$$

Let

$$C' \sqsubseteq D' = \begin{cases} E_1 \sqcup C_2 \sqsubseteq D \text{ if } E_1 \neq \lambda \\ C_1 \sqcup E_2 \sqsubseteq D \text{ if } E_1 = \lambda \neq E_2 \\ \lambda \qquad\qquad\qquad \text{otherwise,} \end{cases}$$

and $\vdash_{\mathbf{T}^{\mathrm{SN}}} \Delta|C_1 \sqcap C_2 \sqsubseteq D \Rightarrow \Delta, C' \sqsubseteq D'$.

Case $C = C_1 \sqcup C_2$. Then, by induction assumption, there are statements $E_1 \preceq C_1, E_2 \preceq C_2$ such that

$$\vdash_{\mathbf{T}^{\mathrm{SN}}} \Delta|C_1 \sqsubseteq D \Rightarrow \Delta, E_1 \sqsubseteq D;$$
$$\vdash_{\mathbf{T}^{\mathrm{SN}}} \Delta, E_1 \sqsubseteq D|C_2 \sqsubseteq D \Rightarrow \Delta, E_1 \sqsubseteq D, E_2 \sqsubseteq D.$$

Let $C' \sqsubseteq D' = E_1 \sqcup E_2 \sqsubseteq D$, and

$$\vdash_{\mathbf{T}^{\mathrm{SN}}} \Delta|C_1 \sqcup C_2 \sqsubseteq D \Rightarrow \Delta, E_1 \sqcup E_2 \sqsubseteq D.$$

Case $D = D_1 \sqcap D_2$. Then, by induction assumption, there are statements $F_1 \preceq D_1$, $F_2 \preceq D_2$ such that

$$\vdash_{\mathbf{T}^{\mathrm{SN}}} \Delta | C \sqsubseteq D_1 \Rightarrow \Delta, C \sqsubseteq F_1;$$
$$\vdash_{\mathbf{T}^{\mathrm{SN}}} \Delta, C \sqsubseteq F_1 | C \sqsubseteq D_2 \Rightarrow \Delta, C \sqsubseteq F_1, C \sqsubseteq F_2.$$

Let $C' \sqsubseteq D' = C \sqsubseteq F_1 \sqcap F_2$, and

$$\vdash_{\mathbf{T}^{\mathrm{SN}}} \Delta | C \sqsubseteq D_1 \sqcap D_2 \Rightarrow \Delta, C \sqsubseteq F_1 \sqcap F_2.$$

Case $D = D_1 \sqcup D_2$. Then, by induction assumption, there are statements $F_1 \preceq D_1$ and $F_2 \preceq D_2$ such that

$$\vdash_{\mathbf{T}^{\mathrm{SN}}} \Delta | C \sqsubseteq D_1 \Rightarrow \Delta, C \sqsubseteq F_1;$$
$$\vdash_{\mathbf{T}^{\mathrm{SN}}} \Delta | C \sqsubseteq D_2 \Rightarrow \Delta, C \sqsubseteq F_2.$$

If $F_1 \neq \lambda$ then by $(T_1^{\sqcup})$,

$$\vdash_{\mathbf{T}^{\mathrm{SN}}} \Delta | C \sqsubseteq D_1 \sqcup D_2 \Rightarrow \Delta, C \sqsubseteq F_1 \sqcup D_2;$$

if $F_1 = \lambda$ and $F_2 \neq \lambda$ then by $(T_2^{\sqcup})$,

$$\vdash_{\mathbf{T}^{\mathrm{SN}}} \Delta | C \sqsubseteq D_1 \sqcup D_2 \Rightarrow \Delta, C \sqsubseteq D_1 \sqcup F_2;$$

and if $F_1 = F_2 = \lambda$ then by $(T_{\sqcup})$,

$$\vdash_{\mathbf{T}^{\mathrm{SN}}} \Delta | C \sqsubseteq D_1 \sqcup D_2 \Rightarrow \Delta.$$

Let

$$C' \sqsubseteq D' = \begin{cases} C \sqsubseteq F_1 \sqcup D_2 \text{ if } F_1 \neq \lambda \\ C \sqsubseteq D_1 \sqcup F_2 \text{ if } F_1 = \lambda \neq F_2 \\ \lambda \qquad\qquad \text{otherwise,} \end{cases}$$

and $\vdash_{\mathbf{T}^{\mathrm{SN}}} \Delta | C \sqsubseteq D_1 \sqcup D_2 \Rightarrow \Delta, C' \sqsubseteq D'$. □

Lemma 11.3.4 *If Θ is a $\preceq$-minimal change of Γ by Δ and Θ' is a $\preceq$-minimal change of $C_{n+1} \sqsubseteq D_{n+1}$ by $\Delta \cup \Theta$ then Θ' is a $\preceq$-minimal change of $D \cup \{C_{n+1} \sqsubseteq D_{n+1}\}$ by Δ.*

Proof Assume that Θ is a $\preceq$-minimal change of D by Δ and Θ' is a $\preceq$-minimal change of $C_{n+1} \sqsubseteq D_{n+1}$ by $\Delta \cup \Theta$. Then, by the definition, Θ' is a $\preceq$-minimal change of $D \cup \{C_{n+1} \sqsubseteq D_{n+1}\}$ by Δ. □

Lemma 11.3.5 *If $C \sqsubseteq F_1$ is a $\preceq$-minimal change of $C \sqsubseteq D_1$ by Δ and $C \sqsubseteq F_2$ is a $\preceq$-minimal change of $C \sqsubseteq D_2$ by $\Delta \cup \{C \sqsubseteq F_1\}$ then $C \sqsubseteq F_1 \sqcap F_2$ is a $\preceq$-minimal change of $C \sqsubseteq D_1 \sqcap D_2$ by Δ.*

Proof Assume that $C \sqsubseteq F_1$ is a $\preceq$-minimal changes of $C \sqsubseteq D_1$ by Δ and $C \sqsubseteq F_2$ of $C \sqsubseteq D_2$ by $\Delta \cup \{C \sqsubseteq F_1\}$.

Then, $F_1 \sqcap F_2 \preceq D_1 \sqcap D_2$, and $\Delta \cup \{C \sqsubseteq F_1 \sqcap F_2\}$ is consistent.

For any G with $F_1 \sqcap F_2 \prec G \preceq D_1 \sqcap D_2$, there are G_1, G_2 such that $G = G_1 \sqcap G_2$, and

$$F_1 \preceq G_1 \preceq D_1,$$
$$F_2 \preceq G_2 \preceq D_2.$$

Then, if $F_1 \prec G_1$ then $\Delta \cup \{C \sqsubseteq G_1\}$ is inconsistent and so is $\Delta \cup \{C \sqsubseteq G_1 \sqcap G_2\}$; if $F_2 \prec G_2$ then $\Delta \cup \{C \sqsubseteq F_2\}$ is inconsistent and so is $\Delta \cup \{C \sqsubseteq F_1 \sqcap F_2\}$. □

Lemma 11.3.6 *If $C \sqsubseteq F_1, C \sqsubseteq F_2$ are $\preceq$-minimal changes of $C \sqsubseteq D_1$ and $C \sqsubseteq D_2$ by Δ, respectively, then $C \sqsubseteq F'$ is a $\preceq$-minimal change of $C \sqsubseteq D_1 \sqcup D_2$ by Δ, where*

$$F' = \begin{cases} F_1 \sqcup D_2 & \text{if } F_1 \neq \lambda \\ D_1 \sqcup F_2 & \text{if } F_1 = \lambda \text{ and } F_2 \neq \lambda \\ \lambda & \text{if } F_1 = F_2 = \lambda \end{cases}$$

Proof It is clear that $F' \preceq D_1 \sqcup D_2$, and $\Delta \cup \{C \sqsubseteq F'\}$ is consistent.

For any G with $F' \prec G \preceq D_1 \sqcup D_2$, there are G_1, G_2 such that

$$G = G_1 \sqcup G_2,$$
$$F_1' \preceq G_1 \preceq D_1,$$
$$F_2' \preceq G_2 \preceq D_2,$$

and either $F_1' \prec G_1$ or $F_2' \prec G_2$, where F_1' is either F_1 or D_1 and F_2' is either F_2 or D_2.

If $F_1' \prec G_1$ then $F_1' = F_1$, $F_2' = F_2$, and by induction assumption, $\Delta \cup \{C \sqsubseteq G_1\}$ is inconsistent and so is $\Delta \cup \{C \sqsubseteq G_1 \sqcup D_2\}$; and if $F_2' \prec G_2$ then $F_2' = F_2$, $F_1' = D_1$, and by induction assumption, $\Delta \cup \{C \sqsubseteq G_2\}$ is inconsistent and so is $\Delta \cup \{C \sqsubseteq D_1 \sqcup G_2\}$. □

Lemma 11.3.7 *If $E_1 \sqsubseteq D$ is a $\preceq$-minimal change of $C_1 \sqsubseteq D$ by Δ and $E_2 \sqsubseteq D$ is a $\preceq$-minimal change of $C_2 \sqsubseteq D$ by $\Delta \cup \{E_1 \sqsubseteq D\}$ then $E_1 \sqcup E_2 \sqsubseteq D$ is a $\preceq$-minimal change of $C_1 \sqcup C_2 \sqsubseteq D$ by Δ.*

Lemma 11.3.8 *If $E_1 \sqsubseteq D, E_2 \sqsubseteq D$ are $\preceq$-minimal changes of $C_1 \sqsubseteq D$ and $C_2 \sqsubseteq D$ by Δ, respectively, then $E' \sqsubseteq D$ is a $\preceq$-minimal change of $C_1 \sqcap C_2 \sqsubseteq D$ by Δ, where*

$$E' = \begin{cases} E_1 \sqcup C_2 & \text{if } E_1 \neq \lambda \\ C_1 \sqcup E_2 & \text{if } E_1 = \lambda \text{ and } E_2 \neq \lambda \\ \lambda & \text{if } E_1 = E_2 = \lambda \end{cases}$$

Theorem 11.3.9 *For any statement set Δ, statements $C \sqsubseteq D, C' \sqsubseteq D'$, if $\Delta | C \sqsubseteq D \Rightarrow \Delta, C' \sqsubseteq D'$ is provable then $C' \sqsubseteq D'$ is a $\preceq$-minimal change of $C \sqsubseteq D$ by Δ. That is,*

$$\vdash_{\mathbf{T}^{\mathrm{SN}}} \Delta|C \sqsubseteq D \Rightarrow \Delta, C' \sqsubseteq D' \textit{ implies } \models_{\mathbf{T}^{\mathrm{SN}}} \Delta|C \sqsubseteq D \Rightarrow \Delta, C' \sqsubseteq D'.$$

Proof We prove the theorem by induction on the structure of C and D. Assume that $\vdash_{\mathbf{T}^{\mathrm{SN}}} \Delta|C \sqsubseteq D \Rightarrow \Delta, C' \sqsubseteq D'$.

Case $C \sqsubseteq D = B \sqsubseteq B'$, a literal statement. Either $\Delta|B \sqsubseteq B' \Rightarrow \Delta, B \sqsubseteq B'$ or $\Delta|B \sqsubseteq B' \Rightarrow \Delta$ is provable. That is, either $C' \sqsubseteq D' = B \sqsubseteq B'$ or $C' \sqsubseteq D' = \lambda$, which is a $\preceq$-minimal change of $C \sqsubseteq D$ by Δ.

Case $C = C_1 \sqcap C_2$. Then, if

$$\vdash_{\mathbf{T}^{\mathrm{SN}}} \Delta|C_1 \sqsubseteq D \Rightarrow \Delta, E_1 \sqsubseteq D$$

then $E_1 \sqcap C_2 \sqsubseteq D$ is a $\preceq$-minimal change of $C_1 \sqcap C_2 \sqsubseteq D$ by Δ; and if

$$\begin{array}{l} \vdash_{\mathbf{T}^{\mathrm{SN}}} \Delta|C_1 \sqsubseteq D \Rightarrow \Delta; \\ \vdash_{\mathbf{T}^{\mathrm{SN}}} \Delta|C_2 \sqsubseteq D \Rightarrow \Delta, E_2 \sqsubseteq D; \end{array}$$

then $C_1 \sqcap E_2 \sqsubseteq D$ is a $\preceq$-minimal change of $C_1 \sqcap C_2 \sqsubseteq D$ by Δ; and if

$$\begin{array}{l} \vdash_{\mathbf{T}^{\mathrm{SN}}} \Delta|C_1 \sqsubseteq D \Rightarrow \Delta; \\ \vdash_{\mathbf{T}^{\mathrm{SN}}} \Delta|C_2 \sqsubseteq D \Rightarrow \Delta \end{array}$$

then λ is a $\preceq$-minimal change of $C_1 \sqcap C_2 \sqsubseteq D$ by Δ.

Case $C = C_1 \sqcup C_2$. Then there are concepts $E_1 \sqsubseteq C_1$, $E_2 \sqsubseteq C_2$ such that

$$\begin{array}{l} \vdash_{\mathbf{T}^{\mathrm{SN}}} \Delta|C_1 \sqsubseteq D \Rightarrow \Delta, E_1 \sqsubseteq D; \\ \vdash_{\mathbf{T}^{\mathrm{SN}}} \Delta, E_1 \sqsubseteq D|C_2 \sqsubseteq D \Rightarrow \Delta, E_1 \sqsubseteq D, E_2 \sqsubseteq D. \end{array}$$

By the induction assumption, $E_1 \sqsubseteq D$ is a $\preceq$-minimal change of $C_1 \sqsubseteq D$ by Δ; and $E_2 \sqsubseteq D$ is a $\preceq$-minimal change of $C_2 \sqsubseteq D$ by $\Delta \cup \{E_1 \sqsubseteq D\}$, and hence, $E_1 \sqcup E_2 \sqsubseteq D$ is a $\preceq$-minimal change of $C_1 \sqcup C_2 \sqsubseteq D$ by Δ.

Case $D = D_1 \sqcap D_2$. Then there are concepts $F_1 \sqsubseteq D_1$, $F_2 \sqsubseteq D_2$ such that

$$\begin{array}{l} \vdash_{\mathbf{T}^{\mathrm{SN}}} \Delta|C \sqsubseteq D_1 \Rightarrow \Delta, C \sqsubseteq F_1; \\ \vdash_{\mathbf{T}^{\mathrm{SN}}} \Delta, C \sqsubseteq F_1|C \sqsubseteq D_2 \Rightarrow \Delta, C \sqsubseteq F_1, C \sqsubseteq F_2. \end{array}$$

By the induction assumption, $C \sqsubseteq F_1$ is a $\preceq$-minimal change of $C \sqsubseteq D_1$ by Δ; and $C \sqsubseteq F_2$ is a $\preceq$-minimal change of $C \sqsubseteq D_2$ by $\Delta \cup \{C \sqsubseteq F_1\}$, and hence, $C \sqsubseteq F_1 \sqcap F_2$ is a $\preceq$-minimal change of $C \sqsubseteq D_1 \sqcap D_2$ by Δ.

Case $D = D_1 \sqcup D_2$. Then, if

$$\vdash_{\mathbf{T}^{\mathrm{SN}}} \Delta|C \sqsubseteq D_1 \Rightarrow \Delta, C \sqsubseteq F_1$$

then $C \sqsubseteq F_1 \sqcup D_2$ is a $\preceq$-minimal change of $C \sqsubseteq D_1 \sqcup D_2$ by Δ; and if

$$\vdash_{\mathbf{T}^{\mathrm{SN}}} \Delta|C \sqsubseteq D_1 \Rightarrow \Delta;$$
$$\vdash_{\mathbf{T}^{\mathrm{SN}}} \Delta|C \sqsubseteq D_2 \Rightarrow \Delta, C \sqsubseteq F_2;$$

then $C \sqsubseteq D_1 \sqcup F_2$ is a $\preceq$-minimal change of $C \sqsubseteq C_1 \sqcup C_2$ by Δ; and if

$$\vdash_{\mathbf{T}^{\mathrm{SN}}} \Delta|C \sqsubseteq D_1 \Rightarrow \Delta;$$
$$\vdash_{\mathbf{T}^{\mathrm{SN}}} \Delta|C \sqsubseteq D_2 \Rightarrow \Delta$$

then λ is a $\preceq$-minimal change of $C \sqsubseteq D_1 \sqcup D_2$ by Δ. □

Theorem 11.3.10 *For any theory Δ and statements $C \sqsubseteq D$, $C' \sqsubseteq D'$, if $C' \sqsubseteq D'$ is a $\preceq$-minimal change of $C \sqsubseteq D$ by Δ then $\Delta|C \sqsubseteq D \Rightarrow \Delta, C' \sqsubseteq D'$ is $\mathbf{T}^{\mathrm{SN}}$-provable. That is,*

$$\models_{\mathbf{T}^{\mathrm{SN}}} \Delta|C \sqsubseteq D \Rightarrow, \Delta, C' \sqsubseteq D' \textit{ implies } \vdash_{\mathbf{T}^{\mathrm{SN}}} \Delta|C \sqsubseteq D \Rightarrow, \Delta, C' \sqsubseteq D'.$$

Proof Let $C' \sqsubseteq D'$ be a $\preceq$-minimal change of $C \sqsubseteq D$ by Δ.

Case $C \sqsubseteq D = B \sqsubseteq B'$. Then $C' \sqsubseteq D' = \lambda$ (if $\Delta, B \sqsubseteq B'$ is inconsistent) or $C' \sqsubseteq D' = B \sqsubseteq B'$ (if $\Delta, B \sqsubseteq B'$ is consistent), and $\Delta|C \sqsubseteq D \Rightarrow \Delta, C' \sqsubseteq D'$ is provable.

Case $C = C_1 \sqcap C_2$. Then there are E_1 and E_2 such that either $E_1 \sqsubseteq D$ or $E_2 \sqsubseteq D$ is a $\preceq$-minimal change of $C_1 \sqsubseteq D$ or $C_2 \sqsubseteq D$ by Δ, respectively. Define

$$C' \sqsubseteq D' = \begin{cases} E_1 \sqcap C_2 \sqsubseteq D \text{ if } E_1 \neq \lambda \\ C_1 \sqcap E_2 \sqsubseteq D \text{ if } E_1 = \lambda \text{ and } E_2 \neq \lambda \\ \lambda \qquad\qquad\qquad \text{if } E_1 = E_2 = \lambda \end{cases}$$

Then, $C' \sqsubseteq D'$ is a $\preceq$-minimal change of $C_1 \sqcap C_2 \sqsubseteq D$ by Δ. By the induction assumption, either (i) $\Delta|C_1 \sqsubseteq D \Rightarrow \Delta, E_1 \sqsubseteq D$ or (ii) $\Delta|C_2 \sqsubseteq D \Rightarrow \Delta, E_2 \sqsubseteq D$, or (iii) $\Delta|C_1 \sqsubseteq D \Rightarrow \Delta$ and $\Delta|C_2 \sqsubseteq D \Rightarrow \Delta$ are provable, and so is $\Delta|C_1 \sqcap C_2 \sqsubseteq D \Rightarrow \Delta, C' \sqsubseteq D'$, where if $E_1 \neq \lambda$ then $\Delta|C_1 \sqcap C_2 \sqsubseteq D \Rightarrow \Delta, E_1 \sqcap C_2 \sqsubseteq D$ is provable; if $E_1 = \lambda$ and $E_2 \neq \lambda$ then $\Delta|C_1 \sqcup C_2 \sqsubseteq D \Rightarrow \Delta, C_1 \sqcup E_2 \sqsubseteq D$ is provable; and if $E_1 = \lambda$ and $E_2 = \lambda$ then $\Delta|C_1 \sqcap C_2 \sqsubseteq D \Rightarrow \Delta$ is provable.

Case $C = C_1 \sqcup C_2$. Then there are E_1, E_2 such that $E = E_1 \sqcup E_2$, and $E_1 \sqsubseteq D$ and $E_2 \sqsubseteq D$ are $\preceq$-minimal changes of $C_1 \sqsubseteq D$ and $C_2 \sqsubseteq D$ by Δ and $\Delta \cup \{E_1 \sqsubseteq D\}$, respectively. Hence, $E_1 \sqcup E_2 \sqsubseteq D$ is a $\preceq$-minimal change of $C_1 \sqcup C_2$ by Δ. By the induction assumption,

$$\vdash_{\mathbf{T}^{\mathrm{SN}}} \Delta|C_1 \sqsubseteq D \Rightarrow \Delta, E_1 \sqsubseteq D;$$
$$\vdash_{\mathbf{T}^{\mathrm{SN}}} \Delta, E_1 \sqsubseteq D|C_2 \sqsubseteq D \Rightarrow \Delta, E_1 \sqsubseteq D, E_2 \sqsubseteq D,$$

and we have that $\Delta|C_1 \sqcup C_2 \sqsubseteq D \Rightarrow \Delta, E_1 \sqcup E_2 \sqsubseteq D$.

Case $D = D_1 \sqcap D_2$. Then there are $F_1 \preceq D_1$, $F_2 \preceq D_2$ such that $F = F_1 \sqcup F_2$, and $C \sqsubseteq F_1$ and $C \sqsubseteq F_2$ are $\preceq$-minimal changes of $C \sqsubseteq D_1$ and $C \sqsubseteq D_2$ by Δ and $\Delta \cup \{C \sqsubseteq F_1\}$, respectively. Hence, $F_1 \sqcap F_2 \sqsubseteq D$ is a $\preceq$-minimal change of $C \sqsubseteq D_1 \sqcap D_2$ by Δ. By the induction assumption,

$$\vdash_{\mathbf{T}^{\mathrm{SN}}} \Delta|C \sqsubseteq D_1 \Rightarrow \Delta, C \sqsubseteq F_1;$$
$$\vdash_{\mathbf{T}^{\mathrm{SN}}} \Delta, C \sqsubseteq F_1|C \sqsubseteq D_2 \Rightarrow \Delta, C \sqsubseteq F_1, C \sqsubseteq F_2,$$

and we have that $\Delta|C \sqsubseteq D_1 \sqcap D_2 \Rightarrow \Delta, C \sqsubseteq F_1 \sqcap F_2$.

Case $D = D_1 \sqcup D_2$. Then there are $F_1 \sqsubseteq D_1$ and $F_2 \sqsubseteq D_2$ such that $C \sqsubseteq F_1$ and $C \sqsubseteq F_2$ are $\preceq$-minimal changes of $C \sqsubseteq D_1$ and $C \sqsubseteq D_2$ by Δ, respectively. Define

$$C' \sqsubseteq D' = \begin{cases} C \sqsubseteq F_1 \sqcup D_2 \text{ if } F_1 \neq \lambda \\ C \sqsubseteq D_1 \sqcup F_2 \text{ if } F_1 = \lambda \text{ and } F_2 \neq \lambda \\ \lambda \qquad\qquad\quad \text{if } F_1 = F_2 = \lambda \end{cases}$$

Then, $C' \sqsubseteq D'$ is a $\preceq$-minimal change of $C \sqsubseteq D_1 \sqcup D_2$ by Δ. By the induction assumption, either (i) $\Delta|C \sqsubseteq D_1 \Rightarrow \Delta, C \sqsubseteq F_1$ or (ii) $\Delta|C \sqsubseteq D_2 \Rightarrow \Delta, C \sqsubseteq F_2$, or (iii) $\Delta|C \sqsubseteq D_1 \Rightarrow \Delta$ and $\Delta|C \sqsubseteq D_2 \Rightarrow \Delta$ are provable, and so is $\Delta|C_1 \sqcup C_2 \sqsubseteq D \Rightarrow \Delta, C' \sqsubseteq D'$, where if $F_1 \neq \lambda$ then $\Delta|C \sqsubseteq D_1 \sqcup D_2 \Rightarrow \Delta, C \sqsubseteq F_1 \sqcup D_2$ is provable; if $F_1 = \lambda$ and $F_2 \neq \lambda$ then $\Delta|C \sqsubseteq D_1 \sqcup D_2 \Rightarrow \Delta, C \sqsubseteq D_1 \sqcup F_2$ is provable; and if $F_1 = \lambda$ and $F_2 = \lambda$ then $\Delta|C \sqsubseteq D_1 \sqcup D_2 \Rightarrow \Delta$ is provable. □

Similarly we have the following

Theorem 11.3.11 *For any statement set Δ, statements $C \not\sqsubseteq D, C' \not\sqsubseteq D'$, if $\Delta|C \not\sqsubseteq D \Rightarrow \Delta, C' \not\sqsubseteq D'$ is provable then $C' \not\sqsubseteq D'$ is a $\preceq$-minimal change of $C \not\sqsubseteq D$ by Δ. That is,*

$$\vdash_{\mathbf{T}^{\mathrm{SN}}} \Delta|C \not\sqsubseteq D \Rightarrow \Delta, C' \not\sqsubseteq D' \textit{ implies } \models_{\mathbf{T}^{\mathrm{SN}}} \Delta|C \not\sqsubseteq D \Rightarrow \Delta, C' \not\sqsubseteq D'.$$

□

Theorem 11.3.12 *For any theory Δ and statements $C \not\sqsubseteq D, C' \not\sqsubseteq D'$, if $C' \not\sqsubseteq D'$ is a $\preceq$-minimal change of $C \not\sqsubseteq D$ by Δ then $\Delta|C \not\sqsubseteq D \Rightarrow \Delta, C' \not\sqsubseteq D'$ is $\mathbf{T}^{\mathrm{SN}}$-provable. That is,*

$$\models_{\mathbf{T}^{\mathrm{SN}}} \Delta|C \not\sqsubseteq D \Rightarrow, \Delta, C' \not\sqsubseteq D' \textit{ implies } \vdash_{\mathbf{T}^{\mathrm{SN}}} \Delta|C \not\sqsubseteq D \Rightarrow, \Delta, C' \not\sqsubseteq D'.$$

□

Generally we have the following

Theorem 11.3.13 (Soundness theorem) *For any theories Θ, Δ and any finite statement set Γ, if $\Delta|\Gamma \Rightarrow \Delta, \Theta$ is provable in $\mathbf{T}^{\mathrm{SN}}$ then Θ is a $\preceq$-minimal change of Γ by Δ. That is,*

$$\vdash_{\mathbf{T}^{\mathrm{SN}}} \Delta|\Gamma \Rightarrow \Delta, \Theta \textit{ implies } \models_{\mathbf{T}^{\mathrm{SN}}} \Delta|\Gamma \Rightarrow \Delta, \Theta.$$

□

Theorem 11.3.14 (Completeness theorem) *For any theories* Θ, Δ *and any finite statement set* Γ, *if* Θ *is a* $\preceq$*-minimal change of* Γ *by* Δ *then* $\Delta|\Gamma \Rightarrow \Delta, \Theta$ *is provable in* $\mathbf{T}^{\mathrm{SN}}$. *That is,*

$$\models_{\mathbf{T}^{\mathrm{SN}}} \Delta|\Gamma \Rightarrow \Delta, \Theta \text{ implies } \vdash_{\mathbf{T}^{\mathrm{SN}}} \Delta|\Gamma \Rightarrow \Delta, \Theta.$$

□

References

1. G. Flouris, D. Plexousakis, G. Antoniou, On applying the AGM theory to DLs and OWL, in *Proceedings of 4th International Conference on Semantic Web* (2005), pp. 216–231
2. C. Halaschek-Wiener, Y. Katz, B. Parsia, Belief base revision for expressive description logics, in *Proceedings of OWL-ED'06* (2006)
3. I. Horrocks, P. Patel-Schneider, Reducing OWL entailment to description logic satisfiability, in *Proceedings of the 2nd International Semantic Web Conference (ISWC)* (2003)
4. I. Horrocks, U. Sattler, Ontology reasoning in the SHOQ(D) description logic, in *Proceedings of the 17th International Joint Conference on Artificial Intelligence* (2001)
5. K. Lee, T. Meyer, A classification of ontology modification, in *Australian Conference on Artificial Intelligence* (2004), pp. 248–258
6. T. Meyer, K. Lee, R. Booth, Knowledge integration for description logics, in *AAAI* (2005), pp. 645–650
7. P. Plessers, O. De Troyer, Resolving inconsistencies in evolving ontologies, in *Proceedings of ESWC'06* (2006), pp. 200–214
8. S. Schlobach, Z. Huang, R. Cornet, F. van Harmelen, Debugging incoherent terminologies. J. Autom. Reason. **39**, 317–349 (2007)
9. L. Stojanovic, *Methods and Tools for Ontology Evolution,* Ph.D. thesis, University of Karlsruhe (2004)
10. J.F. Sowa, Semantic networks, in *Encyclopedia of Artificial Intelligence*, ed. by S.C. Shapiro (1987)

Index

W. Li and Y. Sui, *R-CALCULUS: A Logic of Belief Revision*, Perspectives in Formal Induction, Revision and Evolution,
https://doi.org/10.1007/978-981-16-2944-0

Printed in the United States
by Baker & Taylor Publisher Services